The Population Ecology of White-Headed Langur

Wenshi Pan

The Population Ecology of White-Headed Langur

Springer
北京大学出版社
PEKING UNIVERSITY PRESS

Wenshi Pan
School of Life Science
Peking University
Beijing, China

ISBN 978-981-33-4117-3 ISBN 978-981-33-4118-0 (eBook)
https://doi.org/10.1007/978-981-33-4118-0

Jointly published with Peking University Press
The print edition is not for sale in China (Mainland). Customers from China (Mainland) please order the print book from: Peking University Press.

This Springer imprint is published by the registered company Springer Nature Singapore Pte Ltd.
The registered company address is: 152 Beach Road, #21-01/04 Gateway East, Singapore 189721, Singapore

Preface

The white-headed langur (*Trachypithecus leucocephalus*), which lives deep in the forests on karst hills at the south bank of Zuo Jiang River, in southwestern China, is an unusual kind of animal. They are very timid and numbered very few. By the end of the twentieth century, there were only 105 white-headed langurs living in the Nongguan mountains. It has become one of the most endangered species on earth and has been listed by the International Union for Conservation of Nature (IUCN) in its Red List of Threatened Species as Critically Endangered.

To both wildlife enthusiasts among the public and wild animal zoologists, the white-headed langur is something like a legend animal. The species lives on craggy precipices engulfed by tropical monsoon rainforests. In general, the best an inquisitive watcher can hope for is to catch a distant glimpse of the colorful and slender animal from the base of a craggy hill. Only a selected few people ever have a chance, at dawning hours or after sunset, with the assistance of field researchers, to get near a white-headed langur community's night shelter and hear the rustling sound of tree leaves as the nimble primates swiftly move through the forest, or catch a silhouette of one climbing up a precipice.

This book is a register of 20 years of arduous field research. We will give an account on the "last natural refuge" which sustains the white-headed langurs, relate the restoration process of the geographic population, discuss topics of the white-headed langur's home range and territories, its behaviors and movement patterns, its diets and forage strategies, its reproduction and population ecology, among other aspects. We will discuss conservation and extrapolate the future fate of white-headed langur through a population viability analysis. At the same time as we are writing this book, the taxonomical rank of the white-headed langur is yet under debate. Not even in the authoritative publication *China Species Red List (2004)* (Wang Sung, Xie Yan. China species red list (2004). Beijing: Higher Education Press; 2004) has its rank been resolutely determined. Is the white-headed langur a geographical subspecies of the black langur? Or is it a distinct species in its own right? We shall start by looking into this contentious issue and deliberate the various arguments that have been put forward.

Beijing, China — Wenshi Pan

Contents

1 Origin of the White-Headed Langur: Discovery of Fossils

Abstract

In a limestone karst cave (Sanhe Cave) in Nongguan Mountains in Chongzuo, Guangxi, China, we unearthed in the sediments 82 mammal fossils in a *Gigantopithecus* fauna of 1.4 mya, including fossils of *Gigantopithecus blacki*, *Ailuropoda milanoleuca wulingshanensis*, *Megatapirus augustus*, *Rhinoceros sinensis*, and *Stegodon preorientalis*, the ancestor species of the white-headed langur, named Chongzuo langur (*Trachypithecus chongzuosis*), and the ancestor species of the Sumatran striped rabbit (*Nesolagus* sp.). The animals were buried during the middle phase of the Early Pleistocene Epoch, which coincided with the first extreme glaciations in the Quaternary Period. The glaciations caused the sea level to drop, and the continental shelf was thus exposed. As a result, the coastal areas surrounding Java Island and Sumatra Island became land, which connected with the Malay Peninsula, Indochina Peninsula, and the coastal lowlands around Beibu Gulf in southern Guangxi. This provided land passages for Javan Langurs (*Trachypithecus* sp.) to disperse northward, enter the karst hill district on the south bank of Zuo Jiang River, and gradually evolve into the present-day white-headed langur (*Trachypithecus leucocephalus*).

Keywords

Sediments · *Gigantopithecus* fauna · Ancestor species · Langur evolution · Chongzuo langur (*Trachypithecus chongzuosis*) · Guangxi · China

There is living evidence that white-headed langurs are presently living on the karst hills at the south bank of Zuo Jiang River, southwest China. But how did they happen to be there? Who were their ancestors? Where did the ancestors live? How did the ancestors make their way from their homeland to settle in a foreign place? When did that begin? In this chapter, we will discuss these questions based on the fossil discovery in Chongzuo Sanhe Cave, Guangxi, China.

1.1 Discovery of Fossils in Sanhe Cave

Sanhe Cave is located at the southeast of Jiangzhou District in Chongzuo City, Guangxi Zhuang Autonomous Region, China (Fig. 1.1). It is situated in between Luobai Township and Banli Township, 26 km away from Chongzuo city capital at the northeast corner of the Nongguan Mountains, which consists of a series of karst hills. The geographical coordinates are latitude 22°16.493′N, longitude 107°30.663′E. At a distance of 300 m on its east is Sanhetun, a small village with only seven households. When we published our first research report on the *Gigantopithecus* fauna discovered there, we named the cave where the fossils were discovered "Sanhe Cave," following the administrative name of the place on maps. The cave is also referred to as Feijichang (FJC) Cave in various chapters in this book. There was once a barrack there, but it has been left derelict for 20 years. When we first arrived and stationed ourselves at the derelict barracks in November 1996, we came to notice that to the west of the derelict barracks was a north-south orientated hilly wetland, and on its north arose a sheer precipice. Halfway up the precipice was a big limestone cave, which was the home of white-headed langurs. The cave opens to a broad, flat, round-shaped wetland. As we were making our drafted map, an amusing idea came to us that the wetland could be used as a landing pad for future research if helicopters were needed, so we named the wetland Feijichang (airstrip). Accordingly, the big molten cave high up on the precipice was named Feijichang (FJC) Cave. Since that day, our research team has always referred to the cave by that name. In this book, when it is related to fossil discovery, the cave is referred to as Sanhe Cave, whereas regarding the langur's night shelters, it is referred to by the name of Feijichang (FJC) Cave. Likewise, Peking University

W. Pan, *The Population Ecology of White-Headed Langur*, https://doi.org/10.1007/978-981-33-4118-0_1

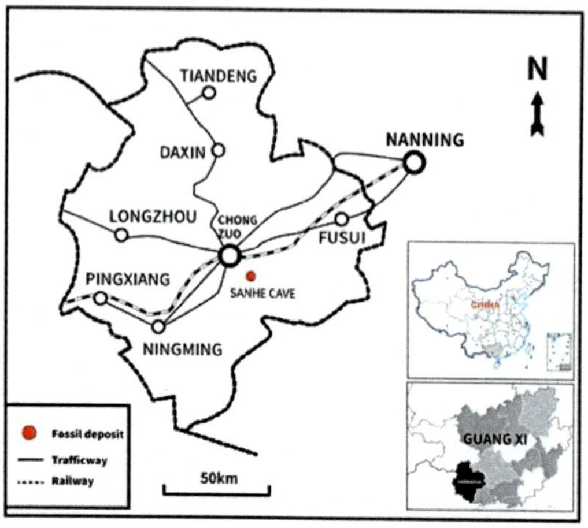

Fig. 1.1 Geographical location of Chongzuo Sanhe Cave

Chongzuo Biodiversity Research Base (shortened to Research Base) and Chongzuo Ecologic Park (shortened to Eco-park) are but different names of the same place. The Research Base was founded in year 1998, and 2 years later (year 2000) the Eco-Park was established at the same place.

1.1.1 Initial Fieldwork

When we arrived at this place in wilderness, we did not have any exact idea of how to start a systematic research, but an accidental discovery led us to unearth many prehistoric fossils with significant value from the sediments in Sanhe Cave.

On November 17, 1996, we set up a temporary fieldwork station at the derelict barracks opposite to Sanhetun. From that day until August 1998, our chief task was to keep observing the movements of the white-headed langur troop living on the precipices of Feijichang Cave and obtain knowledge of their behavior modes and social system. Once as Pan Wenshi was tracking the flotsam on the water toward the entrance to the underground river under the cliff of FJC Cave, he came up with the idea that FJC Cave might be an entrance of the underground river in the early years (geological age).

In 2003, when our observation entered into the sixth year, we found that the members of female langurs remained the same in FJC Cave, while the resident alpha male was constantly being replaced. Despite that, the rhythm of life remains the same, breeding offspring, leaving night shelter in early mornings, and returning in late evenings (Fig. 1.2). We wondered whether they follow the same way of living generations after generations. How many lives did it cost for them to find out that they must climb up sheer precipices to find a cave as a shelter at night? Would they slip and fall off and become fossils eventually?

We decided that we should look into the extant white-headed langur's evolutionary history to have a deeper understanding of their ecology. A strong inner force drove us to enter the FJC Cave to have a look (Fig. 1.3).

1.1.2 In-Depth Research of Sanhe Cave

On December 21, 2003, with the assistance of ropes, fire-torches, and some simple digging tools, we climbed up the precipice and went inside the pitch-dark Feijichang Cave. Under the light of fire-torches, the molten cave at last unveiled itself. Water that flowed since ancient times had permeated the limestone mountain and shaped the molten cave. We conjectured that Feijichang Cave in ancient times was not situated on the mid-hill as it is today, but was a hole on the foot of the FJC hilly wetland which connected to a subterranean river. When it rained, the rainwater would flow from the higher grounds in the southwest, infiltrate the limestone mountain, and come out from divided outlets at the north and northeast, respectively. Geologically, this place is the work of the second phase of the Himalayan Orogeny in history; the Nongguan Mountains came to shape in that orogeny. The craggy peaks in the Nongguan Mountains have ever since kept developing, and the mountain mass is still getting taller today. The once subterranean river was gradually raised above ground. It lost its drainage function and became the duct-shaped cave as it appears today. Today, the once subterranean river inlet appears 70–80 m up on the precipice, which overlooks the hilly wetland.

On December 26, 2003, we installed electric lights at the digging area inside the cave (Fig. 1.4). It became clear and bright, which allowed us to look closely at the cave terrain and the traces left behind by flowing water centuries ago. As we searched along the flowing direction of the bygone subterranean river, we were imagining a scene that the flowing water was blocked by a precipice and was forced to turn around and create downward-drawing swirls on its way to the exit. We eventually found the reflux spot, around which we guessed the swirling water had brought about animal bones, sand, and gravel.

On March 2, 2004, at a conference in Beijing, when we presented several fossil teeth unearthed in Chongzuo Sanhe Cave (Fig. 1.5) to Academician Qiu Zhangxiang, the Director of Institute of Vertebrate Paleontology and Paleoanthropology, Chinese Academy of Sciences (CAS), he immediately wrote on a conference memo slip a list of scientific names (Fig. 1.6). When we saw names of gigantic mammals like *Gigantopithecus Blacki*, *Hylobates* sp., *Rhinoceros Sinensis*, *Ailuropoda Micrta*, *Macaca* sp., *Megalovies* sp., we were excited and realized that there must still be many precious pieces lying buried in the muddy sediments inside Chongzuo

Fig. 1.2 When the dusk is low, they come to this qintan (*Pteroceltis tatarinowii*) and form a harmonious whole; in the dim light, before I can count their number, they jump to the cliff one after another (Photo by Pan Wenshi)

Sanhe Cave. Thus, we decided to establish a joint team to start long-term in-depth research.

In May 2004, we started the in-depth research on the Quaternary mammalian fauna discovered in Chongzuo Sanhe Cave. The Nongguan Mountains where Chongzuo Sanhe Cave is situated has a northern tropical dry–wet alternating climate and exposed limestone geology with karst peak-cluster depression and peak forest valley topography. White-headed langurs live on deep-scarped precipices bordering the hilly wetlands and have their night shelters high up on those karst precipices. Figure 1.7 shows the external view of Chongzuo Sanhe Cave. The limestone hill where the Cave is located can be divided vertically into six horizontal layers. The first and bottom layer is 130 m above sea level and is still developing. The sixth layer at the top is at 270 m above sea level. The fifth layer is about 200 m above sea level, and it was mostly in this sediment layer inside the Cave that fossils of the "Chongzuo langur" and other sympatric animals were buried (Jin Changzhu et al. 2009).

The cave chamber at approximately 30 m inside the Sanhe Cave from the southwest entry was decided to be the site for digging. The chamber lays out roughly on a north-south axis (Fig. 1.8). The extant white-headed langurs live all year round on the nooks at the loft of the southwest entry to the chamber. The cross-section of the sediments in the digging chamber insides the Sanhe Cave reveals two major sedimentary layers, which can be divided top-down into seven sediment strata (Fig. 1.9).

Fig. 1.3 Prof. Pan Wenshi built a wood and bamboo plank route up the 70 m tall sheer precipice... He held a strong belief that there would be dead ancient animals buried in the sediments inside Feijichang Cave (Photo by Liang Zuhong)

Fig. 1.4 December 26, 2003. Prof. Pan Wenshi (leftmost) holding a fluorescent lamp and field assistants Lin Jinghua (second left), Cen Shidong (standing on the ladder), Nong Lihua (on top of Cen), and Xiao Lu (rightmost) were digging the sediments in the Cave for fossils (Photo by Qin Dagong)

1.2 *Gigantopithecus* Fauna and the Fossil of White-Headed Langur's Ancestor

Altogether, fossils of 82 mammalian species living in the middle phase of the Early Pleistocene Epoch comprising 10 orders, 28 families, and 66 genera had been unearthed by the Joint Research Team (Jin Changzhu et al. 2009). Since *Gigantopithecus* fossils were also discovered, we named the animals living in those years "Chongzuo Sanhe Cave *Gigantopithecus* Fauna."

To obtain evidence of the natural evolution of the white-headed langur, we must study the natural history of the fossil langur that appeared in Sanhe Cave, which lived in the middle phase of Early Pleistocene Epoch. We put together the seven *Gigantopithecus* faunas that lived in Chang Jiang (Yangtze River) drainage and Zhu Jiang (Pearl River) drainage to contrast their occurring epochal years and compare the zoogeographical attributes of the composing animals of the seven *Gigantopithecus* faunas, to find out the living conditions of the white-headed langur's ancestor.

1.2.1 Zoogeographical Traits of the Chongzuo Sanhe Cave *Gigantopithecus* Fauna

By comparing the occurring epochal years and the traits of the composing animals in the various *Gigantopithecus* faunas, we found that the ancestor species of white-headed langurs and the Sumatran stripped rabbit appeared only in Chongzuo Sanhe Cave *Gigantopithecus* fauna. Therefore, we surmised that those two species appeared at the south bank of Zuo Jiang River at a time of distinct geology in the Pleistocene Epoch, during which time ancient langurs began to evolve at the south bank of Zuo Jiang River.

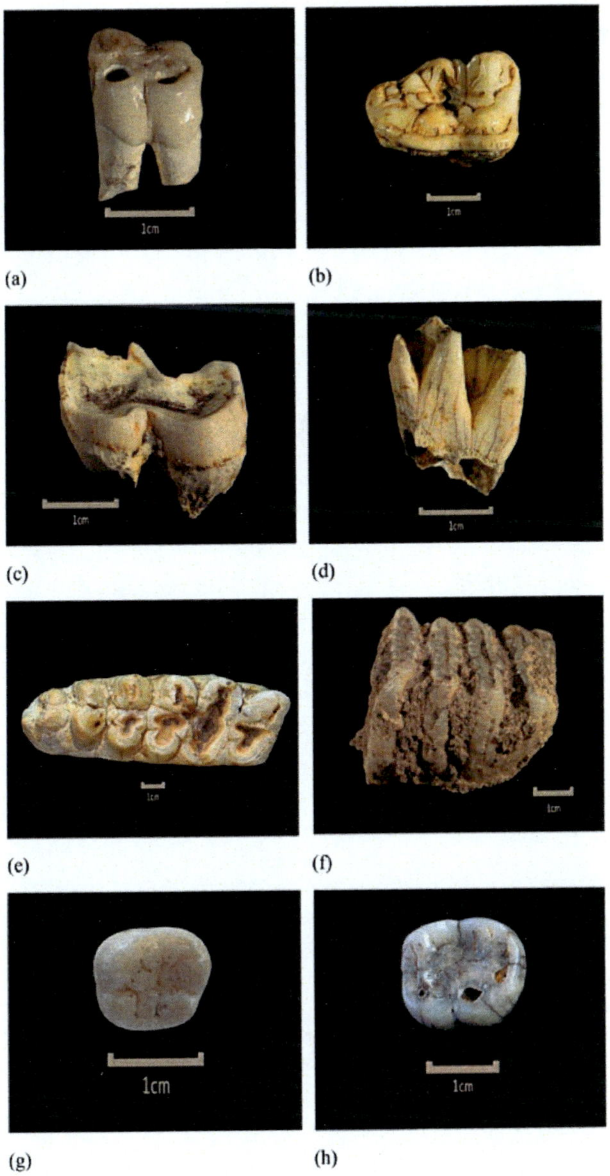

Fig. 1.5 Fossil teeth of some of the mammals among Chongzuo Sanhe Cave *Gigantopithecus* fauna: (**a**) Fossil tooth of *Gigantopithecus blacki*, (**b**) fossil tooth of *Ailuropoda milanoleuca wulingshanensis* (**c**) fossil tooth of *Megatapirus augustus*, (**d**) fossil tooth of *Rhinoceros sinensis*, (**e**) fossil tooth of *Sinomastodon yangziensis*, (**f**) fossil tooth of *Stegodon orientalis*, (**g**) fossil tooth of cf. *Hominidae*, (**h**) fossil tooth of *Pongo* sp

Zoogeographers divide the landmasses on the earth into several large zoogeographic regions reflecting the synergistic outcomes of historical geologic developments, which were different in each region, and the present-day ecologies of the regions. In drawing up the land animal zoogeographic regions, small-sized mammals were generally not considered as criteria, since their activities cover only little areas, and a small living environment would have satisfied their survival needs. Their appearances or disappearances do not reflect accurately on the entirety of general conditions of a large zoogeographic region. The criteria for drawing up the land animal zoogeographic regions therefore lay on the medium and large mammals and birds, for these animals have strong adaptation capabilities, and they disperse over broad areas. The gigantic animals buried underground since ancient years were the ancestors of the animals living on that piece of land today. If we would relate their remains to the distribution and living practice of relative extant species as well as trace the historic years they were living in, we may get to know the zoogeographical traits of regions in which they had lived.

In Table 1.1, we list the composing animals in the seven *Gigantopithecus* faunas which occurred at different places from the south of Chang Jiang to the upper drainages of Zhu Jiang and at different times between the early phase and middle phase of Early Pleistocene. The seven *Gigantopithecus* faunas in sequence of their occurrence geologic periods are Chongqing Wushan Longgupo *Gigantopithecus* fauna (Huang Wanpo 1991), Guangxi Liucheng *Gigantopithecus* fauna (Pei Wenzhong 1987; Han Defan 1987), Guangxi Baise Tiendong Mehui Cave *Gigantopithecus* fauna (Wang Wei et al. 2005), Hubei Jianshi Longgu Cave *Gigantopithecus* fauna (Zheng Shaohua 2004), Guangxi Chongzuo Sanhe Cave *Gigantopithecus* fauna (Jin Changzhu et al. 2009), Guangxi Daxinhei Cave *Gigantopithecus* fauna (Han Defan 1982), and Guangxi Bama Nongmoshan *Gigantopithecus* fauna (Zhang Yinyun et al. 1975).

The data in Table 1.1 shows that as a result of occurring at different geologic periods and at different geographical locations, the seven *Gigantopithecus* faunas had each revealed a different zoogeographic characteristic. The most notable differences are:

- Only Chongqing *G* fauna and Jianshi *G* fauna had Palaearctic animals in their faunas. Those two *Gigantopithecus* faunas were distributed at relatively northerly geographical locations. Northern species like *Hesperotherium* and *Leptobos* appeared in their faunas and lived among southern species, which results in particular north-south overlapping characteristics in the two *Gigantopithecus* faunas. Chongqing *G* fauna occurred in the early phase of the Early Pleistocene with Indo-Malaya species accounting for only 55.5% of the fauna, whereas Jianshi *G* fauna occurred at a slightly later period nearer to the middle phase of Early Pleistocene with more Indo-Malaya species making up 63.9% of the total.
- All the four *Gigantopithecus* faunas which occurred at different places in southwestern Guangxi, namely Tiendong *G* fauna, Chongzuo *G* fauna, Daxinhei *G* fauna, and Bama *G* fauna had no Palaearctic species. Insofar as discovered in southern China they were the

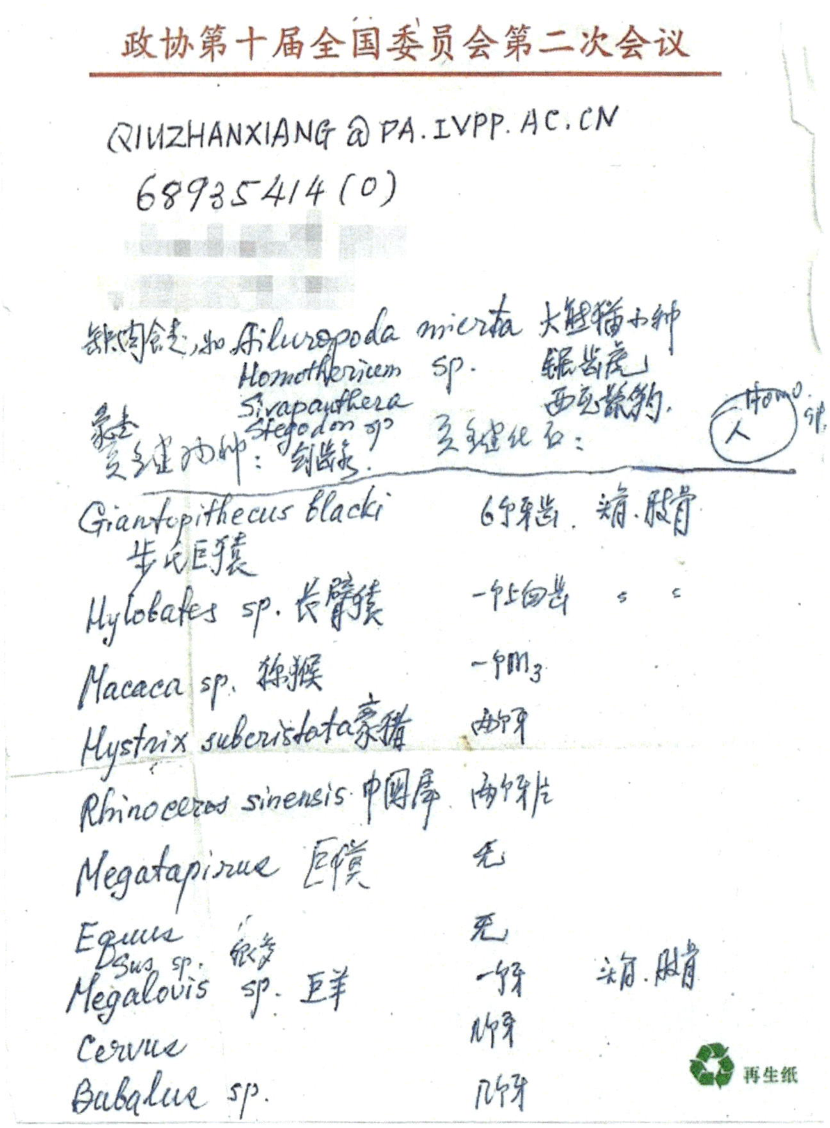
政协第十届全国委员会第二次会议

QIUZHANXIANG @ PA.IVPP.AC.CN
68935414 (0)

Ailuropoda microta 大熊猫小种
Homotherium sp. 锯齿虎
Sivapanthera 西瓦猎豹
Stegodon sp. 剑齿象
人 Homo sp.

Gigantopithecus blacki 步氏巨猿 6个牙齿
Hylobates sp. 长臂猿 一个上臼齿
Macaca sp. 猕猴 一个M3
Hystrix subcristata 豪猪 两个牙
Rhinoceros sinensis 中国犀 两个牙片
Megatapirus 巨貘 无
Equus 无
Sus sp. 猪
Megalovis sp. 巨羊 一个牙
Cervus 几个牙
Bubalus sp. 几个牙
再生纸

Fig. 1.6 Academician Qiu identified the fossil species and wrote their names on the conference memo slip

most typical tropic-subtropical forest mammalian faunas in the Quaternary Period; this could be a consequence of their southerly geographic locations. Indo-Malaya species took up as high percentages as 82.4–92.9% in their faunas.

- Chongzuo *G* fauna had one distinct feature that distinguished it from the other six *Gigantopithecus* faunas: it had the Chongzuo Langur, *Nesolagus*, *Melogale*, *Erictis*, and *Procynocephalus* sp. in its composition, and fossils unearthed in Sanhe Cave of these animals took up 14.7% of the big- and medium-sized fossil animals. Since those five species were not present in the other six *Gigantopithecus* faunas regardless of their occurring before or after Chongzuo *G* fauna, we therefore conjecture those five species had come to the south bank of Zuo Jiang River at an exceptional geologic period that happened in a limited time span.

1.2.2 Fossil Tooth of a Chongzuo Langur (*Trachypithecus Chongzuosis*)

The fossil mandibular third molar of an ancient langur (Fig. 1.10) unearthed from the sediments in Sanhe Cave is monumental in the origin and evolutionary history of white-headed langurs. We named its owner "Chongzuo langur" (*Trachypithecus Chongzuosis*). The Latin name begins with *Trachypithecus* which means it belongs in the continental type langur genus, the species name *Chongzuosis* means

Fig. 1.7 Elevation view of Chongzuo Sanhe Cave (Jin Changzhu et al. 2009)

that it is originated in Chongzuo. Although there was only one molar of ancient langurs discovered in the silted up muddy sand inside Chongzuo Sanhe Cave, it provides sufficient evidence that ancient langurs did once live in that area.

1.3 Origination and Evolution of Langur Species

Fossils buried in sediments provide the most straightforward evidence for researching the origin and evolution of an animal species. As there are yet very few fossil findings of ancient langurs, their originating place still cannot be ascertained. However, it is commonly accepted that the ancestor species of langurs came from Africa. So far, there is only one extinct langur species discovered in Asia and that is *Presbytis sivalensis* (Ayer 1948; Delson 1975), which appeared in the Indo-subcontinent at the end of Miocene epoch about 5.5–8 mya. It is conjectured that it could have emigrated there from Africa.

After the ancient langurs had come into Asia, they adapted themselves to the climatic radiations there and successfully established themselves in many climatic and vegetation zones in south and southeast Asia, as well as on the various islands there. It was estimated that the genus of ancient langurs (*Presbytis* genus) had diversified during the ages from around the end of the Pliocene Epoch until the early phase of the Pleistocene Epoch (3.6 mya to 2.18 mya). Speciation could have happened during the transition period from the late Early Pleistocene Epoch to the Mid-Pleistocene Epoch (1.9 mya to 0.8 mya). It then further diversified into various subspecies in Southeast Asia and the Malay Archipelagos (Zhou Mingzhen 1964) in around the rainforest periods, which happened after several glaciations in the Pleistocene Epoch. Langur species are therefore endemic to tropical Asia. It has adapted to competing for survival in the southern parts of tropical Asia and is a high-

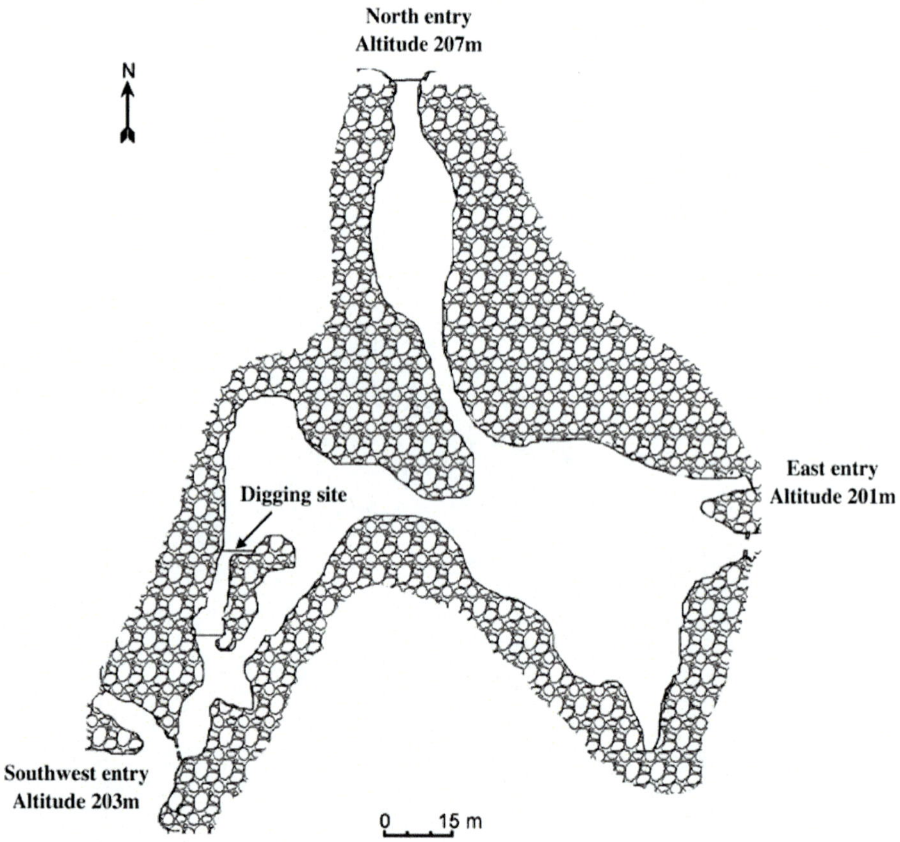

Fig. 1.8 Layout diagram of Chongzuo Sanhe Cave (Jin Changzhu et al. 2009)

order primate species, which only in the latest geologic period diversified and rapidly developed. Table 1.2 lists the langur fossils or DNA evidences found.

We are not going to discuss the origination and evolution of every langur species but shall focus on the questions "Where did the Chongzuo langur come from?" and "How did the white-headed langur come into existence?"

1.3.1 Ancestors of the White-Headed Langur and Their Emigration

Looking back at the occurring times of the fossil animals listed in Table 1.1, we come to note that Sundaland animals like the Chongzuo langur, *Nesolagus*, *Melogale*, etc. did not appear among faunas that occurred during the transition period from the early phase of Early Pleistocene to the middle phase of Early Pleistocene (2.58 mya to 1.6 mya), which are the Chongqing *G* fauna, Jianshi *G* fauna, Liucheng *G* fauna, and Tiendong *G* fauna. We conjecture it was due to the warm weather in the Northern Hemisphere during the one million years transiting from the early phase of Early Pleistocene to the middle phase of Early Pleistocene. In those years, the seas stayed at high levels and the West Pacific continental shelf was submerged below sea level. Isolated islands were formed, and there were no land bridges for animals to disperse northwards to China. As a result, Chongzuo langurs, *Nesolagus, and Melogale* did not appear in Chongqing *G* fauna, Jianshi *G* fauna, Liucheng *G* fauna, and Tiendong *G* fauna.

Also, it could be noted in the most recent faunas in Table 1.1, Daxinhei *G* fauna and Bama *G* fauna, there were also no Chongzuo Langurs, *Nesolagus*, *Melogale*, or other Sundaland animals among their composing animals. We conjecture that changes of the sea level were the factor behind this. The two faunas occurred during the period of transit from the late phase of Early Pleistocene to Mid-Pleistocene (1.2 mya to 0.8 mya), by then, the first glaciations had ended and the Northern Hemisphere entered into a warm interglacial period. Glaciers melted, the sea had returned to high level and the West Pacific continental shelf was once again submerged. As a result, Sundaland animals once again had

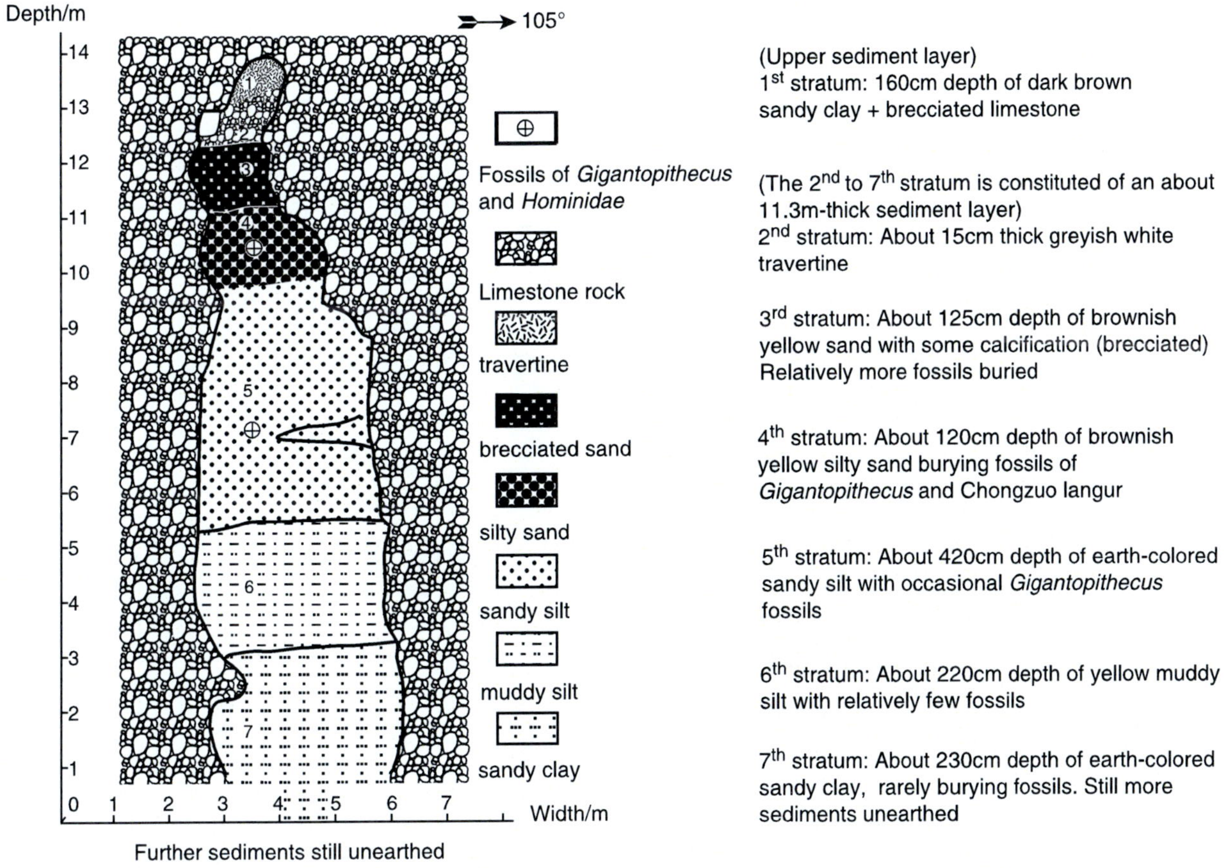

Fig. 1.9 Cross-section diagram of Sanhe Cave digging site (Jin Changzhu et al. 2009)

no land bridge to disperse northwards. Of all the seven *G* faunas, only Chongzuo *G* fauna had animals like Chongzuo langurs, *Nesolagus*, *Melogale*, etc. that came from Java Island, Borneo, and Sumatra. And the time period that those animals appeared in the south bank of Zuo Jiang River was in the middle phase of Early Pleistocene Epoch (1.6 mya to 1.2 mya), which coincided with the first extreme glaciations in Pleistocene Epoch. That easily leads us to associate the origination of the Chongzuo langur and the way it had arrived at Zuo Jiang River basin with the descended sea level, which exposed the West Pacific continental shelf chaining up the islands with land passages during the extreme glaciations.

It would be a significant event in the evolutionary history of whatever species to depart its ancestral homeland for long distance emigration. The lineal ancestors of white-headed langurs should be a species of ancient langur living in Java Island. Jablonski et al. in their research report (Jablonski and Tyler 1999) supposed that a fossil langur species, which lived 1.9 mya in central Java Island, was very much likened to *Trachypithecus* today. We find that the silvered langur (*T. a. auratus*) distributing over Java Island has its body size, physique, and hair color of its pups very similar to those of white-headed langurs living in Nongguan Mountains. Hence we conjecture that the ancient Javan Langurs (*Trachypithecus* sp.) living in Java Island in the early phase of Early Pleistocene Epoch could very possibly be the lineal ancestors of Chongzuo langurs. Questions then arise, how did that long distance emigration come to happen? How did they make the long distance from their ancestral homeland to arrive at the banks of Zuo Jiang River? We believe there were three interrelated factors:

1.3.1.1 The Particular Period in History

The particular period in history was the most critical one among the three factors. Chongzuo *G* fauna with Chongzuo langurs among its composition occurred from 1.6 mya to 1.2 mya and the median, i.e., 1.4 mya, coincided with the peak year of the first extreme glaciations in the Quaternary Period. The ancient Javan langurs would have dispersed

Table 1.1 Comparing the zoogeographic attributes of the chief composing animals in the seven *Gigantopithecus* faunas respectively occurred in southern China in Early Pleistocene Epoch

No.	Animal found in the *Gigantopithecus* fauna	Zoogeographic attribute	Early phase to middle phase of Early Pleistocene (2.6 mya to 1.6 mya)				Middle phase of Early Pleistocene (1.6 mya to 1.2 mya)	Middle phase to late phase of Early Pleistocene (1.2–0.7 mya)	
		Palaearctic	Chongqing *G* fauna	Jianshi *G* fauna	Liucheng *G* fauna	Tiendong *G* fauna	Chongzuo *G* fauna	Daxinhei *G* fauna	Bama *G* fauna
1	中国黄昏兽 *Hesperotherium sinensis*		◎ ★	◎					
2	丽牛 Leptobos sp.			◎ ★					
3	化石马 *Equus* sp.		◎	◎	◎				
4	更新猎豹 *Sivapanthera pleistocaenicus*		◎ ★	◎ ★	◎ ★				
5	巨额剑齿虎 *Megantereon* sp.			◎					
	Number of Palaearctic species in the respective *Gigantopithecus* faunas		3	5	2	0	0	0	0
	Percentage of Palaearctic species respective to the *Gigantopithecus* faunas		8.6	13.9	5.9	0.0	0.0	0.0	0.0
6	崇左叶猴 *Trachypithecus Chongzuosis*	Indo-Malaya (also known as Oriental)					◎		
7	山獾 *Melogale* sp.						◎		
8	苏口答腊兔 *Nesolagus* sp.						◎		
9	巨貘 *Megatapirus augustus*						◎	◎ ★	
10	古豺 *C. antiquus*						◎ ★	◎ ★	◎
11	山原貘 *Tapirus sanyuanensis*		◎ ★	◎ ★		◎ ★			
12	步氏巨猿 *Gigantopithecus blacki*		◎ ★	◎ ★	◎ ★	◎ ★	◎ ★	◎ ★	◎ ★
13	似人似猿 *cf. Hominidae*		□	□	□	□	□		
14	猕猴 *Macaca* sp.		◎	◎	◎	◎	◎	◎	◎
15	疣猴 *Colobinae indet*		◎						
16	猩猩 *Pongo* sp.				◎	◎	◎	◎	◎
17	长臂猿 *Hylobates* sp.					◎	◎	◎	◎
18	金丝猴 *Rhinopithecus* sp.			◎		◎	◎	◎	
19	原狒*Procynocephalus* sp.						◎		
20	小种大熊猫 *Ailuropoda microta*		◎ ★		◎ ★	◎ ★			
21	武陵山大熊猫 *Ailuropoda milanoleuca wulingshanensis*			◎ ★			◎ ★		
22	巴氏大熊猫 *A.m. baconi*							◎ ★	◎ ★
23	拟豺 *Cuon dubius*		◎ ★	◎ ★	◎ ★	◎ ★			
24	爪哇豺 *C. javanicus*							◎	◎
25	中国貘 *T. sinensis*			◎ ★	◎		◎ ★	◎	◎
26	貘 *Tapirus* sp.								◎
27	最后双齿尖河猪 *Dicoryphochoerus ultimus*		◎ ★		◎ ★		◎ ★	◎ ★	
28	小猪 *Sus xiaozhu*		◎ ★	◎ ★	◎ ★	◎ ★	◎ ★	◎	◎ ★

(continued)

Table 1.1 (continued)

No.	Animal found in the *Gigantopithecus* fauna	Zoogeographic attribute	Early phase to middle phase of Early Pleistocene (2.6 mya to 1.6 mya)				Middle phase of Early Pleistocene (1.6 mya to 1.2 mya)	Middle phase to late phase of Early Pleistocene (1.2–0.7 mya)	
29	裴氏猪 *S. peii*		◎★	◎★	◎★	◎★	◎★	◎	◎
30	嵌齿象 *Gomphotherium serridentoides*				◎★				
31	乳齿象 *Sinomastodon* sp.			◎					
32	扬子中华乳齿象 *Sinomastodon yangziensis*		◎★	◎	◎★	◎★	◎★		
33	先东方剑齿象*Stegodon preorientalis*		◎★	◎★	◎★	◎★	◎★	◎	◎
34	中国犀 *Rhinoceros sinensis*		◎★	◎★	◎★	◎	◎★	◎★	◎★
35	犀 *Rhinoceros* sp.								◎
36	麂 *Muntiacus* sp.			◎				◎	
37	湖麂 *Muntiacus lacustris*		◎★	◎	◎★	◎	◎★	◎	
38	凤岐祖鹿 *Cervavitus fenqii*		◎	◎★	◎★	◎★	◎★	◎	
39	云南水鹿 *Cervus (Rusa) yunnanensis*		◎	◎★	◎★	◎★	◎★	◎	
40	丘齿鼷鹿 *Docabune liuchengsis*				◎★				
41	广西巨羊 *Megalovis guangxiensis*		◎★		◎★	◎★	◎★	◎★	
42	山羊 *Caprinae*				◎		◎★		
43	建始鬣羚 *Capricornis jianshiensis*			◎★	◎★				
44	裴氏转角羚 *Spirocerus peii*		◎★						
45	楬牛 *Bibos* sp.		◎		◎		◎	◎	
46	羚牛 Budorcas sp.		◎★	◎★	◎★			◎★	
47	野水牛 Bovinae indet							◎★	
48	大灵猫 *Viverra* sp.			◎	◎		◎★		
49	小灵猫 *Vivericula* sp.							◎	
50	灵狸 *Prionodon* sp.			◎★					
51	果子狸 *Paguma lavata*				◎★			◎★	
52	南方猪獾 *Arctonyx collaris*			◎	◎★	◎★	◎★	◎★	
	Number of Indo-Malaya species in the respective *Gigantopithecus* faunas		20	23	26	19	28	26	14
	Percentage of Indo-Malaya species respective to the *Gigantopithecus* faunas		57.1	63.9	76.5	82.6	82.4	92.9	82.4
53	硕貂鼬 *Erictis* sp.	Widely dispersed					◎		
54	似中国黑熊*Ursus cf. thibetanus*		◎★	◎★	◎★	◎★	◎★	◎★	◎
55	熊 *Ursus* sp.			◎					◎
56	似中国貉 *Nyctereutes cf. sinensis*		◎★						
57	桑氏硕鬣狗 *Pachycrocuta perrieri*		◎★						

(continued)

Table 1.1 (continued)

No.	Animal found in the *Gigantopithecus* fauna	Zoogeographic attribute	Early phase to middle phase of Early Pleistocene (2.6 mya to 1.6 mya)				Middle phase of Early Pleistocene (1.6 mya to 1.2 mya)	Middle phase to late phase of Early Pleistocene (1.2–0.7 mya)	
58	硕鬣狗 *P. licentui*			◎	◎				
59	鹿 *Cervus* sp.		◎					◎	
60	麝 Moshus moschiferus plicodom		◎ ★						
61	豹 *Panthera pardus*		◎ ★	◎ ★	◎ ★	◎	◎ ★		
62	似剑齿虎 *Homotherium cf. palanderi*		◎ ★						
63	剑齿虎 *Homotherium* sp.			◎					
64	德氏猫 *Felis teilhardi*		◎ ★		◎ ★	◎ ★	◎ ★		
65	野猫 *Felis* sp.		◎	◎	◎		◎		◎
66	小野猫 *F. microtis*		◎ ★						
67	貂 *Martes* sp.			◎		◎	◎		
68	狗獾 *Meles cf. M. chiai*		◎ ★						
69	猪獾 *Arctonyx minor*		◎ ★		◎ ★				
70	水獭 *Lutra* sp.			◎ ★					
	Number of widely dispersed species in the respective *Gigantopithecus* faunas		12	8	6	4	6	2	3
	Percentage of widely dispersed species respective to the *Gigantopithecus* faunas:		34.3	22.2	17.4	17.4	17.6	7.1	17.6

Note: ◎ of the same genus, ★ of the same species, □ Hominoidea
Designations of the seven *Gigantopithecus* faunas are shortened as below:
Chongqing Wushan Longgupo *Gigantopithecus* fauna: Chongqing *G* fauna
Guangxi Liucheng *Gigantopithecus* fauna: Liucheng *G* fauna
Guangxi Baise Tiendong Mehui Cave *Gigantopithecus* fauna: Tiendong *G* fauna
Hubei Jianshi Longgu Cave *Gigantopithecus* fauna: Jianshi *G* fauna
Guangxi Chongzuo Sanhe Cave *Gigantopithecus* fauna: Chongzuo *G* fauna
Guangxi Daxinhei Cave *Gigantopithecus* fauna: Daxinhei *G* fauna
Guangxi Bama Nongmoshan *Gigantopithecus* fauna: Bama *G* fauna

northwards from Java Island in those years. That particular glaciation period lasted about 100,000 years, during which, more than a quarter of the earth's surface on the northern part of Northern Hemisphere was covered by ice and snow (Wenshi et al. 1988). The extreme cold had enormously impacted the distribution of animals and vegetation in the Northern Hemisphere, compelling the animals in the north to migrate southwards. However, the extreme cold impact on the south was minimal, particularly in the tropical climate regions. Rain water that got frozen on land by glaciations in the Northern Hemisphere caused the sea level to drop, and the continental shelf was thus exposed above sea becoming land passages for animals in the south to disperse northward.

1.3.1.2 The Particular Geographical Location

Research into the last glaciations in the Quaternary Period had revealed that in the coldest period (18,000 years ago to 11,000 years ago), the sea level in the West Pacific Ocean had descended by 165 m, desiccating the shallow seas of Beibu Gulf, Bay of Siam, and the seas around Sumatra and Java Island, exposing the continental shelf above the sea. Vast expanses of seashores surrounding the Sundaland Archipelagos, Kalimantan, Sumatra, and Java Island became dry land and linked up with the coastal lowlands (formerly the continental shelf) around the Malayan Peninsula, Indo-China and Beibu Gulf in Guangxi (Ying 2012). Practically it would take only 100–200 years for the exposed continental shelf to become natural habitats clustered with

Fig. 1.10 We say white-headed langur is a species endemic to Chongzuo, because its ancestors came into existence from evolution at Chongzuo, and the descendants have been until the present living only in Chongzuo and no other places in the world (The mandibular third molar (M_3) fossil of a Chongzuo Langur)

vegetation and forests, providing a very nice green passage for animals to disperse or emigrate from the south to the north. The interglacial period after the first extreme glaciations lasted no less than 100,000 years. Taking into account that it would take 5 years to produce a new generation of langur, and the increasing number of langurs from every new generation would bring up the population density and induce a tendency toward dispersing to distant land. In addition, 100,000 years would have given ancient Javan langurs 20,000 attempts to emigrate.

1.3.1.3 There Had to Be a Relating Ancestor Species

About 3.6–3.4 mya, the sea level had once descended 50 m, and the ancestor species of langurs, *Presbytis sivalensis,* took the opportunity and dispersed southwards along the Indian Ocean coast. About 2.8 mya, the earth had come to a severe cold glaciations period, during which the sea level had descended more than 100 m, and the ancestor species *Presbytis sivalensis* in the Indian Ocean coastland further dispersed to the ancient landmass of Sundaland. Until 1.9 mya (Early Pleistocene Epoch), *Trachypithecus* sp. came into being in central Java Island. It had a larger body size and was already morphologically very much alike *Trachypithecus a. auratus*, which happened later in the Mid-Pleistocene Epoch. We conjecture that *Trachypithecus* sp. had multiplied in numbers in the tropical paradise of Java Island by the time the first extreme glaciations in the Quaternary Period set in. They then discovered the new coastal forest belt springing up on the former continental shelf between the former islands with abundant food, water, resources, and few competitor species. It had given the ancient Javan langurs (*Trachypithecus* sp.) plenty of room for rapid population growth. With the increase in population, ancient Javan langurs dispersed north along the coast, and the belt of newly grown tropical rainforests became the speedy highway for their northward emigration. We suppose it must be much easier to disperse along the coastal forest belt than to disperse through twists and turns into the Java Island inland. Yet, some of them had dispersed inland, adapted to the new environment there, and developed into other species of *Trachypithecus.* We speculate that a batch had arrived at Ha Long Bay of Vietnam, in Beibu Gulf, and adapted to living there on the karst hills jutting out of the sea. And we may think of the extant golden-headed langurs (*Trachypithecus piliocephalus*) in Cat Ba Island to be a population of ancient Javan langurs that stayed behind on their emigration road to the north. Morphologically, white-headed langurs and golden-headed langurs are very much alike. Even the DNA properties of the two are very approximate to each other, to the extent that some people consider the white-headed langur to be a subspecies of the golden-headed langur. Since in the days of the Early Pleistocene Epoch, there were karst hills in rolling columns covered with luxuriant vegetation on Cat Ba Island, and before long, the island became the last stop for the Javan Langurs on their emigration en route to Guangxi, China. The Javan Langurs that left Cat Ba Island moved along the meandering Shiwanda Mountain, passing through the low hills and river valleys in Sifangling and without much difficulty reached at the south bank of Zuo Jiang River. They might have first settled in the areas around Nonglin Mountains and Jiuwanshan, and then dispersed gradually to the direction of Nongdou Mountains and Longfeng Mountains, as well as along the banks of Zuo Jiang River toward Bazhiqi Mountains and Nongrui Mountains. It appears that the white-headed langurs in the Nongguan Mountains had been descended from a few family troops coming from Nonglin Mountains and Jiuwanshan in a relatively recent time in history. As there are no distinct genetic variations among the populations of extant white-headed langurs in Nongguan Mountains, the mitochondria DNA diversity between them is also very low.

1.3.2 Allied Species

Nesolagus has been living in Sumatra Island since ancient times. Yet fossils unearthed from Chongzuo Sanhe Cave

Table 1.2 Fossil evidence or DNA evidence relative to langur species (Ayer 1948; Delson 1975; Harrison et al. 2006; Changzhu et al. 2009)

Occurring period (mya)	Geographic location	Evidence	Species name
5.5–8	Indo-subcontinent	Fossil mandible (complete with third molar)	*Prebytis sivalensis*, the initial langur species
2.8	Sumatra	Fossil and DNA evidences	*Trachpithecus* sp., the earliest diversified continental langur genus
1.9	Central Java	Fossil and DNA evidences	*Trachpithecus* sp. (continental langur)
1.4	Chongzuo, Guangxi, China	Fossil mandibular third mola	*Trachypithecus Chongzuosis* (Chongzuo langur)
Extant	Chongzuo, Guangxi, China	Fossil mandibular third molar + DNA evidence	*Trachypithecus leucocephalus Tan* (white-headed langur)

Fig. 1.11 Wild Sumatran stripped rabbit has lustrous fur on its body

Fig. 1.12 Distribution of present day Sumatran stripped rabbits (red shaded area) (Drawn by Liu Chuan)

have indicated that it appeared alongside white-headed langurs in Chongzuo *G* fauna while they were both absent from the other six *G* faunas. That leads us to surmise that the two species had at the same time taken the opportunity allowed in the first glaciations in the Quaternary Period to disperse from the south to the north. Studying the present distributions (Fig. 1.11) of *Nesolagus* (Fig. 1.12) and associating with the conditions of its appearance in Early Pleistocene Epoch in Zuo Jiang River Basin (Changzhu et al. 2010), we conjecture that it had followed the same migration route as that of Javan langurs. They both had taken the shortest route, i.e., the route along the coastline, to arrive at the south bank of Zuo Jiang River.

References

Ayer AA. The anatomy of *Semnopithecus entellus*. Madras: The Indian Publishing House Ltd.; 1948. p. 182.

Changzhu J, Dagong Q, Wenshi P, et al. Gigantopithecus fauna discovered in Chongzuo Cave and Characteristics of the Fauna (广西崇左三合大洞新发现的巨猿动物群及其性质). Science China Press. 2009;6:765–73.

Changzhu J, Yukimitsu T, Yuan W, et al. The first discoveries of fossil Nesolagus Leporidae in Southeast Asia (苏□答腊兔化石种类在东南亚地区的首次发现). Sci China Press Earth Sci. 2010;8:1014–21.

Defan H. Fossils of mammals from Daxinhei Cave, Guangxi (广西大新黑洞哺乳动物化石). Vertebr Paleontol Acad J. 1982;1982:1.

Defan H. Fossils of Artiodactyla from Guangxi Liucheng Gigantopithecus Cave (广西柳城巨猿洞偶蹄目化石). In: Collected articles of Institute of Vertebrate Paleontology and Paleoanthropology CAS, vol. 18. Beijing: Science Press; 1987. p. 135–208.

Delson E. Evolutionary history of the Cercopithecidae. Contrib Primatol. 1975;5:167–217.

Harrison T, Krigbaum J, Manser J. Primate biogeography and ecology on the Sunda Shelf Islands: a paleontological and zooarchaeological perspective. Primate biogeography. New York, NY: Springer US; 2006. p. 331–72.

Jablonski NG, Tyler DE. *Trachypithecus auratus sangiranensis*, A new fossil monkey from Sangiran, Central Java, Indonesia. Int J Primatol. 1999;20(3):319–26.

Mingzhen Z. Zoogeography transformations in China in the quaternary period (中国第四纪动物区系的演变). Chin J Zool. 1964;6:274–8.

Shaohua Z. Relics of Homo erectus Jianshi (建始人遗址). Beijing: Science Press; 2004.

Wang Wei J, Potts R, Yamei H, et al. Human fossils of the early-pleistocene discovered in Mehui Cave in Bubing Basin, Guangxi

(广西布兵盆地么会洞新发现的早更新世人类化石). Sci China Press. 2005;50(17):1879–83.
Wanpo H. The Wushan Ape-man Relic (巫山猿人遗址). Beijing: Ocean Press; 1991.
Wenshi P, Zhengsheng G, Zhi L. Natural Refuge of Giant Panda in Qinling Mountains (秦岭大熊猫的自然庇护所). Beijing: Peking University Press; 1988.
Wenzhong P. Fossils of Carnivora, Proboscidea and Rodentia in Guangxi Liucheng Gigantopithecus Cave and other caves (广西柳城巨猿洞及其他山洞之食肉目、长鼻目和啮齿目化石). In: Collected articles of Institute of Vertebrate Paleontology and Paleoanthropology CAS, vol. 18. Beijing: Science Press; 1987. p. 5–134.
Ying W. Regional oceanography of Chinese seas: marine geomorphology (中国区域海洋学:海洋地貌学). Beijing: Ocean Press; 2012.
Yinyun Z, Linghong W, Mingren D, et al. Fossil teeth discover in Bama, Guangxi (广西巴马发现的巨猿牙齿化石). Vertebr Paleontol Acad J. 1975;13:3.

2 Evolution of the White-Headed Langur

Abstract

Combining lines of evidence of paleontology and zoogeography of Asian langurs, we here offer a hypothesis of ancestral origin and evolution of China's white-headed langur, divided into four phases. In the initial phase, during the extreme glaciation period in the Early Pleistocene Epoch, some ancient Javan langurs took the opportunity to emigrate northwards and entered the small land mass sandwiched between Zuo Jiang and Ming Jiang Rivers in Southwest China. Barricaded by the rivers, the Javan langur ancestors evolved independently into Chongzuo langurs (*Trachypithecus chongzuosis*). In the second phase, under the pressure of predacious animals, the Chongzuo langurs were forced to abandon their exclusive arboreal life and find natural refuges on the numerous precipices of Guangxi's karst topography. In the third phase, on that extensive infertile land with a scant human population, the early white-headed langurs (*T. leucocephalus*) found sufficient suitable habitat to further evolve locally, adapting to coexist with humans. In a short, anthropogenic phase 4, from the 1950s to the year 2000, trees were brutally chopped and forests were recklessly exploited, which devastated the white-headed langurs' natural habitat. Along with the rampant poaching activities, white-headed langurs were driven to an evolutionary bottleneck, at the brink of extinction.

Keywords

Paleontology · Zoogeography · Pleistocene · *Trachypithecus* spp. · Javan langur · Chongzuo langur · White-headed langur · Guangxi

We shall divide the white-headed langur's evolutionary history into four phases.

2.1 Phase 1: Origination from the Ancient Javan Langur

Phase 1 lasted about 100,000 years, from 1.45 mya to 1.35 mya.

The first extreme glaciations in the Pleistocene Epoch lasted for a period of about 100,000 years from 1.45 mya till 1.35 mya. As was discussed in Chap. 1, during that period, some ancient Javan langurs dispersed north and came to the small land area sandwiched between Zuo Jiang River and Ming Jiang River in Guangxi. This period ended with an interglacial period, when the Pacific Ocean sea level rose and submerged the continental shelf again. Those ancient Javan langurs hence were detached from their ancestors far in the south. They were barricaded by the two rivers from mingling with other langur populations (if there were any). We believe that those ancient Javan langurs had successfully established themselves at the south bank of Zuo Jiang River and gradually evolved into Chongzuo langurs.

2.2 Phase 2: "Garden of Eden" Period for the Chongzuo Langur

Phase 2 lasted about 1.35 million years, from 1.35 mya to 2000 years ago. This was the "Garden of Eden" period for Chongzuo langurs.

From perspectives of global biodiversity evolution, paleontologists termed the period when biodiversity developed without restraint as the "Garden of Eden period." It was postulated that the "Garden of Eden period" began 450 mya when the first influx of organisms appeared at the beginning of the Phanerozoic Eon and ended about 50,000–10,000 years ago. It was at the time around the advent of the Upper Paleolithic Age and the Neolithic Age in human history. Human species in the Paleolithic Age were

W. Pan, *The Population Ecology of White-Headed Langur*, https://doi.org/10.1007/978-981-33-4118-0_2

still at a relatively primitive state of development with small populations and were using only crude rudimentary tools. Hence, human activities in this period posed no threat to the survival of animals.

Zuo Jiang River Basin was a part of tropical Asia zoogeography. The hominids living and evolving in the Zuo Jiang River drainage areas were indigenous to the tropical zoogeography in the Zuo Jiang River Basin. The *early Homo sapiens* could have appeared in southern Zuo Jiang River at about the same time as the ancient Javan langurs. The evolutionary histories of the two should have involved with each other. Still, human species was in the *Homo erectus* or *early Homo sapiens* evolutionary stage, and the population was small and the tools were rudimentary. They were then hunter-gatherers in the south of Zuo Jiang River and were constantly in alert of being preyed upon by predacious animals. Therefore, early humans were not endangering Chongzuo langurs and the ecosystem as a whole. Even when *early Homo sapiens* started using fire, the damage to nature was minimal and could be restored rapidly. Those were the "Garden of Eden" days for Chongzuo langurs, a time when they were developing without hinderance. The only strain came from finding for themselves ecological niche in the competition with other species, from competition for food and from the danger of falling into prey. These are the issues we are going to discuss in this chapter.

2.2.1 Chongzuo Langur's Habitat in the "Garden of Eden"

When the ancient Javan langurs left their homeland and arrived at the Zuo Jiang River Basin, they are faced with the following problems: First, they must find an endurable habitat. The climate of the place (e.g., the varying range of temperature and humidity) must be bearable for them to adapt. Second, the new environment must possess sufficient potential for them to live and reproduce. Third, as newcomers, ancient Javan langurs should be able to compete with other species that already exist there and be able to hide away from predators.

2.2.1.1 Climate of Zuo Jiang River Basin in the "Garden of Eden" Period

Geologic evidence has shown that climate changes in the Pleistocene Epoch had an enormous effect on the present zoogeography of Northern Hemisphere land animals. The overall climate in the Quaternary Period gradually turned colder as time went by. Glaciations in the Quaternary period started about 2 mya, and the last glacial period had ended just 10,000 years ago. During this period, the earth's surface experienced 16–22 fluctuations between cold and warm temperatures as glaciations and interglacial periods occurred right after each other. The northern parts of China encountered four striking temperature drops over extensive areas in the Pleistocene Epoch (Quaternary), which caused the fauna distributions to expand and contract over high and low latitudes. The ancestor species of the white-headed langur appeared and evolved in the first extreme glaciations in the Quaternary Period. They emigrated from low latitude areas (ancient Sundaland at the southern tip of Asia tropic) northwards to high latitude areas (Zuo Jiang River drainage at the northernmost of Asia tropic).

Evidence of changes in climate can be obtained by studying the morphologic characteristics of fossil pollens or fossil spores from the form of their stomata, swellings, and epidermal patterns. Under favorable natural conditions, exines of pollens and spores are resistant to high acids, strong bases, intense heat, and high pressures and could be preserved for thousands of years without disintegration. Palynology is therefore able to provide information on climates and floras of past geologic eras. Relatively little research had been done on the climate of southern China in the Quaternary Period, but we found a research report by Wang Kaifa et al. (Table 2.1) analyzing the assemblage of Quaternary Period spores found in the coastal areas along the Beibu Gulf of China. It provided us great insight in our research of the region's climate in ancient geologic eras. Pollen and spores can be dispersed by air and will sink when fallen on water. Without disturbances, they will be deposited in layers in successive sequence over the years. The analysis of the spore assemblage had shown that the region had a tropical and subtropical climate from the Early Pleistocene through the Late Pleistocene to the Holocene Epoch.

Table 2.1 tabulates the analysis of the spore assemblage of ancient flora along the coast of Beibu Gulf from the Pleistocene Epoch to the Holocene Epoch. The results reflect the climates relative to the geologic eras (Wang Kaifa et al. 1977).

In the Minguo 26th year edition of Chongshan County Archive (2011), it was recorded that "Taiping (present day *Jiangzhou District* of Chongzuo City) is situated at a blistering, hot, desolate and remote area where the climate is entirely different from that of central China. The weather starts getting warm from the lunar months of January and February, and in March, it gets as hot as mid-summer in central China. From April to June, the weather is so sweltering that it scalds. It gets even more humid in autumn and there is not even a breeze of cool air. It is mostly sunny and warm in winter and there is never frost or snow withering the vegetation. Yet the wind and rain over the nights bring cold temperatures... ." In the book *Taiping Prefecture Archive*, it was recorded that "the weather is summer-like for all four seasons, but once it rains, the temperature drops to that of winter. Rainfall will turn a summer day into winter, and sunny skies will turn a winter day into summer." The two

Table 2.1 The tropical Asia characteristics of the ancient flora and ancient climate as reflected by the spore compositions

Epoch		Major floras	Vegetation cover	Climate
Early Pleistocene[a]	Early phase	Composed chiefly of *Moraceae, Lauraceae, Proteaceae, Sapindaceae*	Tropical and subtropical forest	Mildly hot and moist
	Late phase	Composed chiefly of *Poaceae, Artemisia, Hicriopteris*, with among them *Proteaceae* and *Sapindaceae*	Tropical and subtropical savanna	Very hot and dry
Mid-Pleistocene	Early phase	Composed chiefly of *Sapindaceae, Cycas, Annonaceae, Arecaceae*, and *Symplocaceae*	Tropical and subtropical forest	Very hot and humid
	Late phase	Composed chiefly of *Poaceae, Artemisia*, Myrtaceae, with among them *Moraceae* and *Sapindaceae*	Tropical and subtropical savanna	Very hot and dry
Late Pleistocene	Early phase	Composed chiefly of *Moraceae, Arecaceae, Cycas, Rutaceae*, and *Annonaceae*	Tropical and subtropical forest	Very hot and humid
	Late phase	Composed chiefly of *Poaceae* and *Hicriopteris*, with among them *Sapindaceae*	Tropical and subtropical savanna	Very hot and dry
Holocene[b]		Composed chiefly of *Poaceae, Hicriopteris*, and *Myrtaceae*, with among them *Sapindaceae*	Tropical and subtropical savanna	Very hot and dry

[a]In the middle phase of Early Pleistocene Epoch, about 1.4 mya, ancient Javan langurs came to the south bank of Zuo Jiang River
[b]There were relatively few specimens of the Holocene Epoch analyzed, and they covered only a part of the period, so they do not stand for the whole Holocene period

archives, written at different times, had described exactly the climate of the Zuo Jiang River basin. The basin is situated at the low latitudes of the Northern Hemisphere, south of the Tropic of Cancer where the climate is warm and moist. The climate of the basin has no essential differences to that of the southern part of present-day tropical Asia.

Since October 2006, a team lead by Prof. Li Xiangdong from the Forestry College of Guangxi University has conducted in-depth research on the plants and vegetation of the habitats of white-headed langurs. Their statistics analysis showed that 24 out of 98 families of plants, accounting for 25.5% of the total, were tropical species. Thirty-two families, accounting for 34.0% of the total, belonged to tropical-subtropical distributing species. The data shows that tropical plant families and tropical and subtropical plant families account for 59.5% of all plant families in the habitat of white-headed langurs. This community structure is consistent with the geographical location of the white-headed langur distribution in the southwestern part of Guangxi, the northern edge of tropical Asia. The team listed the plant composition of the genus *Polygonaceae*, *Sapindus*, *Cycadaceae*, *Annona*, *Palmaceae*, *Moraceae*, etc. in the transition period from the Early Pleistocene to the Middle Pleistocene, indicating that from the time of the Chongzuo langur to the white-headed langur, the habitat climate has always been a tropical-subtropical humid monsoon type, with a tropical type.

The ancient Javan langurs may have dispersed northwards in search of a more comfortable place, when they arrived at the southern bank of the Zuo Jiang River and could not cross it, they found that the climate here was not worse than the place they left, which also fit their living standards.

2.2.1.2 Food Resources in the "Garden of Eden"

The ancient Javan langurs went into the evolutionary course of developing a new habitat when they dispersed to the small area surrounded by Zuo Jiang River and Ming Jiang River. Many other herbivores live there at the same time. The competition between them must be fierce.

In the collection of fossils unearthed in Chongzuo Sanhe Cave, we found 82 mammalian species coexisting with the Chongzuo langur. Among them, 12 were large- to medium-sized herbivores. They were competitors of the Chongzuo langurs, as their ecological niches in foraging were overlapping or partially overlapping. Nevertheless, nature would always justly reconcile competitions between species and keep developments balanced through compromises.

By comparing the geographical distributions and habitats of the various genera of fossil mammals with those of corresponding extant species (Table 2.2), we can get to know the sort of natural environments in the south of Zuo Jiang River at the time of the Chongzuo langur, as well as the foraging niches of the competing species.

We refer to the research approach by Zheng Shaohua (2004), and most data on extant animals are with references to published materials from *Distribution of Terrestrial Animals in China* (陆生中国哺乳动物分布) (Nowak and Paradiso 1983; Zhang Rongzu et al. 1997) and from *Picture Guide of World Animals* (世界动物图鉴) (Zhu Yaoyi 1995).

The data in Table 2.2 has revealed that the herbivorous mammals in Chongzuo *Gigantopithecus* fauna included species of three basic types of habitats, namely the tropical rainforest habitat, the tropic-subtropic forest habitat, and the tropic-subtropic-temperate forest habitat. Though all the 12 herbivorous mammals listed in the table, which are the *Pongo* sp., *Hylobates* sp., *Presbytis* sp. (Chongzuo langur), *Tapirus sinensis, Rhinoceros sinensis, Cervus* sp. (*Rusa*

Table 2.2 The habitat and foraging niches of Chongzuo langur and coexisting animals in Early Pleistocene Epoch as against those of extant animals of corresponding genera in the present

Extant species corresponding to the genus of the fossil animal unearthed in Chongzuo Sanhe Cave	Geographical distribution of the extant species	Habitat and foraging niche of the extant species	The fossil species unearthed from Chongzuo Sanhe Cave	Habitat and foraging niche of the fossil species in Early Pleistocene Epoch
Bornean orangutan (*Pongo pygmaeus*)	Borneo, Sumatra	*Habitat*: tropical rainforest *Forage niche*: fruits, tree sprouts, leaves, bird eggs	Orangutan (*Pongo* sp.)	*Habitat*: middle stratum of tropical rainforest *Forage niche*: fruits, tree sprouts, leaves, bird eggs
Lar gibbon (*Hylobates lar*)	Java, Borneo, Sumatra, SE Asia, Southern China	*Habitat*: tropical rainforest *Forage niche*: fruits, leaves, tree sprouts, flowers, insects, bird eggs	Gibbon (*Hylobates* sp.)	*Habitat*: middle and upper strata of tropical rainforest *Forage niche*: fruits, leaves, tree sprouts, flowers, insects, bird eggs
Silvered langur (*T.a. auratus*)	Sunda archipelagos, Malay Peninsula, Indochina	*Habitat*: mmangroves and broadleaf forests *Forage niche*: leaves, flowers, tree sprouts, barks	Chongzuo langur (*Presbytis* sp. = *Trachypithecus* sp.)	*Habitat*: middle and upper strata of tropical rainforest—karst hill forests *Forage niche*: leaves, flowers, tree sprouts, barks
Malayan tapir *(Tapirus indicus)*	Southern Burma, Thailand, Malay Peninsula, Sumatra	*Habitat*: tropical rainforest *Forage niche*: grass, leaves, fruits	Chinese tapir (*Tapirus sinensis*)	*Habitat*: bottom stratum of tropical rainforest *Forage niche*: grass, leaves, fruits
Javan rhinoceros (*Rhinoceros sondaicus*)	Java, Sumatra, Malay Peninsula	*Habitat*: tropical rainforest *Forage niche*: reeds, leaves, bamboos	Chinese Rhinoceros (*Rhinoceros sinensis*)	*Habitat*: bottom stratum of tropical rainforest *Forage niche*: reeds, leaves, bamboos
Sambar deer (*Cervus unicolor*)	Malay archipelagos, Malay Peninsula Indochina, Mountainous regions south of Qinling Mt., Southern India	*Habitat*: Subtropical lowland shrub bushes and mid-mountain temperate forests *Forage niche*: twigs, leaves, barks, fruits, grass	Yunnan deer (*Cervus* sp.) (*Rusa yunnanensis*)	*Habitat*: bottom stratum of tropic-subtropical forest *Forage niche*: twigs, leaves, barks, fruits, grass
Asian elephant (*Elephas maximus*)	Borneo, Sumatra, Malay Peninsula, Indochina, Yunnan in China, Burma, Southern India, Sri Lanka	*Habitat*: tropic forest, from plains to sub-alpine forests *Forage niche*: leaves, tree sprouts, barks, fruits, grass	Yangtze Sinomastodon (*Sinomastodon* sp., *yangziensis*)	*Habitat*: bottom stratum of tropic-subtropical forest *Forage niche*: leaves, tree sprouts, barks, fruits, grass
			Pre-orient Stegodon (*Stegodon* sp., *preorientalias*)	*Habitat*: bottom stratum of tropic-subtropical forest *Forage niche*: leaves, tree sprouts, barks, fruits, grass
Extinct species	Fossil grounds in Java, Northern Vietnam, Guangxi of China, South bank of Chang Jiang	*Habitat*: tropical forest *Forage niche*: omnivores on forest	Great ape (*Gigantopithecus blacki*)	*Habitat*: bottom stratum of tropic-subtropical forest

(continued)

Table 2.2 (continued)

Extant species corresponding to the genus of the fossil animal unearthed in Chongzuo Sanhe Cave	Geographical distribution of the extant species	Habitat and foraging niche of the extant species	The fossil species unearthed from Chongzuo Sanhe Cave	Habitat and foraging niche of the fossil species in Early Pleistocene Epoch
		floor competing with herbivores		*Forage niche*: (forest floor omnivores competing with herbivores)
			Hominini (*cf. Hominidae*)	*Habitat*: bottom stratum of tropic-subtropical forest *Forage niche*: (forest floor omnivores competing with herbivores)
Vietnam snub-nosed monkey (*Rhinopithecus* sp.)	High grounds of median and low mountains in Northern Vietnam	*Habitat*: subtropical-temperate forests *Forage niche*: leaves, tree sprouts, fruits	Snub-nosed monkey (*Rhinopithecus* sp.)	*Habitat*: middle and upper strata of subtropical-temperate forests *Forage niche*: leaves, tree sprouts, fruits
Macaque (*Macaca*)	Widely dispersed in low-latitude areas in Northern Hemisphere	*Habitat*: Tropic-subtropical plains and mountainous forests *Forage niche*: chiefly herbivorous but also striking small animals	Macaca (*Macaca* sp.)	*Habitat*: middle stratum of tropic-subtropical-temperate forest *Forage niche*: chiefly herbivorous but also striking small animals

yunnanensis), *Sinomastodon* sp. (*yangziensis*), *Stegodon* sp. (*preorientalias*), *Gigantopithecus blacki*, *cf. Hominidae*, *Rhinopithecus* sp., and *Macaca* sp., lived on forest vegetation in the south of Zuo Jiang River, they had different foraging niches which could be divided into three different categories:

- Seven species had their foraging niches in the bottom stratum of tropical rainforests. They were *Tapirus sinensis*, *Rhinoceros sinensis*, *Cervus* sp. ((*Rusa*) *yunnanensis*), *Sinomastodon* sp. (*yangziensis*), *Stegodon* sp. (*preorientalias*), *Gigantopithecus blacki*, and *cf. Hominidae*, and they fed on short grasses, leaves, barks, tuberous roots, fruits, bamboo, etc.
- Three species had their foraging niches in the middle and upper strata of tropical rainforests. They were *Hylobates* sp., *Rhinopithecus* sp. and *Presbytis* sp. (Chongzuo langur), and they fed on fruits, tree sprouts, leaves, flowers, insects, bird eggs, etc.
- *Pongo* sp. and *Macaca* sp. had their foraging niches in the middle and lower strata of tropic-subtropical forests, and they fed on twigs, barks, grasses, fruits, insects, and various small animals.

During the Pleistocene, there were multifarious forest environments on the terrain south of Zuo Jiang River, including tropical rainforests on low-lying hilly wetlands, swampy valleys, and tropic-temperate deciduous forests at 450 m up on mountains. Different environments allowed the foraging niches of the different animals in the region to be separated, and therefore, they could forage for different resources in different ways in the same forests (Fig. 2.1). Chongzuo langurs, which were new to the environment, had to compete for food with two other primate species, the snub-nosed monkey (*Rhinopithecus* sp.) and the gibbon (*Hylobates* sp.), which also had their foraging niches at the middle and upper strata of forests. By advantage of their smaller size and lighter weight, Chongzuo langurs were able to forage for leaves from taller trees. There were other factors compelling them to evolve to live in distinct habitats in karst hills. There are also two other points to note:

- Langur species are indiscriminate in their vegetative diets and are opportunistic. Sprouts and fruits of nearly every tree species may serve their diet, so there was an abundant and endless supply for them.
- The various fruits and seeds that have gone through the digestive guts of langurs will be laid in hill clefts or on soil on the bottom of the hills, which keeps the ecosystem of the karst hills thriving.

(a) (b)

Fig. 2.1 (**a**) Reproduced image of habitat I; (**b**) Reproduced image of habitat II (drawn by Gu Tieliu)

2.2.1.3 Places to Hide from Predators in the "Garden of Eden"

The term "predators" used here is limited to the animal species that strike, hurt, kill, and prey on other animals. Table 2.3 lists nine predatory species, which appeared in Chongzuo *Gigantopithecus* fauna in the early phase of the Early Pleistocene. Their existence was definitely a threat to the survival of Chongzuo langurs. Seven of the nine predators were ferocious carnivores, and among them the most threatening was *Cuon* sp. (*antiques*), which foraged on the forest floor. *Cuon* sp. (*antiques*) gathered up in packs to hunt various mammals. Even fierce *Panthera pardus* and massive buffalos would fall prey to them. *Panthera pardus* and *Erictis* sp. were capable of climbing trees and they hunted by using ambushes. In addition, there were giant pythons lurking on trees that ambushed mammals, while high-order primates *cf. Hominidae* and *Gigantopithecus blacki*, which had well-developed brains, would hunt by tricks and snares. It is found that at the very beginning, Chongzuo langurs led risky lives with predators everywhere, but finally they survived.

Among the nine predators illustrated in Table 2.3, at least seven of them were larger than Chongzuo langurs, and they were all hunting on the forest floor. That made Chongzuo langurs afraid to forage on the forest floor. But on trees, there were still *Martes flavigula* sp., *Martes* sp., and *Erictis* sp. preying on them. Fortunately, there were columns of karst hills on the Zuo Jiang River Basin, providing natural refuge to them. The danger in the environment had altered the evolutionary course of the Chongzuo langur and made it abandon the arboreal life to rest on the precipices. The strain of being preyed upon had entered deep into their genes, which became an instinct and was passed on from generation to generation.

2.2.2 Chongzuo Langur's Evolution Course

The morphology of the third molar on the mandible (the lower third molar) of the white-headed langur now living in the small area surrounded by Zuo Jiang River and Ming Jiang River is very similar to that of the Chongzuo langur, which lived in the same small area 1.4 mya. This allows us to associate its ancestors to the Javan langurs, which lived in central Java in the ancient.

Emigration is a way of evolution, while fortuitousness in taking hold of the right time and the right opportunity is essential to the evolution. Extant white-headed langur families are female-centered, although the gene-flow follows that of adult males. The system of its society and dispersal implies that population emigration must be slow, and it would take a longer time and more detours to establish a new habitat. Nonetheless, the founding ancestors had emigrated successfully, arriving at the south bank of Zuo Jiang River and taking root on the karst hills. We conjecture that the ancient Javan langurs had taken the opportunity of the first extreme glaciations in the Early Pleistocene Epoch through the time span of that 100,000 years, and after hundreds of thousands of northward emigration attempts, the population gradually accumulated to a group that can reproduce effectively and finally rooted on the "isolated island" surrounded by rivers.

Table 2.3 The habitats and foraging niches of the predatory species preying on Chongzuo langurs in the Early Pleistocene Epoch compared to those of extant animals of corresponding genera in the present

Extant species corresponding to the genus of the fossil animal unearthed in Chongzuo Sanhe Cave	Geographical distribution of the extant species	Habitat and foraging niche of the extant species	The fossil species unearthed from Chongzuo Sanhe Cave	Habitat and foraging niche of the fossil species in Early Pleistocene Epoch
Dhole (*Cuon alpines*)	From eastern and southern China all the way to Indochina, India, Burma	Wide disperse species in tropic-subtropics mountainous forest habitats Preying ferociously in packs on various animals on forest floor	Ancient dhole (*Cuon* sp. (*antiques*))	Wide disperse species in tropic-subtropics mountainous forest habitats Ground predator Preying ferociously in packs on various animals on forest floor
Asian black bear (*Selenarctos thibetanus*)	From eastern and southern China all the way to Indochina, India, Burma	Wide disperse species in tropic-subtropics-temperate mountainous forest habitats Omnivorous, capable of preying in karst hill forests	*Ursus* sp. (*c.f. thibetamus*)	Wide disperse species in tropic-subtropics forest habitats Ground predator Omnivorous, capable of preying in karst hill forests
Leopard (*Panthera pardus*)	Extensively distributed in the whole of tropic Asia. From Java to Malay Peninsula, Indochina, eastern and southern China, and through India into Africa	Wide disperse species in tropic-subtropics-temperate-subarctic forest habitats Preying on medium to large size mammals (including primates) on forest floor and those up on trees on karst hills	*Panthera pardus*	Wide disperse species in tropic-subtropics forest habitats. Ground predator Preying on medium to large size mammals (including primates) on forest floor and those up on trees on karst hills
Leopard cat (*Prionailurus bengalensis)*	Extensively distributed in the whole of tropic Asia. From Malay Archipelagos to Indochina, southern China and all the way to southern India	Wide disperse species in forest on plains and on mountain habitats Preying on various small animals on forest floor and those up on the trees	*Felis teilhard*	Wide disperse species living in tropic-subtropics forest habitats Preying on various small animals on forest floor and those up on the trees
Asian golden cat (*Catopuma temminckii*)	Distributing over the southern slopes of Qinling Mt., China, southwestern China, Indochina, and Sumatra	Wide disperse species in mountainous forest habitats Preying on medium to small size animals (including monkeys) on forest floor on karst hills	Wild cat (*Felis* sp.)	Wide disperse species in tropic-subtropics forest habitats Ground predator Preying on medium to small size animals (including monkeys) on karst hill forest floors
Yellow-throated manen (*Martes flavigula*)	Extensively distributed in China	Wide disperse species randomly inhabiting on plains, in mountains, forests, and lakes Distributing still in Nongguan Mountain Territory Arboreal living Omnivorous with broad diet range. Strikes various small animals including small langurs	Martes (*Martes* sp.)	Wide disperse random inhabiting species Arboreal living Terrestrial predator. Broad diet range. Strikes various small animals including small langurs
			Erictis sp.	Wide disperse random inhabiting species. Arboreal living Terrestrial predator. Broad diet range. Strike various small animals including small langurs
Extinct species	In places in Java, Guangxi, and the south bank of Chang Jiang in China, where fossils had been discovered		*Cf. Hominidae*	In tropic-subtropics forest habitats. Mostly preying on forest floors
			Gigantopithecus blacki	In tropic-subtropics forests. Mostly preying on forest floors

Most often, pre-adaptations are essential to evolution. We conjecture that ancient Javan langurs had the same living habits as the extant silvered langurs in Java Island: forage on leaves, sprouts, and fruits in mangroves and in broadleaf forests, and assemble at night, sleeping high up on trees to stay away from ferocious predators including crocodiles and giant pythons. When they moved along the belts of new forests on the exposed West Pacific continental shelf and gradually arrived at the karst hills south of Zuo Jiang River, they found that the warm tropical climate there was no different from that of where they came from. It was just as suitable for living, although it was not as moist as their old habitat in the south. There are also diverse natural habitats and vegetation covers in the new habitat. The huge biomass with various plants with leaves, sprouts, flowers, fruits, and bark provided them food all year round. In the competition for food with the many other herbivorous animals there, the ancient Javan langurs developed a diet indiscriminate of plant species. In field research, we identified no less than 80 plant species in the diets of extant white-headed langurs. Studies also found that they readily explore the leaves and fruits of newly found exotic plants for food. This opportunistic and versatile eating habit of langurs would have helped them survive in the "isolated island" in the south bank of Zuo Jiang River.

In the Pleistocene Period, Javan langurs in the south bank of Zuo Jiang River met the same carnivorous predators that their ancestors were familiar with in the old homeland in central Java. Among them were *Cuon* sp. (*antiques*), *Felix* sp., and *Ursus* sp. (*cf. thibetamus*), which stalked the forest floor and would readily vanquish other species which dared appear on the bottom stratum of forest, and *Panthera pardus* crouching on trees ready to prey on animals on both the forest floor and the trees. There were also *Martes flavigula* sp., *Martes* sp., *Erictis* sp., and *Macaca* sp., which were capable of climbing to the upper branches of trees to prey on small animals like squirrels and infant langurs. Primates like *Gigantopithecus blacki* and *cf. Hominidae* with developed brains would hunt in all areas of the habitat and had the skill to snare all kinds of animals. In addition, there were always giant pythons lurking noiselessly in dark corners waiting for their prey.

Faced with such a precarious environment, the ancient Javan langur had only a narrow path to survival, with no other choice but to evolve. They gathered in the day to feed together, to move together, helping each other. By being together, they could be more cautious of the predators and flee rapidly all at the same time. They might gather to sleep high up on tall trees, yet the giant pythons could still approach them without making a sound. Thus, they were forced to take the columns of karst hills on the Zuo Jiang River Basin as natural refuge. Every night, they would climb up the precipitous precipices to sleep there together. This was more than a night shelter for them; it was also a delivering room for pregnant female langurs. They chose stark slippery precipices with no trees or even grass around, so that no predators could reach them. In this way, they kept their safety and enhanced the survival rate of their offspring.

2.3 Phase 3: Coevolution of the Chongzuo Langur and Chongzuo Aborigines

Phase 3 lasted about 500,000 years; from 1 mya to 50,000 years ago.

Worldwide, the use of fire by *late Homo Sapiens* dates back to 60,000 years ago. At that time, with a small population and a mobile hunter-gatherer lifestyle, fire may have been used at first to drive away animals hidden in the forest or to change the growth of grasslands to lure prey. Therefore, fires were small in scale and had little impact on the environment. However, after the last glaciations (1 mya), people around the world began to use fire to clear forests, develop agriculture, and build villages. To cope with the ever increasing population, more and more land had to be cleared for planting, leading to rapid changes in the natural environment, which forced many species out of survival.

In the year 2009, a joint team led by Peking University and the Institute of Vertebrate Paleontology and Paleoanthropology CAS unearthed a mandible of a *late Homo sapiens* in the Nongguan Mountains (Fig. 2.2). It was verified to be living 111,000 years ago (Jin et al. 2009). There is no fossil evidence or records that there were extensive forest clearing in the karst hill regions at Zuo Jiang River Basin in those years like what had happened in the Eurasian plains. The reason might be that the soil in karst hill regions was infertile and that the hilly terrain made it difficult for the *late Homo sapiens* to find room for expansion. In fact, the scale of agriculture and population size in the Nongguan Mountains had for a long time hardly been developed.

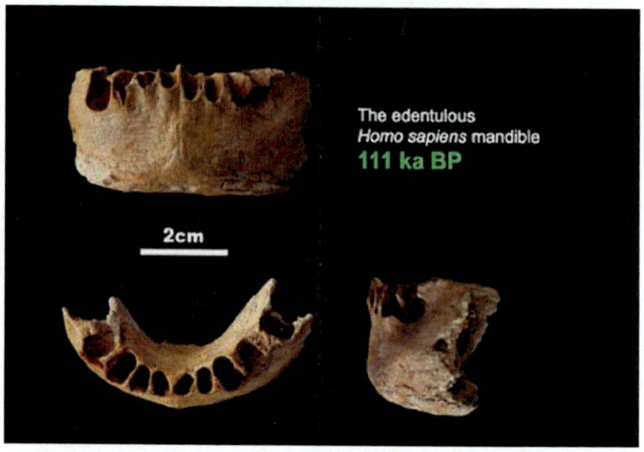

Fig. 2.2 The mandible of *late Homo sapiens* in Mulan mountain unearthed in Nongguan mountain, Chongzuo, Guangxi

It was recorded in history that in the year 2 AD, when the aristocrats of the West Han Dynasty were practicing sophisticated farming and silk-worm cultivations on the banks of Wei He River and were expanding the strong feudal dynasty, Guangxi was still a savage land with only about 200,000 people in the entire province, and the population density was merely 1 km^{-2}. It could only be even less populated in Chongzuo region at the middle and upstream parts of Zuo Jiang River and still less in the karst hilly region, which was described as "mountains after mountains of bamboo forests" where "routes were precarious and lost in dense forests." There it would be difficult to develop agriculture with scant labor. It was also recorded in history that in the times of the Song Dynasty and Yuan Dynasty, the agricultural practice in the region was still very primitive. It was said to be "sowing seeds without plowing the field," "simply let the seeds grow where they fall, no re-lining of seedlings, not introducing water upon drought and not draining the fields upon flood, not applying manure, no cultivating but just letting nature takes its course." Not until 300 years ago in Ming and Qing Dynasty did the Chongzuo County (it was then known as Chongshan County) reach a population of 7860, which amounted to less than 5 people/km^2. While there were improvements in agricultural practices, the soil was so thin and unfit for crop growing that farmers' livings were still below subsistence levels. The region was still branded as "untamed territory" where "the mountains are high and people are savage and ruthless." Hundred years ago (the second year in *Xuantong* calendar), census showed that the population in Chongshan County had grown to 35,029. The average population density was then 22 people/km^2, but agricultural cultivation remained in primitive, using slash and burn techniques.

On that extensive infertile region with scant human populations living threadbare lives, the white-headed langur had sufficient room and time to mutate genetically, adapting to coexistence with humans. Everywhere on the banks of Ming Jiang River, or in the mountain caves in Fusui, or on the sugarcane field divides at Chongzuo, we hear the same beautiful folk legend:

> Long time ago, there was a small, impoverished village. Villagers often suffered from hunger because food was insufficient. One day, an aged man passed away and the villagers buried him in the forest according to their customs. At the funeral procession, adults and children wore the same white cotton bands on their heads and waists to lament the deceased. In the forest, the children found many edible wild fruits and leaves and ate them to satisfy their hunger. After the funeral, the children decided to stay in the mountain rather than going home. They played and picked plants for food in the forest during the day and spent the nights in caves to stay away from possible dangers. Time passed and the lives in the village became easier as the village population decreased. The adults, missing their children, went to the mountain, wishing to bring them home. But the children had adapted to living in the mountain and they said to the adults, "We are not going home. At home we have to take our food from the share of our parents and grandparents. We are doing well here in the mountain so please let us stay here." Hence the children grew to be adults in the mountain, had children of their own and lived happily in their new home for generations and generations. As time went by, the white cotton bands they wore on their heads to express lament wore out but white hairs developed, and the white cotton band on their waists wore out and was replaced by a long white tail. Wearing this costume, they expressed their yearning for their parents and village peers.

That is the story of the origin of white-headed langurs. In the old days, *Zhuang* people had no writing system, but they have ever since conveyed their history by telling stories orally. Their stories were participatory materials, stored and organized in the memories of people from generation to generation. Most significant of all is the harmonious relationship between the Zuo Jiang River people and the white-headed langurs. The story is a folkloric myth and also a piece of white-headed langur epic.

2.4 Phase 4: White-Headed Langurs' Endangered Status by the Late Twentieth Century

When the human population on earth was small, the regenerative cycle of the earth was able to fulfill the needs of all mankind. But ever since the beginning of the twentieth century, conditions have drastically changed. Along with the rapid growth of the global population, the devastation of natural habitats has been more and more rampant, which has led to rapid decline of biospecies and even extinction of some. This is often referred to as "The 20th Century Phenomenon." In less than 100 years, the world population had seen exponential growth. If there are no checks on population growth, "The power of population is indefinitely greater than the power in the earth to produce subsistence for man." (British Reverend Thomas Roberts Malthus was taken aback by his own presage.)

In the 1950s, the President of Peking University Ma (1979), an economist himself, propounded the need to curb population growth, which could hinder China's developments if unrestrained. However, the policymakers of the time did not heed Ma's advice; instead, they stripped Ma of the presidency of Peking University and put forward a slogan "More people, more discussions, extra enthusiasm, extra vigor" to encourage population growth. As a result, the population in China grew from 450 million in year 1949 to the extent that, in the span of 30 years, an "extra nation" was reproduced. The figure reached one billion in 1982, and it now stands at 1.4 billion. Despite the government's utmost efforts in checking population growth afterwards, the enormous population that had already been produced has been affecting the nation's continual development on education,

Fig. 2.3 (a) Blasting mountains; (b) Burning wasteland for reclamation (Photo by Pan Wenshi)

medical, and health services, as well as hampering the growth of industries and agriculture. We take Chongzuo County as an example. In 1910, there was a population of 35,029 with density of 22 people/km^2. In 1947, Chongshan County and Chongzuo County had an aggregate population of 94,945 and the population density was at around 60 people/km^2. In 1982, the population in Chongzuo County reached 281,187 and the population density rose to 97 people / km^2. In 1985, the population rose to 297,314 and the population density became 102.5 people/km^2. And if we look at Luobai Township, one of the two only regions with extant white-headed langur distributions, we see that it had a population of 4785 in the year 1937 with a density of 30.3 people/km^2. In 1997, its population reached 24,591 and the population density rose to 155.6 people/km^2. The figures reveal that the population in Luobai Township had increased more than fivefold in the last 60 years, and to sustain the larger population, more resources have necessarily been exhausted. Take the case of the eight villages closely bordering the north of the Nongguan Mountains, there were 3131 people (year 2000 statistic) engaged in agriculture, and they get their fuel by cutting the trees in the forest. If 1 kg of firewood is necessary for cooking for one person/day (this is the lowest possible estimation, the actual figure should be far greater), 1142.82 tons of tree would have been cut each year. *Leizhai* is an example: the cantonment had 1500 *mu* (Chinese acre) of forest covered areas in the year 1951, but there remained only 40 *mu* in 1999. The forested area had been reduced by 97.4% in 48 years. When we came to Nongguan Mountains in November 1997, we came across one farming family cutting trees for firewood.

> "Why are you cutting even the small sapling?" I asked.
>
> "We feel no less pity than you do. With every chop our hearts pain as if we're culling a piglet, butchering it before its time." The farmer replied.
>
> "Can't you refrain from cutting saplings?"
>
> "The old and young members of the family won't live if there isn't any firewood in the house," the farmer continued, "Look everywhere around you, where can you find a full-grown tree?"

In 1958, the slogan calling for the "Big Leap Forward, Big Efforts for Socialism" was propagandized and the "Backyard Furnaces" campaign rolled out projecting the annual production target at 1070 tons of steel in order to meet the unrealistic goal of "development at the pace of one day equaling 20 years of senescent Britain in steel production." In just one day, "People's Communes" was set up in all villages. All farmers were exhorted to cast steel in patios. Hundreds of millions of trees were chopped, sun-dried, and put into furnaces for fuel. The so-called steel was produced, ton after ton, but was subsequently discarded, ton after ton, as it comprised of high impurity contents. Yet the whole nation was immersed in joy, and people were "casting steel" day and night. Farms were left untended with overgrown weeds and grasses, and poor harvests took place over extensive areas. Deforestation, land reclamation, mountain blasting, and other acts of destroying the natural environment were pushing the white-headed langur to the edge of extinction (Fig. 2.3). No sooner than that, famine struck nationwide. Countless wild animals were hunted for food, depleting China from being the world's biggest wild animal trove. A militia captain living in the white-headed langur distribution region during those years recalled with regret and helplessness, "The village was so impoverished that practically nothing was available. Every time a senior came to inspect this area, I had to pick up my rifle to hunt a langur in order to have some meat... ."

Quan et al. (1981) had pointed out in their book *Classification and Distribution of Primates in Our Country* (我国灵长类动物的分类与分布) that "The habitat environment of primates is being altered at excessive speed. The continuous excessive forest clearings are rapidly curtailing primate distributions. As large expanses of forests have been exploited, these precious species have already withdrawn to narrow stripes of land at the most remote regions." It goes on further to state that "the sphere of excessive hunting has come to an extreme; nobody oversees producing tonics with wild animals. Regulations are never rigorously enforced and the number of excessive trappings and reckless killings are shocking." It has also been recorded in *Chongzuo County Archives* (Nong Biqiang 1994) that "in the year 1979, the County Forestry Department had purchased 39 white-headed langurs for the Nanning Zoo." Lu and Huang (1993) in their paper *White-Headed Langurs Population Surveys and Researches* (1993) (白头叶猴种群的调查研究) had pointed out that "Random incidents of hunting white-headed langurs are still taking place."

In short, all the data collected by professionals and academics from field surveys have shown that incidents of environmental devastation and the hunting of white-headed langurs took place frequently. No doubt, these were the two fundamental causes leading to the losses of white-headed langur habitats and the decrease in population. Such had brought the disappearances of the four populations in the Longrui Mountains (east of Ming Jiang River in Longgang Natural Reserve), the Bazhiqi Mountains (Renling District), the Longfeng Mountains, and the Daling Mountains. The diminution of a natural habitat reduces the number of species the habitat can sustain. The situation in the white-headed langur habitat has already reached an alarming state as told by the "Law of Area–Species Ratio."

References

Biqiang N. Chongzuo County archive (崇左县志). Nanning: Guangxi Renmin; 1994.

Chongshan County Archive. Chongshan County archive (Minguo 26th year edition) (photo-copy version). Nanning: Guangxi Renmin Publisher; 2011.

Jin CZ, Pan WS, Zhang YQ, et al. The Homo sapiens Cave Hominin Site of Mulan Mountain, Jiangzhou District, Chongzuo, Guangxi with emphasis on its age. Chin Sci Bull. 2009;54(21):3848–56.

Lu LR, Huang CM. Investigations and researches on white-headed langurs population (白头叶猴种群的调查研究). Acta Theriol Sinica. 1993;13(1):11–5.

Ma YC. New populations theory –discourse paper on the 4th National Peoples' Congress (新人口论——在第一届全国人民代表大会第四次会议上的书面发言). Archive. 1979;02:34–59.

Nowak RM, Paradiso JL. Walker's mammals of the world. 4th ed. London: The John Hopkins University Press; 1983.

Quan GQ, Wang S, Zhang RZ. Classification and distribution of primates in our country (我国灵长类动物的分类与分布). Chin J Wildl. 1981;1981:3.

Rongzu Z, et al. Distribution of terrestrial animals in China (中国哺乳动物分布). Beijing: China Forestry Publishing; 1997.

Wang KF, Zhang YL, Ye ZH, et al. The discovery of composition of quaternary spores on the coast of Beibu Gulf in our country and the ancient climate of the place (我国北部湾沿岸第四系孢粉组合的发现及其古气候). Beijing: Science China Press; 1977.

Zheng SH. Relics of Homo erectus Jianshi (建始人遗址). Beijing: Science Press; 2004.

Zhu YZ. Picture guide of world animals (世界动物图鉴). Beijing: Dolphin Books; 1995.

3 The White-Headed Langur Is an Endemic Species of Chongzuo

Abstract

More than 60 years ago, based on a black and white piebald pelt, a formerly unknown primate species, the "Piebald langur" *Presbytis leucocephalus* (Tan 谭邦杰 1955) was described from the Chongzuo area, in China's Guangxi Province. It was later reassigned to the genus *Trachypithecus*, but further controversies arose regarding its systematics. Some propose that it is a geographic subspecies of the black langur *T. francoisi*, while others hold the view that the white-headed langur is a distinct species.

Our research found that geographic isolation due to Zuojiang and Mingjiang Rivers had effectively barred the gene flow between the white-headed langur and the black-headed langur. We also found that the fine-scale morphology of the mandibular third molar of the Chongzuo langur, in particular the occlusal configuration of cusps and ridges, is very similar to that of its ancestor *P. sivalensis* from Pakistan and is almost the same as those of extant white-headed langurs. While the mandibular third molar of black-headed langur does not have these same attributes. We conclude that the white-headed langur is an endemic species of Chongzuo, Guangxi.

Keywords

White-headed langur · Mandibular third molar · Resident species · Endemic species · Primate · China

3.1 The Distinct Appearance of White-Headed Langur

Of all the 30–40 langur species in the world, the white-headed langur perhaps has the most distinct appearance. The black and white pied pelage of adult individuals is well-described, yet primatologists still know very little of other characteristic external features, in particular that of the appearance of the pups. The white-headed langur has a sleek black and white contrasting pelage, an exquisitely slender body shape with long, slim limbs, and an even longer slim tail, which gives it a well-balanced and supple physique. When a troop of white-headed langurs are sitting motionlessly on trees or on karst hills, the long white tails hanging in midair are typically what observers see at the first glance. Its tawny, friendly looking eyes, its graceful movements, and generally mellow manner do not inspire aggression.

The scientific name of the white-headed langur, *Trachypithecus leucocephalus,* literally means continental-type langur species with a white head (Fig. 3.1a, b). From age 1 through adulthood, the white hair on a white-headed langur's head grows upright and toward the center, forming the characteristic steepled hair crown. The head, neck, shoulder, and most of the tail are covered with white hair as well. On the chests of mature adults, white hair mixes with black hair, making it appear to be light brown. The rest of the body is covered with sleek black hair with the longest strands reaching 10–12 cm on the ribcage to keep the body from being soaked on rainy days. Underneath is a layer of velvety hair to maintain body temperature on cold winter nights. The hair growing on the back of palms, on the upside of feet, and on the tail are the shortest with a length between 1 and 2.5 cm, which may be white or black in color. The black and white colors do not make the white-headed langur stand out on the mottled craggy precipices; on the contrary, it camouflages the animal.

The adult male white-headed langur usually sits high up on treetops or on projecting rocks, prominently displaying its gleaming white hair on crown and shoulder. The particular way he sits with legs spreading wide apart showing his male genitals clearly sends the message that he is the protector of the family.

The adult female white-headed langur would sometimes sit the way the adult male does, high up on treetop spreading legs, revealing the triangular patch of creamy white skin on her groin. When a female is in estrus, the triangular patch of

W. Pan, *The Population Ecology of White-Headed Langur*, https://doi.org/10.1007/978-981-33-4118-0_3

(a) (b) (c)

Fig. 3.1 (**a**) Adult male; (**b**) Adult female; (**c**) 7-day-old newborn (Photo by Liang Zuhong)

skin would turn moist and red. As in females of other primate species, the changing color of the genital skin sends out signals of being receptive for sex. There are black blotches on that triangular patch in every female white-headed langur whose outline is different for each individual and remains unchanged throughout life. This individualistic feature has become the definitive mark upon which we identify female individuals.

The newborn white-headed langur has glossy golden yellow hair, with two round eyes on a delicate pastel-colored face. It has a round head that does not yet have the steeple hair crown (Fig. 3.1c). The skin on its face, ears, and limbs has a clean pastel flesh color. It weighs about 350 g, has a body length of 15 cm with a tail of about 30 cm, similar to a kitten. The hair and skin color on a pup's face rapidly change as it grows. In the big family of langurs, the hair color of pups is a criterion in classifying the genus. Golden-headed langurs and silvered langurs, which are consanguine with the white-headed langur also bear golden yellow–haired pups.

At the age of 1 year, a white-headed langur pup would be capable of moving around independently of its mother but would still live with its parents, aunts, and siblings in one harmonious family. They will be living together for 4 years (6 years in few cases) until the lineal father is defeated in a contest for mating rights by an outlier adult male invading the family. The family would then split, and the lineal father would take along all its sons (in some cases also take along a few immature daughters) leaving the family, while nearly all the females including the lineal mother, aunts, and mature and immature sisters would stay behind and form a new family with the incoming adult male in the original family territory. The white-headed langur population propagates over many rounds of such family splits. Family territory is inherited matrilineally, and the alpha male would be an incoming outlier. Female white-headed langurs will be able to reproduce at the age of 5 years. It is estimated that male white-headed langurs physically mature at the age of 5 years. But at this age, it could only be following its defeated father roving about together with its immature brothers. When it reaches around age 6–7 years, it may be strong enough to invade a family and make that its own.

It is indeed difficult to assess the relationship between the growth and weight of white-headed langurs in the wild. However, our research team has learnt from 20 years of field experience that we can tell the age of a white-headed langur by the black and white color distributions on its body. By placing a weighing scale on the langurs' usual passage and by improvising a measuring scale from a stem when the opportunity comes, we have found the white-headed langurs'

Table 3.1 Age–weight relationship for individuals in "Yintangxiaotu all-males troop"

Age (years)	Mean weight (kg)	No. of weights taken
7.2	10.2	1
6.7	10.0	6
5.3	8.5	18
4.8	8.0	4
4.3	7.4	5
3.7	6.6	5
2.3	4.5	2
0.1	0.4	2

Table 3.2 Age–tail length (a) and age–body length (b) relationships for individuals in "Yintangxiaotu all-males troop"

(a) **Age–tail length**		
Age (years)	*Mean tail length (cm)*	*No. of measures taken*
13.0	85.0	1
7.0	96.0	1
3.7	90.5	4
3.8	83.5	2
2.8	73.0	1
0.1	32.0	1
(b) **Age–body length**		
Age (years)	*Body length (cm)*	*No. of measures taken*
13.0	45	1
6.8	44	1
6.2	42	1
5.0	40	1
4.5	38	1
3.5	37	1
0.1	17	1

weights and measurements of the various parts of their bodies.

Tables 3.1 and 3.2 list statistics for three physical parameters for the male white-headed langurs of different ages in the "Yintangxiaotu all-males troop." It can be seen from the growth curve graph in Fig. 3.2 that body weight, body length, and tail length of male white-headed langurs grow in synchrony of each other. These growths stop at the age of 5–6 years.

Scientists conjecture that the ancient colobus, native to Africa, was the ancestor species of langurs. However, the extant langur finger is structurally different from the colobus finger. Langurs have five fingers with the short thumb opposing the other four fingers, enabling it to nimbly pluck sprouts, leaves, and fruits and to hold on to branches in order to scramble among trees. The rear foot palms are sturdy and flat which allow for walking, and the toes are markedly elongated with the big toe opposing the other four toes, allowing a langur to hold on to tree branches (Fig. 3.3). The langur has thus evolved to live a combination of arboreal and saxatile life (Fig. 3.4). It is capable of scrambling among vines and trees as well as leaping and sprinting among karst hills with its long and sturdy tail functioning as a balancing rod.

3.2 Controversies Over the Systematics of the White-Headed Langur

In 1952, Mr. Tan Bangjie, the first director of the Peking Zoo discovered a black and white piebald pelt at a native products trading post in Fuisui County of the Chongzuo District. After researching, Tan named the species 'Piebald Langur.' Then in 1955, the species was officially named "*Presbytis leucocephalus* Tan 谭邦杰 1955." The article was published in *Zoo Life* (Tan, 1957). Since then, the name "piebald langur," which was later renamed as the white-headed langur, was entered into Linnaean taxonomy.

At that time, little was known of the biology of the piebald langur, so it was natural that there were doubts on the correctness, of its systematics. Anatomists, ethologists, and molecular biologists were enthused to collect data to support their views. Hence, the discussion on the systematics of the piebald or white-headed langur is filled with controversies and has drawn much attention.

3.2.1 White-Headed Langur as a Subspecies of the Black Langur?

Twenty-seven years after the discovery of the "piebald langur" by Tan, its classification as a distinct species was challenged by two researchers from Guangxi, namely Shen Lantian and Li Hanhua (1982). They held that the morphology, habits, and geographical distribution of the "piebald langur" were not distinct from other langur species, but instead were very much alike those of the black langur. In addition, they took into consideration the manifold hair color changes in black langurs and hence determined that the white-headed langur should not be classified as a distinct species, but instead a subspecies of the black langur.

Thirty-six years after Tan's discovery, renowned zoogeographer Mr. Zhang Yongzu led a large team of Chinese mammalian specialists and co-authored the authoritative book *Distributions of Mammals in China* (中国哺乳动物分布) edited by The Endangered Species Import and Export Management Office of The Peoples' Republic of China (1997). In that profoundly influential book, white-headed langurs were still not classified as distinct species but listed as a geographic subspecies of the black langur *Presbytis francoisi leucocephalus* Tan 1957, from Fusui, Guangxi, distributing in the region south of the Zuo Jiang River.

From the mid- and late 1990s, molecular genetics research began to contribute new evidence to the systematics of the white-headed langur.

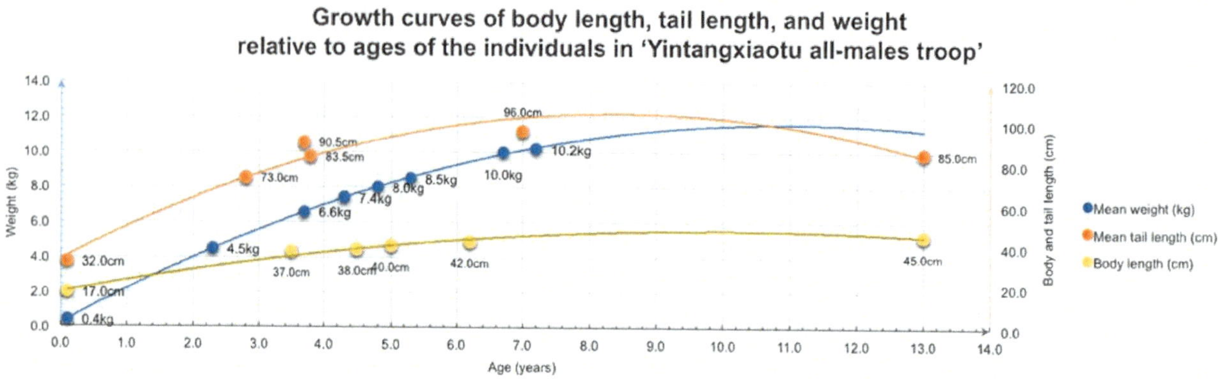

Fig. 3.2 Growth curves of body length, tail length, and weight relative to ages of the individuals in "Yintangxiaotu all-males troop"

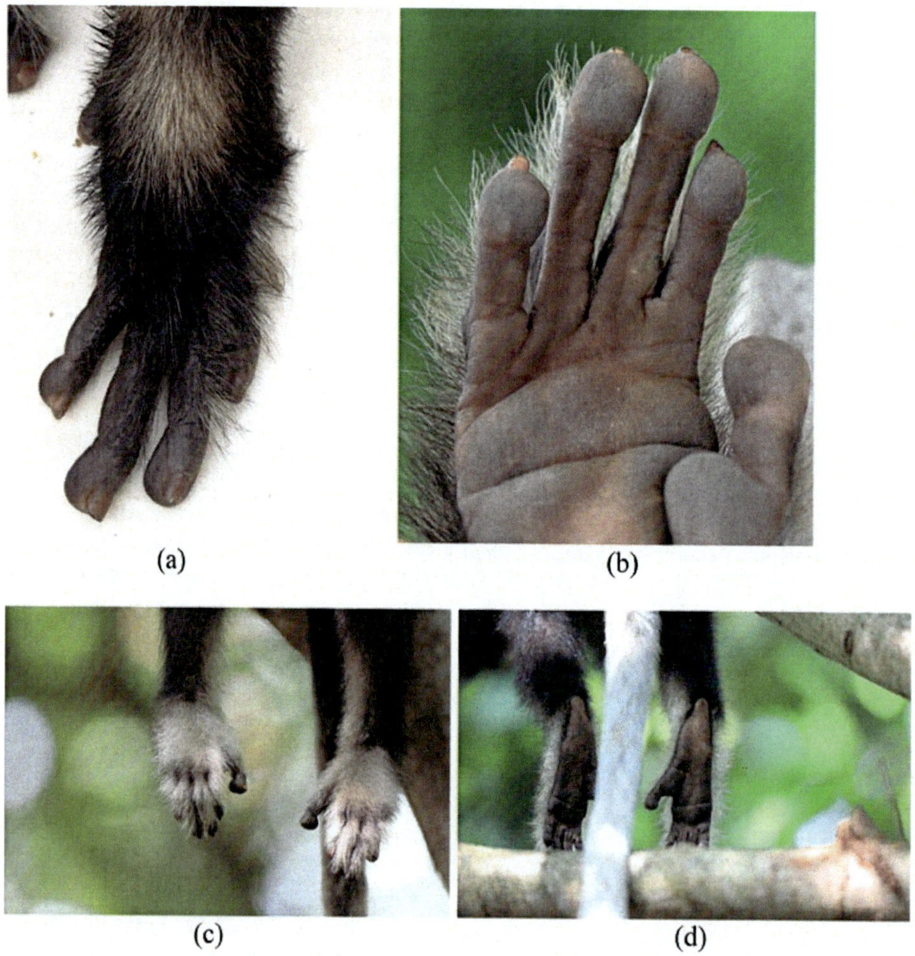

Fig. 3.3 The arboreal cum epigeal adaptive hand and foot of white-headed langur (Photo by Liang Zuhong). (**a**) Back of left hand; (**b**) Palm of left hand; (**c**) Instep; (**d**) Palm of left foot

Liu Zimin et al. (1997) made comparisons between the mitochondrial ND4 genome, base variances, and genetic variances in the mitochondrial D-loop sequencing between white-headed langurs (*T. francoisi leucocephalus*), black langurs (*T. francoisi*), and Phayre's langur (*T. phayrei*). The team found that the white-headed langur is but an

Fig. 3.4 (**a**) Arboreal; (**b**) Saxatile (Photo by Liang Zuhong)

Evolutionary Significant Unit (ESU) of the black langur, and so they also advocated white-headed langurs to be a geographic subspecies of black langurs.

Ding Po et al. (1999), by way of random amplified polymorphic DNA, analyzed the phylogenetic relationship between purple-faced langurs (*T. vetulus*), long-tailed langurs (*S. entellus*), Phayre's langur, black langurs, and white-headed langurs and found that the genetic variances between white-headed langur and black langur are no more significant than those found within the black langur species. Therefore, they deemed that gene flow may have happened between the two species in recent times.

Hu Yanling et al. (2004) and Chen Yiping (1989) had both come up with reports on successful reproduction by outbreeding white-headed langurs with black langurs in captivity. It was a very convincing experiment and was often quoted as proof that no reproduction barrier exists between white-headed langurs and black langurs to support the proposition that the white-headed langur is a subspecies of the black langur.

3.2.2 Arguments Supporting the White-Headed Langur as a Distinct Species

In 1980, Mr. Tan Bangjie in his paper "How I Discovered the White-Headed Langur" (我怎样发现白头叶猴) first portrayed some biological characteristics of white-headed langurs. He wrote that "this langur species (white-headed langur) distributes almost in the same regions as black langurs, but they keep living in separate territories, staying within their own populations, and never intermingle." Moreover, there are cranial differences between the two species: (1) "Piebald langur" has a longer nasal bone than that of the black langur; (2) The eye socket width of white-headed langur extends further than the width of the neurocranium; (3) Black langurs have a plumper posterior portion of the neurocranium.

Scientists who firmly advocate the white-headed langur as a distinct species have all based their arguments on research findings either on the origin and evolution of langur species or on their morphology and ethology.

Brandon-Jones (1984) upon researching the evolutionary history of the Asian colobus suggested classifying the white-headed langur as a distinct species.

Zhou Mingzhen (1964) held that the Pleistocene Epoch was a time when the langur genus rapidly diversified, and it was at that time many contemporary species came into existence, including the white-headed langur. Eudey (1987) has put forward further evidence from comparative morphology substantiating the idea that white-headed langurs should be classified as a distinct species.

Liren and Zhaoyuan (1991) deemed that in the view of morphologic, ecologic, and ethologic evidences, and on grounds of species conservation, white-headed langurs should be separated from black langurs. They considered the "in-between species" with "a white head and a black tail" discovered by Li Xisiang et al. (1980) was in fact just one varied individual in the white-headed langur population.

Roos et al. (2007), through analyzing the base variances and genetic variances on a 576 bp fragment of cytochrome b genes in the mitochondrial DNA of several langur species, accounted that although the genetic distance between black langurs and golden-headed langurs (*T. poliocephalus*) is small, there are relatively broad variances in the morphologies, vocalizations, and ethologies between the two. So he suggested recognizing the golden-headed langur as a distinct species and considered the white-headed langur to be a subspecies of the golden-headed langur.

Groves (2001), based on morphological variations and differences in geographic distribution patterns, bracketed all langur species that reproduce yellow-haired pups into one genus and created the new genus *Trachypithecus*. At the same time, he classified white-headed langurs (*Trachypithecus leucocephalus*) as a distinct species, on the same hierarchy as black langurs (*T. francoisi*), golden-headed langurs (*T. poliocephalus*), Delacour's langur (*T. delacouri*), Hatin langurs(*T. hatinhensis*), and Laotum langurs (*T. laotum*). Since then, the new genus *Trachypithecus* comprises of *Presbytis cristata* and *Presbytis obscure* and includes six continental types of langur species with distributions ranging from India to Southeast Asia and southern China for all of those that reproduce pups with brown or golden yellow hairs (Delsons 1997; Fleagle 1988).

3.2.3 Focus of the Argument

Is the white-headed langur a geographic subspecies of the black langur or is it a distinct species? Arguments on this subject have been going on for 60 years. But, what is their focus? Classifying species was difficult enough in the days when comparative morphology was the only approach. But molecular biology brought even more questions and controversies in classifying species, for it still requires evidence from all aspects and a new series of complex protocols and statistical applications to do the assessment. In our opinion, it is necessary to assess the species in its natural environment. Carefully study its ecology and biology and estimate the geologic era when the species appeared, in order to understand its evolutionary course and its natural distribution in the past and present. Then, the arguments would lead to a more authentic taxonomic classification of white-headed langurs.

3.2.3.1 Where Does the Boundary Between Species Lie?

The diversity boundary between species is in fact clearly defined in nature. Nature has from the start put many reproductive barriers between species to ensure their distinctness. In addition, nature would also make offspring from cross-species mating sterile. As a result, in the realm of sexual reproductive species, every individual is a reproductive unit, and the diversity boundary between species is in a way drawn by the feasibility of natural and effective reproduction between individuals.

To tell species apart, taxonomists have to ascertain whether natural and effective reproduction would happen between individuals by looking at the issue from various aspects.

First, to verify that the geographical distribution isolation between the species is an actuality, for actual geographical distribution, isolation would engender reproductive isolation.

Second, it must be confirmed that reproduction isolation is the result of different mating predispositions that have evolved after a long period of time at different population distributions and that there are also differences in the timing of estrus and mating behaviors.

Third, it must also be noted that if mating occurred between different species under certain unusual conditions, it would be difficult to undergo reproduction and likewise difficult to produce healthy offspring.

Fourth, even if hybrid offspring were reproduced, it would be difficult for them to pass on their genes in nature.

So how should we regard the cases of viable offspring reproduced by white-headed langurs and black langurs mating in captivity as observed by Hu Yanling and Chen Yiping (1989)? It appeared that the reproduction barrier between the two could be erased and that the two animals seemed to belong to the same species. We find that this conclusion is not valid, because the mating of the white-headed langur and the black langur was induced through the artificial conditions of forced co-habitation and captivity. In the wild the distribution ranges of the two animals are isolated from each other by the Zuo Jiang and Ming Jiang Rivers. They would not even have a chance to encounter each other, let alone mate.

In the history of the domestication of fowls and other animals by mankind, there have been cases of arranged sexual mating between different species with remotely related ancestries, but only in a select few cases was reproductive offspring produced. Mating horses with donkeys result in sterile mules. Cross-mating the Muscovy duck and the Peking duck produces the Moulard that is sterile. Pheasant species developed the characteristic feature of rich colored feathers after generations of hybridization. With today's technology, the boundary between species can be altered to a large extent by humans. In nature, there are also cases of reproduction between species that produce viable but usually infertile offspring. There are two species of pheasants (red-breasted pheasant and white-breasted pheasant) inhabiting in the same habitat in southwestern China. Researchers can find hybrid individuals from mating between the two species in their overlapping distribution, but the hybrids were infertile. Mismatches happen as well to the well-known Darwin's finches in the Galapagos Islands. As positions of the different species of finches in the phylogenetic tree are very close, their behaviors are all very similar. Songs by the males of different finch species are so alike that a few females during love season failed to distinguish and mated with the wrong species. Nevertheless, field data has

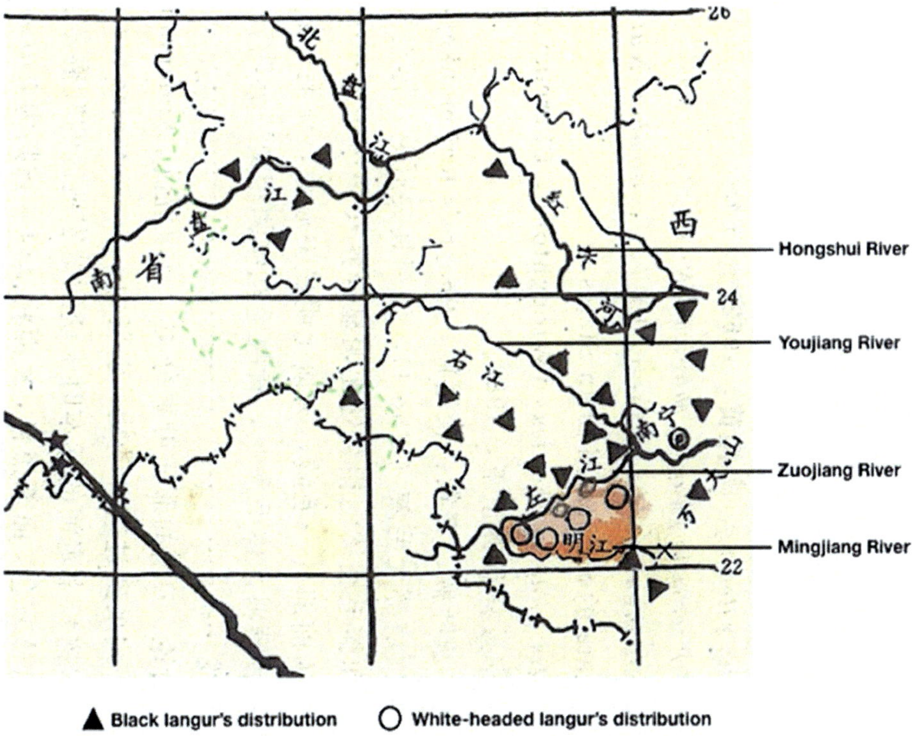

Fig. 3.5 The "small land area" where white-headed langurs distribute (orange-colored area) (map modified from "*Langur Biology*" (叶猴生物学) Ye Zhizhang et al. (1993))

indicated that the rate of interspecies mating is very low, only about 1/50 of normal mating. And research into their natural history has concluded that the distinctness of the various finch species has been maintained in the natural ecosystem for thousands of years. Scientists have found that the hands of natural selection would mercilessly eliminate hybrid progenies from cross-species mating, and to halt abnormal mating, it would mercilessly exterminate hybrid individuals that are not fit for evolution. In the end, various species of Darwin's finches are still sharing the Galapagos Island without blending.

3.3 The Critical Encircling Rivers

The most straightforward method for researchers to differentiate between a land-locked species and a subspecies is by examining its natural habitat and distribution. The distributions of the white-headed langur and the black langur are effectively isolated from each other by two encircling rivers. Reproduction isolation between them has existed since ancient times. These facts alone make it suffice to say that they are two distinct species.

3.3.1 The White-Headed Langur Distribution on the "Small Land Area" Isolated by Rivers

In 20 years of field research, we have found that the distribution of the white-headed langurs remains limited to a "small land area" (Fig. 3.5) surrounded by the Zuo Jiang and Ming Jiang Rivers. Like moats, the two rivers barricaded the white-headed langurs from moving outward and dispersing. Although the two rivers are narrow, they posed real barricades between the black- and the white-headed langurs, thus effectively barring geneflow between the two species. This conclusion is in line with the field research findings by Tan Bangjie (1952–1957) as well as with the research report by Ye Zhizhang et al. (1993).

3.3.2 The Ancestor Species of the White-Headed Langur, the Chongzuo Langur, Was Distributed on the Same "Small Land Area"

From the time of existence and geographical distributions (Fig. 3.6) of the five *Gigantopithecus* Faunas in Guangxi as

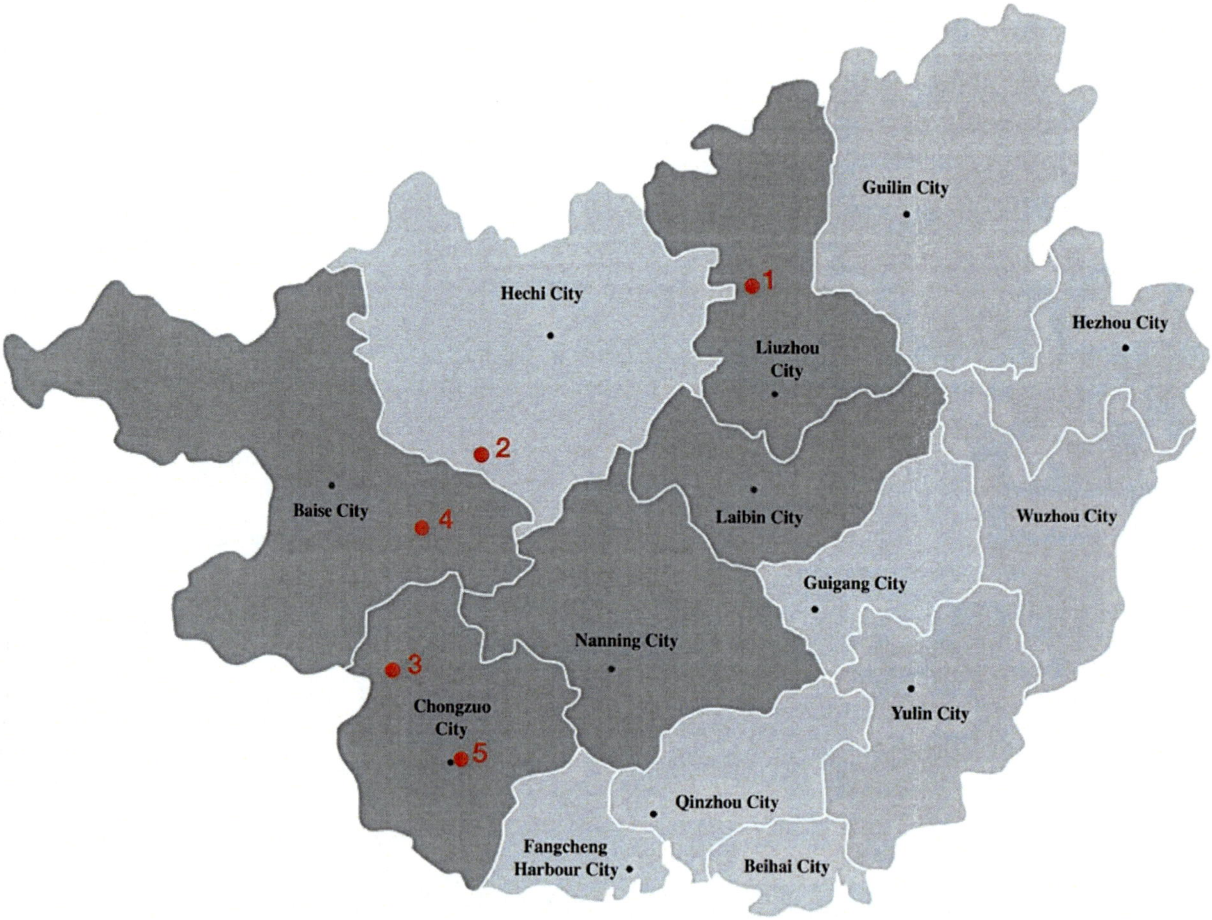

Fig. 3.6 Distributions of the five *Gigantopithecus* faunas. (1) Liucheng *Gigantopithecus* fauna; (2) Bama *Gigantopithecus* fauna; (3) Daxinhei *Gigantopithecus* fauna; (4) Tiendong *Gigantopithecus* fauna; (5) Chongzuo *Gigantopithecus* fauna

illustrated in Table 1.1, we noted the following relevance between time, geography, and the fauna composition:

1. The timing between 1.6 and 1.2 mya coincided with the first glaciations of the Quaternary Period in the Northern Hemisphere, when enormous ice sheets covered a large part of the Northern Hemisphere, causing the sea level to descend and expose the Western Pacific continental shelf above the sea. The new exposed coastland provided a green alleyway, which allowed species to disperse and migrate. Ancient langurs at the time living in Java Island also made use of the "speedy highway," dispersed north into the region south of Zuo Jiang, and became one of the member animals in the Chongzuo *Gigantopithecus* Fauna. At the same time, there were other tropical animals like the Sumatran rabbit (*Nesolagus* sp.), etc. also dispersing north on the same alleyway. The Sumatran rabbit also arrived at the tropical forests south of Zuo Jiang. Its appearance along with ancient langurs in the Chongzuo *Gigantopithecus* Fauna substantiates the time and route of the northward migration of the ancient Javan langurs.
2. The Tiendong *Gigantopithecus* Fauna and the Liucheng *Gigantopithecus* Fauna occurred in the early phase of the Early Pleistocene Epoch between 2.58 and 1.6 million years ago (mya). They both had no langurs and Sumatran rabbits (*Nesolagus* sp.) in their fauna composition. In other words, those two species had not yet appeared in Tiendong and Liucheng in the early phase of the Early Pleistocene Epoch. During this time, the earth enjoyed a warm interglacial period with lots of rain. The sea level was constantly high, and most of the continental shelf was submerged, providing no alleyway to ancient langurs and Sumatran rabbits to cross the vast ocean and migrate northward to arrive at the Zuo Jiang drainage. The event indicates that the northward dispersal of ancient langurs and Sumatran rabbits could only have taken place in the years during the first glaciations in Northern Hemisphere in the middle phase of the early Pleistocene.

3. The first extreme glaciations in the Quaternary Period took place 1.4 mya, and it is conjectured to have ended 1.3 mya. When earth warmed again at about 1.2 mya, the sea level rose to its previous height, and the continental shelf was once again submerged below the sea. The soaring waves blocked the ancient Javan langur and the Sumatran rabbit from returning to Java Island so they had to make the south bank of Zuo Jiang their homeland. In order to survive in their new homeland at the south bank of Zuo Jiang, they had to adapt to the natural environments there and continue adjusting their survival strategies.
4. Why were there no ancient Javan langurs or Sumatran rabbits appearing in the Daxinhei *Gigantopithecus* Fauna and the Bama *Gigantopithecus* fauna? The reason could only be that Zuo Jiang and Ming Jiang Rivers had barricaded the two species from dispersing outward. Geology had revealed that the entire drainage system of Ming Jiang, Zuo Jiang, and Zhu Jiang was the result of tectonics 2.6 mya during the second phase of Himalayan orogeny. Ming Jiang and Zuo Jiang were natural barriers barricading the migration and dispersal of many species long before the arrival of the Javan langur (*Trachypithecus* sp.) and the Sumatran rabbit on the south bank of Zuo Jiang.

3.4 The White-Headed Langur Is a Recent Lineage

Putting together fossil findings and analysis of DNA traces may inspire insight and provide unequivocal evidence. The fossil species *P. sivalensis*, the oldest of the langur family, which existed in Pakistan in the Miocene Epoch of the Tertiary Period (between 8 and 5.5 mya), is generally regarded as the ancestor species of the langur family and might have come from Africa. Fossils of langurs (*Presbytis* spp.) were widely found across India and Pakistan in subsequent years. It is therefore conjectured that the Indian subcontinent was the location of origin for langur species. During the Pliocene Epoch (3.6 and 3.4 mya) when the sea level dropped about 50 m, ancient langurs started to disperse toward the east of India and reached as far as Myanmar (Burma) and Southeast Asia. In the late Miocene Epoch (about 2.8 mya), the earth once again entered a glaciation period and the sea level at that time dropped by more than 100 m. A troop of ancient langurs (*Presbytis* sp.) took the opportunity to disperse to ancient Sundaland. Researchers reckoned from DNA traces that diversification of ancient langurs occurred on Sundaland during the transition period between the late Miocene Epoch and the Early Pleistocene Epoch (about 2.1 mya). Jablonski and Tylor (1999) noticed that fossils of the ancient langurs which appeared in central Java 1.9 mya were already very much alike the large continental type langur (*Trachypithecus*) which appeared in the late phase of the Early Pleistocene Epoch. We also found that the morphology of the mandibular third molar of Chongzuo langur (*Trachypithecus Chongzuosis*) (see Fig. 1.10), specifically the number and the shape of its cusp fine-scale morphology, is very similar to those of its ancestor *P. sivalensis*. It is also very similar to the mandibular third molar of the extant white-headed langur. These morphological similarities provided evidence that the extant white-headed langur, its lineal ancestor the Chongzuo langur, and the ancestor species *P. sivalensis,* all belong to one and the same clade.

The tooth of the fossil species of the Chongzuo langur was buried in the sediments from 1.6 to 1.2 mya in the Chongzuo Sanhe Cave. It turns out to be monumental evidence concerning the evolutionary and migration history of the langur family. It provides us hints that trace back to central Java, which leads us on to Myanmar (Burma) and eastern India, and continues westward to Pakistan where ancestors of the Chongzuo langur appeared. It gives us a clear outline of the origin of the langur family and the journey their evolutionary history had covered (Fig. 3.7).

New species often appear when a branch of an ancient species, which the new species is lineal to, becomes spatially isolated and is compelled through natural selection to adjust morphologically and ethologically, until it completely adapts to the environment physically and biologically. White-headed langurs have since ancient times lived only in the karst hills on the south bank of the Zuo Jiang River. On that "small land area," they stood through trials by natural selection. We could not be sure of the exact time taken for Chongzuo langurs to evolve into white-headed langurs, but we can be sure that span of time would not be longer than 1.4 million years. In comparison to the minimum known duration of eight million years of evolutionary history of langurs (Fig. 3.7), white-headed langurs are to be considered a recent lineage.

3.5 The White-Headed Langur Is a Resident, Endemic Species of Chongzuo

All animal species come into existence at a specific location, which remains inherently linked to its evolutionary history and is often central to its distributional range. If the members stayed there since the species' origin and also do not migrate seasonally, they are considered a resident species. Take the orangutans, the Sumatran rhinoceros, and the Malayan tapirs for example, which originated in the tropical rainforests of ancient Sundaland (present-day Kalimantan, Sumatra, and

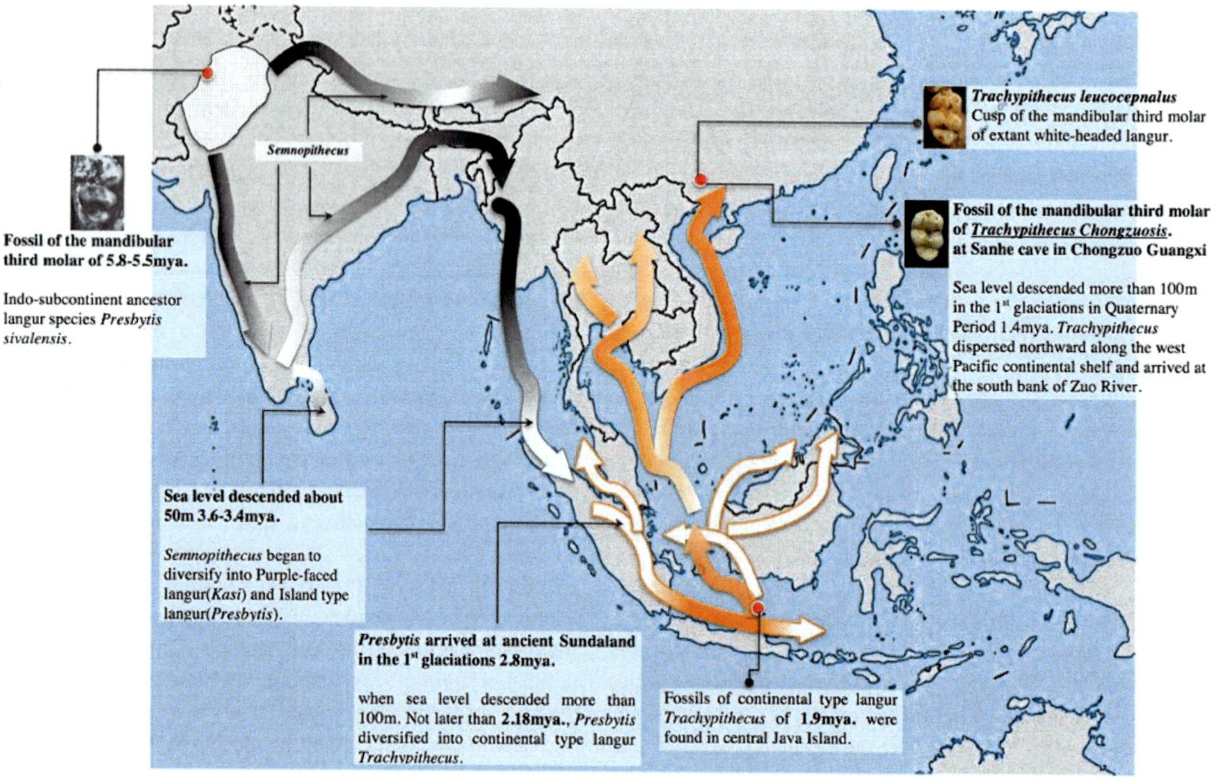

Fig. 3.7 The fossil-found locations on the map take us through time and space to get an overview of the eight million years of evolutionary history of langurs (Drawn by Pan Yue)

Malaysia) and kept inhabiting this region ever since, now known as the Malayan Archipelago. If the distribution of a species is restricted to a single specific limited area, then that species is called endemic. For example, the very first giant pandas (*Ailuropoda melanoleuca*) arose in Guangxi (very likely at Chongzuo), then expanded their distribution and dispersed to the north. Once their distribution covered vast expanses in southeastern China, but in the last hundreds of years, they lost many of their habitats in low mountains and river valleys and were compelled to withdraw to the high mountain gorges on the eastern edge of the Tibetan Plateau and to the high mountain peaks on the southern slopes of the Qinling Mountains. The giant panda occurs nowhere else; hence, it is an endemic species of southern China.

For over a million years, white-headed langurs and their lineal ancestor, the Chongzuo langur, have lived only in the south bank of Zuo Jiang River and only in the areas with a particular landscape of karst hills. This region was and is the core range of the white-headed langur. So, the white-headed langur is both a resident and endemic species of Chongzuo, in the Guangxi Province of China.

The existence of the spectacular white-headed langur is gratifying to the people of Chongzuo and provides them with a sense of pride for their splendid natural heritage. We have reason to believe that the white-headed langur will continue to thrive thanks to its steadfast adaptation abilities that have enabled it to prosper on the karst hills, which are their home.

References

Brandon-Jones D. The colobus and leaf-monkeys. In: MacDonald DA, editor. All the world's animals. New York, NY: Torstar Books; 1984. p. 102–12.

Delson E, Tattersall I. Primates. In: Encyclopedia of human biology, vol. 7. New York, NY: Academic; 1997.

Eudey AA. Action plan for Asia primate conservation. Gland: IUCN – The World Conservation Union; 1987.

Fleagle JG. Old world monkeys. In: Primate adaptation & evolution. San Diego, CA: Academic; 1988. p. 159–201.

Groves CP. Primate taxonomy. Washington, DC: Smithsonian Institution Press; 2001.

Jablonski NG, Tylor DE. Trachypithecus auratus sangiranensis, A New Fossil Monkey from Sangiran, Central Java, Indonesia. Int J Primatol. 1999;20:319–326.

Lantian S, Hanhua L. White-headed langur in Guangxi (广西的白头叶猴). Guangxi Normal Univ J Nat Sci. 1982;1:27–33.

Liren L, Zhaoyuan L. Discussions on white-headed langur systematics and idea exchanges with Ma Shilai (论白头叶猴的分类:兼与马世来商榷). Guangxi Normal Univ J Nat Sci. 1991;02:67–70.

Mingzhen Z. Quaternary zoogeological evolvements in China (中国第四纪动物区系的演变). Chin J Zool. 1964;6:274–8.

Po D, Yaping Z, Ziming L, et al. RAPD analysis and probes on white-headed langur taxonomic position (RAPD分析与白头叶猴分类地位探讨). Zool Res. 1999;20(1):1–6.

Roos C, Thanh VN, Walter L, et al. Molecular systematics of Indochinese primates. Vietnam J Primatol. 2007;1:41–53.

Tan PC. Rare catches by Chinese animal collectors. Zoo Life. 1957;12:61–3.

Yanling H, Pengcheng Q, Chengming H, et al. Probes on white-headed langur taxonomic position (关于白头叶猴分类地位的探讨). Chin J Zool. 2004;39(04):109–11.

Yiping C. Successful outbreeding of white-headed langur with black langur (白头叶猴与黑叶猴的杂交成功), Chinese Zoos, compiled by the China Zoos Editorial Board. Garden J Publ. 1989;1989:12–3.

Zhixiang L, Shilai M. Rectifications on the systematics of white-headed langur (白头叶猴的分类订正). Zool Syst. 1980;1980:4.

Zhizhang Y. Langur biology (叶猴生物学). Kunming: Yunnan Technology Publishing House; 1993.

Ziming L, Yi W, Xiuzhen M, et al. Mitochondria ND4 genome and D-loop sequencing in white-headed langur and probes on its taxonomic position (白头叶猴线粒体ND4基因和D-环区的序列及其分类地位初探). Guangxi Sci. 1997;1:64–71.

Check for updates

4 Poaching of White-Headed Langurs by the Late Twentieth and Early Twenty-First Century

Abstract

Poets in Ming Dynasty and Qing Dynasty portrayed the white-headed langur in the Longrui Mountains during the fifteenth–nineteenth century as "*Thousands of monkey hoots fill the forsaken land.*" But the late twentieth century and the early twenty-first century had become a catastrophe for the white-headed langur. In 1977, three Guangxi researchers pointed out that the white-headed langurs were distributed in six karst mountain districts on the south bank of Zuo Jiang River with a total area of 500–600 km^2. In 1988, Pan Wenshi and his research team ventured into all the white-headed langur habitats and found a still rich biodiversity. They estimated the number of white-headed langurs between 1000 and 1200. In November 1996, Pan Wenshi set up the Peking University Biodiversity Research Base in the Nongguan Mountains. At that time, we witnessed rampant poaching activities in Nongguan Mountains, which shattered the white-headed langur families. From 1999 to 2002, Ran Wenzhong penetrated deep into and surveyed the known distribution range of white-headed langur, spending a total of 2359 h studying the vastly reduced langur communities while interviewing forest guards and local residents. He discovered that four out of six regional populations had been virtually wiped out. Apart from blasting mountains and burning wasteland for agricultural reclamation, it was rampant poaching that had decimated the species in just a few decades. The enormous demand for tonics extracted from langur organs and for langur bushmeat put the white-headed langur on the road to extinction.

Keywords

Surveys · Population abundance · Poaching · Habitat destruction · Over-exploitation

As was related in earlier chapters, it took more than a million years for the white-headed langur to evolve into one of the most unique primates, adapted to living in the karst hills on the southern bank of the Zuo Jiang River. When the anthropoid has not yet evolved into human, the white-headed langur has already been living in the southern bank of the Zuo Jiang River in the jungle. Since the emergence of *early Homo sapiens*, the fates of the two have been intertwined: they could coevolve in early times, but in the mid-1950, massive cutting of trees in the tropical monsoon forest in the Zuo Jiang River basin forced white-headed langur to retreat into six pieces of discontinuous distributions. They might still be able to survive on these karst hills, but the rampant poaching at the turn of the century has driven two-thirds of their population to extinction in just 5 years.

We should look into the state of the white-headed langurs in the past to have an accurate view and a realistic approach to its present and future.

4.1 The State of White-Headed Langurs Before 1980s

4.1.1 White-Headed Langurs in History Books

The earliest texts on white-headed langurs appeared in a fifteenth-century poem by litterateur Xia Yin (1482–1548) of the Ming Dynasty (History and Chronicles Office 2000). In the poem, he rhapsodized about the bustling animals living at a bend of Ming Jiang River (present-day Longrui Mountainous Region in Ningming County):

> Like a phoenix opening its wings, the mountain in Longzhou unfurls
> Round the canton, swerving Ming Jiang River flows
> Don't be scared off by the monkey hoots in the forsaken land

W. Pan, *The Population Ecology of White-Headed Langur*, https://doi.org/10.1007/978-981-33-4118-0_4

There General Fubo[1] had copper pillars forever stand.

The "hooting monkey" in the poem was the white-headed langur. Until today, native inhabitants everywhere in Chongzuo still call the white-headed langur the "white-headed darkish monkey." The "swerving Ming Jiang River" referred to the bend where Ming Jiang River joins Zuo Jiang River. The canton where Ming Jiang River swerved round is the present-day Longrui Mountains at the county border between Ningming and Longzhou.

Many expedition groups from Western countries had come to China in the mid-nineteenth century. They were impressed by various exotic vegetation and animals of China and during their expeditions collected many species that were not found in the West. Yet the white-headed langurs living at the south bank of Zuo Jiang River remained out of their sight. In the same period (Qing Dynasty), a *Zhuang* poet from the Chongshan County by the name of Xie Lan portrayed the white-headed langurs in the Longrui Mountains in the first poem in his manuscript "*8 Zhuzhi-arrangement Poems of Lijiang*" (Zhang Jiong 1997). It reads:

Where terrain of Ming Jiang River and Longzhou adjoin
The two rivers from apart mingle into one
Thousands of monkey hoots fill the forsaken land
Where General Fubo had copper pillars forever stand.

Throughout the 400 years from the fifteenth to the nineteenth century, poets could hear the hoots of the "white-headed darkish monkeys" in Longrui Mountains. Was there any more of their hoots to enter people's ears in the twentieth century?

4.1.2 A 1977 Survey of White-Headed Langur Distributions

In the months June through August, and then again in December 1977, researchers Shen Lantian, Li Hanhua, Wu Mingchuan, et al. from the Guangxi Precious Animal Resources Survey Team conducted ground-breaking and foundational research on the geographical distributions of white-headed langurs and their numbers. They came up with a survey report that was significant for its time. The treatise on their research not only contained accounts of the actual conditions of white-headed langurs in those days but also laid the grounds for the research of white-headed langur ecology in the future.

- Excerpt from "*White-headed langurs of Guangxi*" (广西的白头叶猴) by Shen Lantian and Li Hanhua (1982):

White-headed langurs are distributed on the piece of very narrow delta sandwiched between Zuo Jiang River and Ming Jiang River in the south-western part of Guangxi situating at approximate longitude 107–108° and latitude 22°06′–22°42′. The region encompasses parts of Xiangshui Commune and Shangjin Commune in Longzhou County, parts of Tingliang Commune and Tuolong Commune in Ningming County, the whole of Luobai Commune, Laituan Commune and Tuolu Commune in Chongzuo County, and parts of the five communes in Fusui County, namely Qujiu, Quli, Bapan, Shanwei, and Dongmen. The total area with white-headed langur distributions is approximately 199.5 km^2.

The researchers used the "absolute value by small location method" and the "relative values from line transect and belt transect method" to find the number of white-headed langurs in the distributing area. They also visited all the experienced hunters in the area and obtained their estimated numbers. The numbers were then matched against each other and cross-checked. The result gave the researchers a preliminary overview of the number of white-headed langurs and locations of their distributions. Shen Lantian and Li Hanhua divided the delta-shaped region containing the white-headed langurs, at the east of Ming Jiang River and south of Zuo Jiang River, according to the natural landscape and the distributions of the white-headed langurs. The region was divided into three segments: western, central, and eastern. The western segment encompassed parts of the communes Xiangshui and Shangjin in Longzhou County and parts of the communes Tingliang and Tuolong in Ningming County, covering a total area of about 68 km^2 with 244 white-headed langurs inhabiting it. The central segment encompassed Luobai Commune, Laituan Commune, and Tuolu Commune in Chongzuo County, covering an area of about 43.5 km^2 with 117 white-headed langurs. The eastern segment encompassed the five communes in Fusui County, i.e., Qujiu, Quli, Bapen, Shanwei, and Dongmen, covering an area of about 88 km^2 with 272 white-headed langurs. The aggregate limestone mountain area with white-headed langur distributions of the four counties mentioned above came to 199.5 km^2 with a total of 633 white-headed langurs, which averaged to 3.18 white-headed langurs/km^2 (see Table 4.1 and Fig. 4.1).

- Excerpt from "*An Overview of Wide Animal Resources in Guangxi*" (广西野生动物资源概况) by Mr. Wu Mingchuan (1983) of Guangxi Forestry Survey and Planning Institute.

The white-headed langur distributes only in Guangxi, China. It has never been found anywhere else in the world. As a rare and precious species, it is classified as a Schedule I protection species in the List of Endangered and Protected Species of China. White-headed langurs are arranged in a stripe of narrow, delta-shaped karst regions where the various mountain tracts of the 12 Communal counties Longzhou, Ningming, Fusui, and Chongzuo meet. It has an area of 500 km^2 and is roughly estimated to have about 400 white-headed langurs living in the region.

[1] Fubo was a general of the East Han Dynasty around 43 AD who erected copper pillars at Jiaozhi, present-day northern Vietnam, to mark the southern border of the East Han Dynasty.

Table 4.1 White-headed langur population in some of the distributions in the 1979 survey (compiled from research report 1982 by Shen Lantian and Li Hanhua)

Segment division	County	Commune	Topography of the area with distributions	Surveyed area (km^2)	No. of white-headed langur	Density (no./km^2)
Western	Longzhou	Xiangshui, Shangjin	Densely forested karst hills in Longrui Mountains	68	244	3.59
	Ningming	Tingliang, Tuolong				
Central	Chongzuo	Luobai, Laituan, Tuolu	Karst hill forests and shrub forests (six distributions)	43.5	117	2.69
Eastern	Fusui	Qujiu, Bapan, Quli, Shanwei, Dongmen	Karst hill forests (nine distributions)	88	272	3.09
Total				199.5	633	3.17

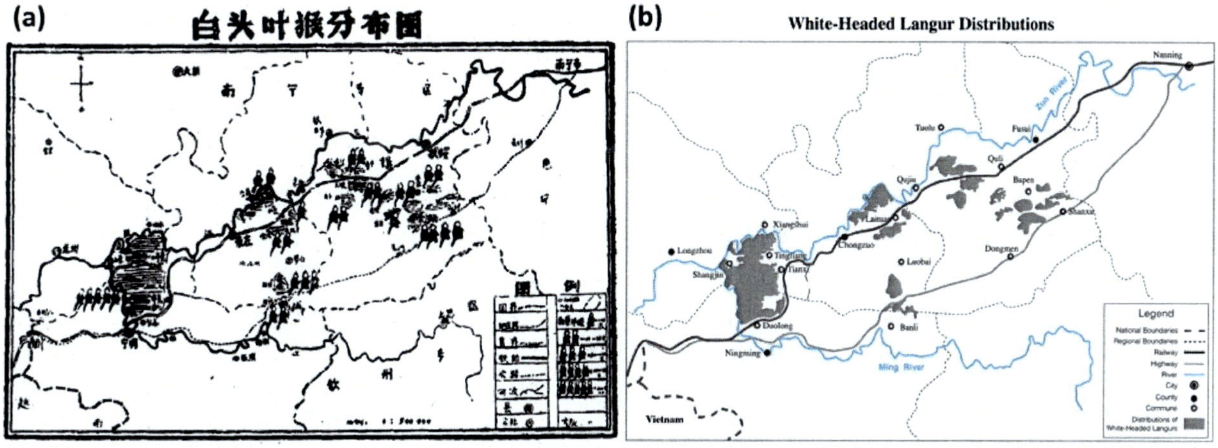

Publisher's note: Springer Nature remains neutral with regard to jurisdictional claims in published maps and institutional affiliations.

Fig. 4.1 (**a**) White-headed langur distributions in 1977 (Shen Lantian 1982) (**b**) Repaired diagrammatic sketch from 4.1 (a)

The survey reports of Shen Lantian and Wu Mingchuan had shown that even though the massive tree cutting during the Big Leap Forward and the calamities of the Cultural Revolution had significantly reduced the size of white-headed langur habitats, the remaining karst hill areas (including the Longrui Mountains) in the early 1980s were still sustaining the survival of white-headed langur.

4.2 White-Headed Langurs Were Still Spotted in the Late 1980s

Following is an excerpt from Pan Wenshi's December 30, 1988, field survey diary:

> Until today, the tropical monsoon rainforest in Longrui still appears as it was 2 million years ago. Various types of original tropical broadleaf evergreen trees are enormous and soaring silently to the sky amidst the karst hills. Lianas, thick and sturdy, are creeping all over the precipices. It remains nothing short of a remote wilderness that has so far escaped human devastation. Its iconic animal, the white-headed langur, lives high up on precipitous precipices. While the huge rocks that landscaped Longrui Nature Reserve had come to be surrounded by farming fields on all sides, the hilly terrain of the Reserve had made agricultural planting there infeasible. As a result, the big trees, shrubs, and the various vine species are left flourishing luxuriantly. Huge trunks and hardy branches are still growing wildly, and the canopy screens the sun just as much as it did in the past. Through the binoculars, I see for the first time tens of tropical frugivorous mega-bats 70 m away, crammed together, hanging their heads on the branches of a shriveled tree, which I fail to discern the name. Among extant mammals, mega-bats can be traced back to the most ancient lineage. As I went into the Longrui Nature Reserve, I came across 35 noisy macaques, and not long after that, I encountered a rare and precious Guangxi red-faced macaque, which came as near as 3 m away from me.
>
> An old patroller of the Nature Reserve, Mr. Huang, said he could lead me to see a troop of 11 lorises. Loris forages at night

and stays at a regular place during the day. Mr. Huang also said that there was still at least 1 female leopard, which was seen leading her 3 cubs to forage in the woods. He saw them once, crouching on a tree, ready to ambush a wild boar. There were also masked palm civets and Siberian weasels, which turned up every night at the village margin. Villagers of the Panlong Village said that several years ago, a tiger prowled about the boundary of the village every evening around 16:00. . .In fact, up until the end of the 1980s, there had still been rich biodiversity.

Pan Wenshi had covered about 20 km^2 in the core of the white-headed langur distribution, which amounted to one-tenth of the karst hill area in the Longrui Nature Reserve (Figs. 4.2 and 4.3). By summing up the number of white-headed langurs in the 13 troops he had come across, he found that there were 119 white-headed langurs: No. 1 + No. 2 + No. 3 + (No. 4–No.10) + No. 11 + No. 12 + No. 13 = 10 + 19 + 5 + 65 + 8 + 1 + 11 = 119. All of them were distributed in the core area, making an average density of 5.95 langurs/km^2.

The population density in a species distribution generally decreases gradually from its core areas toward its perimeter areas. If we set the density at the core area as 1 and decrease it gradually toward the perimeter in three concentric rings, the density in the next ring about 120 km^2 outside the core ring would be halved, and it would be halved again in the outermost ring 60 km^2 further still from the core. Then our calculations are:

- Number of white-headed langurs in the core area is 119 langurs, where the average population density is 5.95 langurs/km^2.
- Number of white-headed langurs in the 120 km^2 outside the core area will be approximately $(5.95 \times 120) \times 0.5 = 357$ langurs.
- Number of white-headed langurs in the yet further extended 60 km^2 will be approximately $(5.89 \times 0.5 \times 60) \times 0.5 = 89.25$ langurs.
- Hence, we got the statistical number of white-headed langurs in Longrui Nature Reserve at the end of December 1988, which was 119 + 357 + 89.25 = 565.25 langurs. The figures may serve as a reference for future researches.

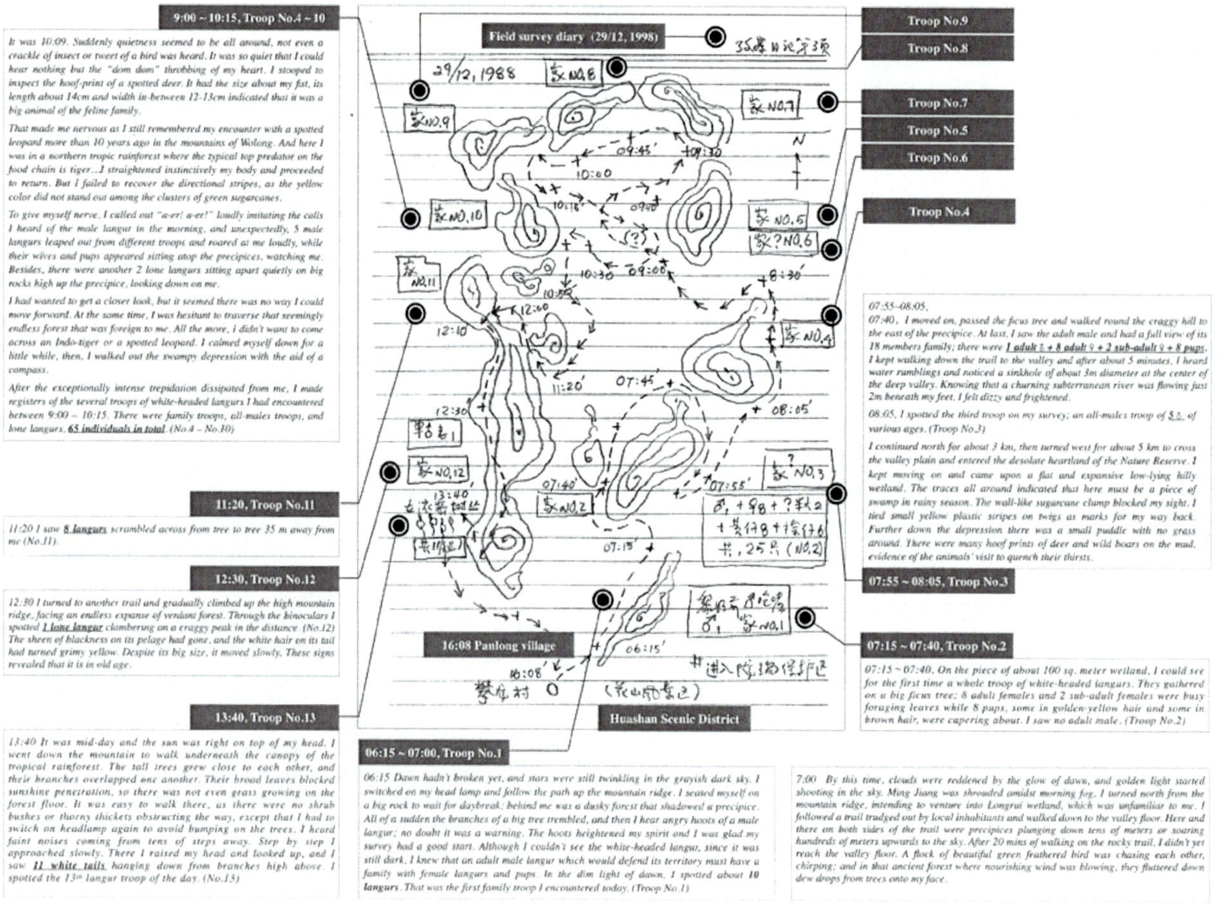

Fig. 4.2 Pan Wenshi's hand-drawn map of Longrui nature reserve at the end of 1988

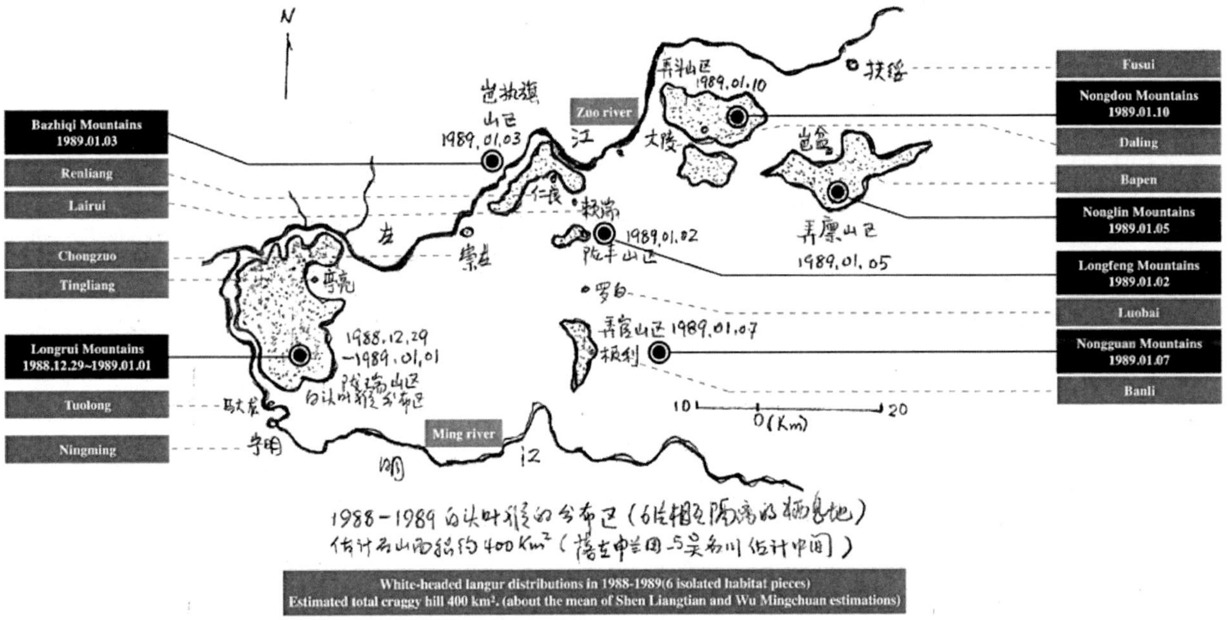

Fig. 4.3 Sketch by Pan Wenshi on his 1988–1989 survey of white-headed langur distributions. The number of white-headed langur was estimated to be 1000–1200

Pan Wenshi's January 10, 1989, diary:

> In the last ten days, I traveled from the stepped topography in the northern bank of Ming Jiang River to the river terraces at the southern bank of Zuo Jiang River. Along the way, I saw a few isolated villages, and all the big trees within reach of the villages were felled, and every piece of cultivable land was cultivated. All the evidence indicated that people there had been farming for a long time. However, the sight was completely different when I stepped into the region of karst hills. I saw precipices and huge rocks of all shapes strewn about. The land was too dry, and the soil was too infertile for cultivation. Nobody inhabited those mountain tracts, but any big trees had all been chopped in the 1950s, and the valleys and hill slopes were laid barren.
>
> Nonetheless, from the skirts of the karst hills towards the peaks, thorny bushes, shrubs, and meso-trees seemed to be able to regenerate, because the columns of karst hills with jagged precipices inhibited humans from entering the region. If there were flat and fertile river terraces, nothing would have remained on the Chongzuo basin. Nature has, in fact, shown us its readiness to sustain the white-headed langurs in the area, and I earnestly pray that is going to last.

4.3 The State of the White-Headed Langurs in the 1990s

In the spring of 1998, 10 years after Pan Wenshi's first solo visit to Longrui Nature Reserve, he led a team of 35 graduates from the Peking University School of Life Science with hopes of spotting the white-headed langurs once again. He took the students on the same familiar mountain trails that he trekked 10 years ago. The tropical forests, the huge Renmianguo and ficus trees, the shrubs by the creeks, and the woody vines creeping on the precipitous precipices all looked the same to me as 10 years ago. The northern tropical climate and environment had not changed a bit, but no more wild animals could be seen. The troops of macaques, the spotted deer, and wild boars disappeared, and so did the red-faced macaque, the spotted leopard and the 11 lorises. Not even one white-headed langur was spotted. Pan Wenshi asked the villagers of Panlong Village why the white-headed langurs vanished and received a wry reply, "We saw many white-headed langurs in your TV program series, *Zuo Jiang River Diary*, produced by CCTV, and we figured the white-headed langurs left here for CCTV." That of course was not true. The villagers all knew the reason behind the disappearance of the white-headed langur but no one dared to say it.

In just 10 years, 500–600 white-headed langurs living in a 200 km^2 karst tropical monsoon rainforest habitat had vanished! What were the reasons behind this disappearance?

4.3.1 Poaching Incidents in the Nongguan Mountains

The poaching of the white-headed langur has become so widespread that they happened even in and around our Peking University Research Base. Below are real incidents Pan Wenshi came across in his fieldwork:

> **Incident 1**: It was a heart-breaking morning on April 3, 1998. I saw a white-headed langur being savagely killed by poachers, an

incident I could never forget. I saw the female white-headed langur's head hanging down the branch of a tree atop the 70 m high Sanhe Cave. I was unable to climb up the precipitous wall to rescue her and remained standing below the cave in disquiet. I recognized she was Heimama, the eldest female of the troop we were studying. Her right leg was clamped by a 'steel cat' (a type of animal trap), and the trap had a chain securely tangled with shrub branches so she couldn't get away. I couldn't take my eyes off her wretched face. After a few feeble winks of her eyes, the fire of life in her was extinguished.

Incident 2: One week after the first incident, at noon on April 10, 1998, when Liang Puzhang and two other farm workers from Leizhai Cantonment were working in the fields in the vicinity of my field research base, they heard the wailings of a white-headed langur. They climbed the hill to the direction of the wail and at the halfway point, found a white-headed langur clamped by a 'steel cat,' unable to get loose. The three kind-hearted farmers carried it down to the research base for help. It was an adult male. I examined its entire body and found no injuries except on its left wrist bones, which had been crushed from the sturdy grip of the clamp. When I bandaged its wound with Yunnan Salve, the langur did not move but blinked its eyes meekly. It then rested on the floor and consumed a little water, and some of the puaeraria leaves we fed. After 2 hours it arched its body and rose, and using its 3 unwounded limbs leaped limply back to the karst hill... Three days later (1998.4.14), we spotted it plucking leaves with its right hand on a tall tree 6 km from the research base. I knew that the male langur, having lost a palm, would not stand a chance when competing for reproduction. It would have no chance to have offspring of its own in its lifetime.

Incidents happened as well beyond the Peking University Research Base vicinity. Poachers laid traps everywhere in the mountain tracts in the Nongguan Mountains. They put rat glue on branches to catch infant langurs, used bamboos to put up large nets at the entrances of caves on precipices, and placed tripping snares, steel clamping traps ("steel cat"), etc. on white-headed langurs' passages (Fig. 4.4). All sorts of tools were used to poach white-headed langurs. We noticed that many "suspects" who were sneaking into the Nongguan Mountains to spy on the usual passages of white-headed langurs were riding motorcycles bearing non-local number plates.

Given the shattering of white-headed langur families and the dwindling population due to rampant poaching, if we did not take emergency actions, in just a few months, the white-headed langurs in Nongguan Mountains could be completely wiped out by poachers.

After that, what would become the fate of the even smaller populations on the other geographical locations? We had to conduct in situ surveys.

4.3.2 Field Surveys by Ran Wenzhong in 2001

By the end of the last century, poachers in the Nongguan Mountains were different from that of individual amateur hunters in history. They had since become occupational, organized, and with classified job duties. The plight in the Nongguan Mountains had alarmed us to also pay attention to the conditions of white-headed langurs in other locations.

From 1999 to 2002, Ran Wenzhong, while studying for his Ph.D. degree at Peking University School of Life Science, had conducted in-depth surveys and research on the distributions and the numbers of the six regional white-headed langur populations spreading over the region surrounded by Zuo Jiang River and Ming Jiang River. He worked in the wild for a total of 581 days, spending a total of 2359 h penetrating deep into the Longrui Nature Reserve, the Longfeng Mountains, and the Bazhiqi Mountains. He

(a) (b)

Fig. 4.4 (**a**) Bamboo frame built by poachers; (**b**) "Steel cat" trap (Photo by Pan Wenshi) interviewed the inhabitants and forest patrollers, collected statistical data of cave shelters currently used by white-headed langurs, abandoned ones with white-headed langur spoors, and found that four regional populations have been wiped out.

The following are excerpts from Ran's field registers of those days (Ran Wenzhong 2003).

4.3.2.1 White-Headed Langur Population Survey Findings in Ningming–Longzhou Region

Ran Wenzhong surveyed nine routes in total in the Longrui and Longshan region at the border between Ningming County and Longzhou County, which spans a total of 55.3 km. The total area with white-headed langurs in the region was 180 km^2, which is more than 30 times that of the Nongguan Mountains, and more than five times the entire area with high-density white-headed langurs in Fusui County. The surveyed area covered by the nine routes amounted to 25.86 km^2, which came to 14.4% of the whole white-headed langur distribution area in the Longrui and Longshan region (Fig. 4.5).

Ran did not even come across one white-headed langur in that survey (Table 4.2), nor did he hear any sounds made by white-headed langurs. That is to say, data collected by way of "line transect" showed that the number of white-headed langurs in the Longrui and Longshan regions at the end of the 1990s was 0. Alternatively, by way of night shelter counting, there was no active night shelter found on all nine surveyed routes. The total of 147 night shelters found were all obsolete ones (Table 4.2). Thus, it is sufficient to say that the number of white-headed langurs in the entire 25.86 km^2 surveyed area was 0.

Historically, the Longrui region had widespread white-headed langur distributions with high densities and numerous populations. Until the end of the 1980s, the situation there still matched the findings from surveys at the end of the 1970s. In the report by Shen Lantian et al. (1982), the region had the most extensive distribution of extant white-headed langurs and was the most populous among all white-headed langur distributions. The number of obsolete night shelters registered by Ran also bore evidence to the white-headed langur boom there once was.

Ran Wenzhong interviewed 11 people, all of whom were forest patrollers in the survey area with the duty of overseeing conditions of white-headed langurs. They patrolled their duty territory no less than once a month. The 11 patrollers belonged to ten different villages, seven of which were in

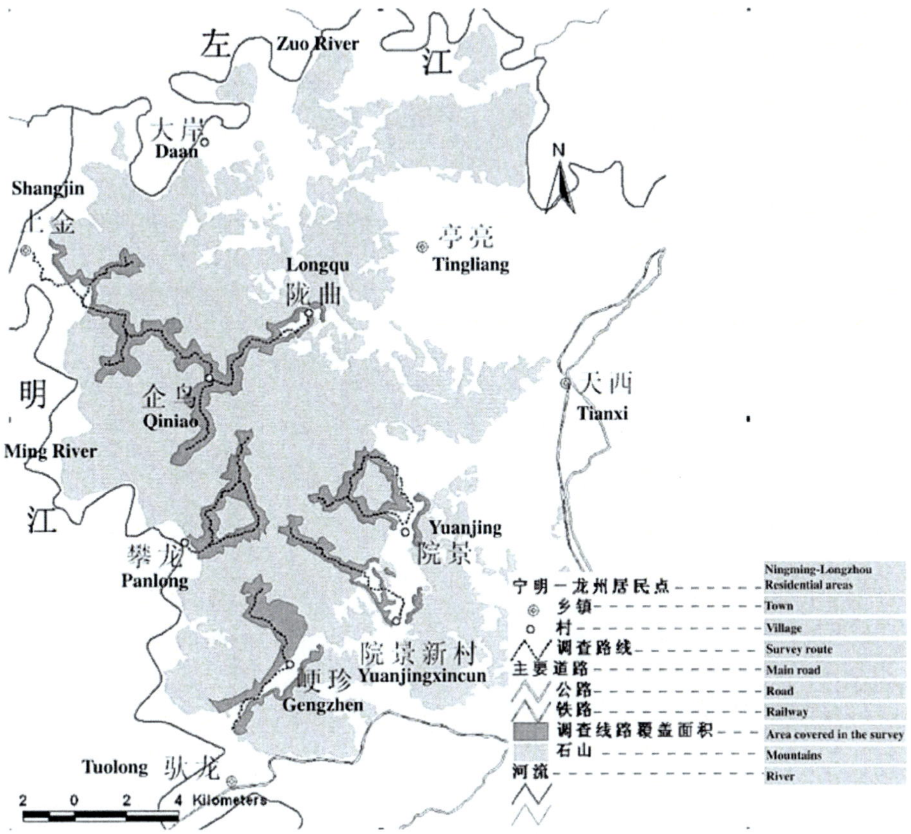

Fig. 4.5 Survey routes taken by Ran Wenzhong in Ningming–Longzhou region in 1999–2000

Table 4.2 Ningming–Longzhou region white-headed langur population survey findings (Quoted from the doctoral thesis by Ran Wenzhong, Ph.D.)

Route	Route length (km)	Survey covered area (km^2)	No. of langur	Inferred no. of langur	In-use night shelter found	No. of langur inferred from in-use night shelter found	Obsolete night shelter found
1	9.4	4.15	0	0	0	0	21
2	6.9	5.08	0	0	0	0	3
3	5.8	2.52	0	0	0	0	14
4	9	3.56	0	0	0	0	24
5	5.2	10.55	0	0	0	0	16
6	3.3		0	0	0	0	22
7	4		0	0	0	0	16
8	7		0	0	0	0	13
9	2.5		0	0	0	0	18
Total	55.3	25.86	0	0	0	0	147

Table 4.3 Findings from forest patroller interviews in Longrui and Longshan regions (no. of patrollers = 11)

Subject		Villages in the survey and findings
Villages		Panlong, Xinlong, Yuanjing New Village (two villages), Yuanjing, Longqu, Qiniao, Longbu, Longmen, Longkou, Shangjin
The years 1980s	No. of white-headed langurs	Many (11)
	Distributing area	On the periphery of the villages (11)
The years 1990s	No. of white-headed langurs in the early 1990s	Many (11)
	No. of white-headed langurs in the late 1990s	Very few (11)
	Distributing area	Within conservation area
The last time spotting white-headed langur		Year 2000 (Jiong 1997), April 2002 (Johanson 2001)
Are there cutting trees and hunting activities?		Yes (Wu 1983), a few (Liu et al. 2006)
The reason responsible for the dwindling of white-headed langur population		Poaching (Stringer 2003), but had not ever witnessed (Johanson 2001), don't know (History and Chronicle Office 2000)

the periphery of white-headed langur habitats, and three were inside the habitats. In the interviews, only one patroller reported seeing one solitary langur on Route 6 in April 2002. The last time the other ten patrollers witnessed a white-headed langur in the Nature Reserve had been 2 or 3 years ago. They were seen within conservation areas (Table 4.3).

All the 11 patrollers interviewed agreed that the number of white-headed langurs in the Longrui Region had kept dwindling since the 1980s. They all remembered that there were many white-headed langurs before, which could be easily spotted on the margins of their habitats near the villages. The obsolete night shelters found have verified that white-headed langurs could still be spotted in the early 1990s, but since 1997, they were hardly seen.

Eight of the interviewed patrollers believed that there were still hunting activities and the chopping of trees for fuel in the conservation area. Animals that were hunted were primarily wild boars, masked palm civets, bamboo rodents, etc., but six of the patrollers believed that hunters would hunt white-headed langurs if they ever came across one (Table 4.3). Hence, poaching was plausibly the cause of the dwindling white-headed langur population.

4.3.2.2 White-Headed Langur Population Survey Findings in the Longfeng Mountains (Laituan) and the Bazhiqi Mountains (Renliang)

At the end of April 2001, Ran Wenzhong surveyed the distribution and population of white-headed langurs in the two townships in Chongzuo County, Laituan Township (Longfeng Mountains) and Renliang Township (Bazhiqi Mountains) and surveyed the habitat conditions there. The survey was conducted principally in two regions: one being the region about 20 km^2 in size lying between Pujiang Village and Weizhou Village in the Longfeng Mountains and the other

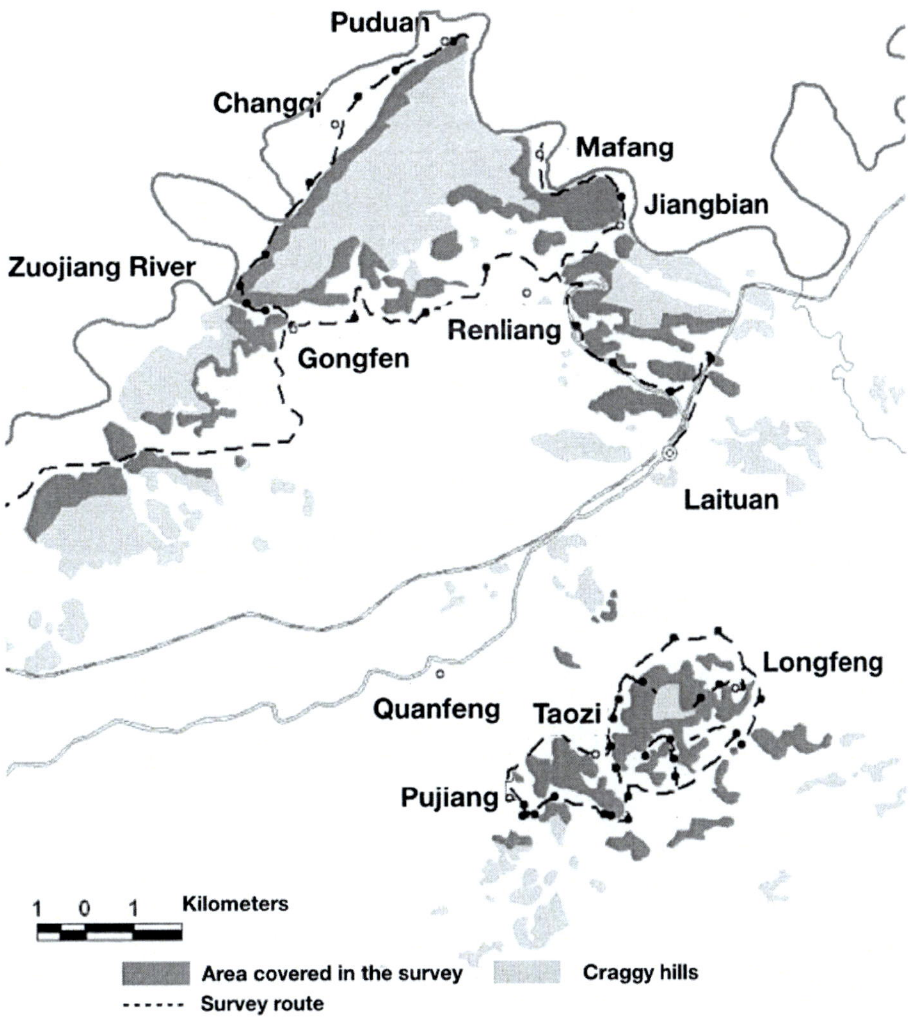

Fig. 4.6 Survey routes in Laituan and Renliang taken by Ran Wenzhong in 2001

region is about 45 km^2 in size in Renliang Village in the Bazhiqi Mountains at the south bank of Zuo Jiang River (Fig. 4.6).

The survey had interviewed 30 people from 12 villages. Seventeen people from Longfeng Territory had spotted white-headed langurs not so long ago, but all of them were seeing the same troop or the same several lone langurs. Thirteen people from Renliang Territory had spotted white-headed langurs, but they were all solitary ones.

All the people interviewed agreed that there were many white-headed langurs before the 1980s, and they could easily be found on the periphery of their habitats in places like in the section of the Longfeng Mountains between Longfeng Village and Taozi Village, in the mountain west of Taozi, in Renliang Mountains, and the villages on the hills beyond Laituan New Road. Their accounts were verified by the distribution of obsolete night shelters found. Nevertheless, since the 1980s, white-headed langur populations in Longfeng Territory and Renliang Territory kept dwindling. They could scarcely be spotted by the end of the 1990s. Most of the people interviewed believed that there was only one white-headed langur troop remaining in the Longfeng Mountains, on a precipice found in the hilly wetland. There were also very few white-headed langurs living in Renliang Territory, staying only in the deep parts of the karst hills (Table 4.4).

4.3.2.3 White-Headed Langur Population Survey Findings in the Nongdou Mountains (Daling)

Routes surveyed by Ran Wenzhong in Daling totaled 6.19 km. Karst hills take up a large area of the region but were less observed because of factors like limited time, etc. Only areas with dense white-headed langur distributions were surveyed. The area occupied by karst hills was 4.7 km^2, while the area covered by the survey routes was

Table 4.4 White-headed langur population survey findings in Longfeng Mountains (Laituan) and Bazhiqi Mountains (Renliang) (Quoted from the doctoral thesis by Ran Wenzhong, Ph.D.)

Location	Route length (km)	Karst hill area (km^2)	Survey covered area (km^2)	No. of langur	Inferred no. of langur	In-use night shelter	No. of langur inferred from in-use night shelter	Obsolete night shelter
Longfeng Territory	21.4	7.58	6.7	0	0	1	3	23
Renliang Territory	28.0	23.27	11.41	0	0	1	5	35
Total	49.4	30.85	18.11	0	0	2	8	58

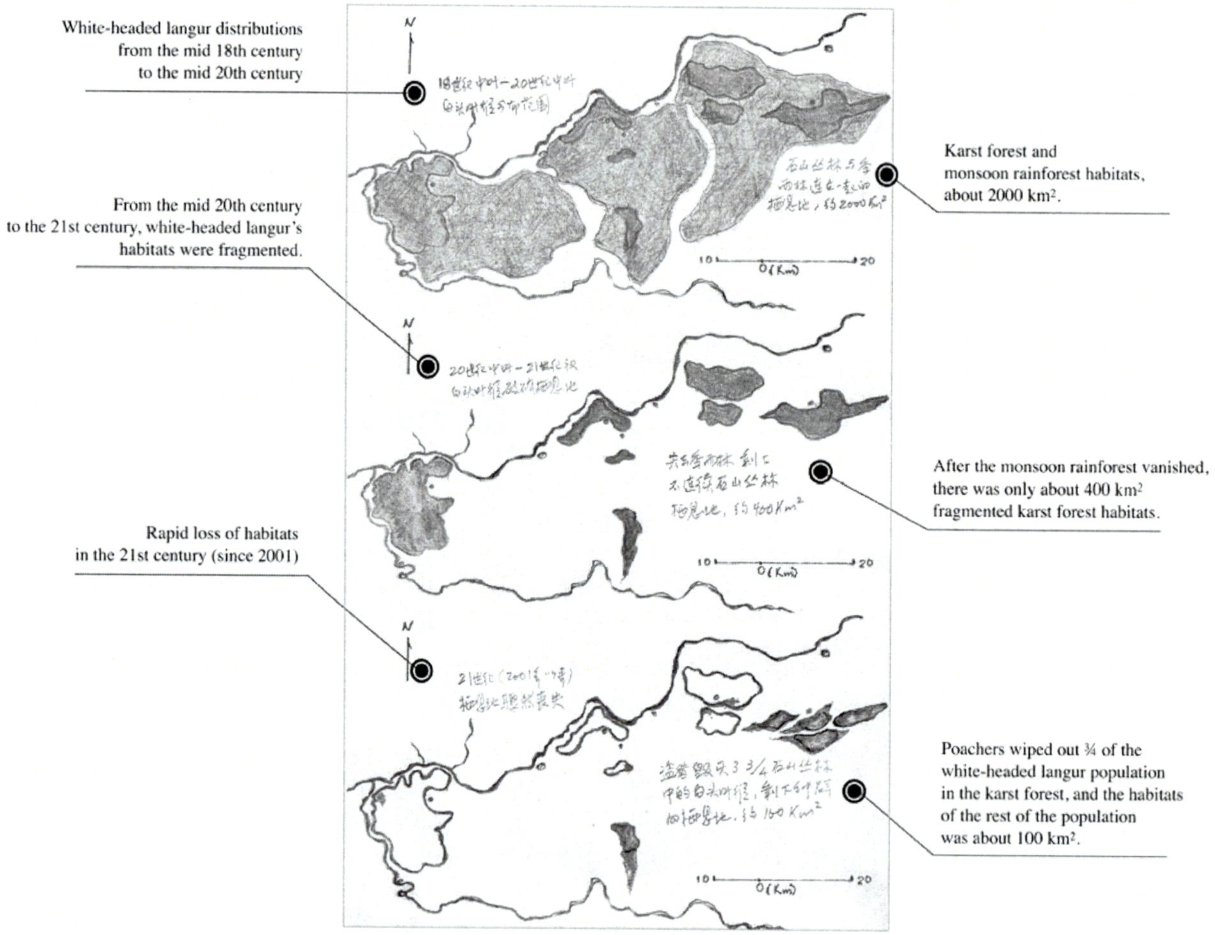

Fig. 4.7 Fragmentation of white-headed langur habitats (black colored areas) over the years (Sketch by Pan Wenshi, December 2001)

2.66 km^2. No white-headed langur had been spotted, nor any langur voice heard in the Daling region, but 25 obsolete night shelters and eight active night shelters were found. From the distribution of the active night shelters, white-headed langur distributions were mostly concentrated in an area of 1.32 km^2, of which 1.3 km^2 was covered intensively by survey routes. By way of "line transect," the area was found to have no white-headed langurs, whereas inferred estimation from random night shelters found that there were approximately 18 white-headed langurs, and thus the density was 13.81 white-headed langurs/km^2. From the distribution and the number of obsolete night shelters, it could be ascertained that the region had many white-headed langurs in the past, which were distributed over extensive areas. Nevertheless, by the time of the survey, the distribution area had significantly been reduced (Fig. 4.7).

4.4 White-Headed Langurs Pushed Toward Extinction by Poachers

In the 1980s and 1990s, there were at least 550 white-headed langurs in the Longrui Nature Reserve. The white-headed langur population there, as noted from the many research reports, was the biggest in earth's history. Moreover, the later surveys by Ran Wenzhong had discovered three more regional populations in the Longfeng Mountains, the Bazhiqi Mountains, and the Nongdou Mountains. By surveys of the abandoned night shelters, it was estimated that each regional population in the 1980s and 1990s had 200–300 white-headed langurs. But the total of more than 1000 white-headed langurs from those four regional populations had all been put to death by the hands of poachers at the dawn of the twenty-first century.

We have known for a long time that in Tuolong Town, just tens of kilometers away from the white-headed langur habitat in Longrui Nature Reserve, there was a rice wine distillery that specialized in processing Darkish Monkey Wine (the local people called white-headed langur "white-headed darkish monkey"). The making of Darkish Monkey Wine is relevant to the extinction of white-headed langur. The flawed understanding of the knowledge of traditional Chinese medicine was a driving force behind poaching. But to a more significant extent, it was the brutal mentality of cruel humans and the greed of the rich and powerful who got into the habit of eating game. Poachers and charlatans together made up lies inducing people to believe that the white-headed langur tonic can cure many diseases, including fevers and headaches, and even increase male sexual potency, thereby leading to the enormous demand of white-headed langurs. It was the thousands of feasts that put white-headed langur on the road to extinction.

References

History and Chronicle Office. Shangsi County archive. Guangxi: History and Chronicle Office. Guangxi Renmin Publishing House; 2000.

Jiong Z. Comprehensive Chinese history, Vol. 5 (中华文学通史:第五卷). Beijing: Hua Yi Publishing House; 1997. p. 751.

Johanson DC. Origins of modern humans: multiregional or out of Africa? Herndon, VA: American Institute of Biological Sciences; 2001.

Lantian S, Hanhua L. White-headed langur in Guangxi (广西的白头叶猴). Guangxi Normal Univ J Nat Sci. 1982;00:27–33.

Liu H, Prugnolle F, Manica A, et al. A geographically explicit genetic model of worldwide human-settlement history. Am J Hum Genet. 2006;79(2):230–7.

Stringer C. Human evolution: out of ethiopia. Nature. 2003;423 (6941):692–5.

Wenzhong R. Analysis on the present status of wild white-headed langur – distribution, population, habitat usage and population sustainability (野生白头叶猴 (*Trachypithecus francoisi leucocephalus*) 的现状:分布、数量、栖息地利用和种群生存力分析(博士论文)). Doctoral thesis. Beijing: Peking University; 2003.

Mingchuan Wu, 1983 An overview of the wild animal resources in Guangxi (广西野生动物资源概况). J South Agric (1)20.

5 The White-Headed Langur's Habitat and Natural Refuge

Abstract

The Nongguan Mountains in Guangxi Province have a landform of peak cluster depression and peak forest valley, which has been formed since the Early Pleistocene. This karst landscape is located at the south of the Tropic of Cancer at the junction of tropical and subtropical monsoon belts. The subtle synchronization of water availability and heat in this area is beneficial to the growth of lush vegetation, many perennial, and provides sufficient food for white-headed langurs year-round. From 1053 AD to 1956 AD, the Nongguan Mountain area was a region with still scant population and rich biodiversity, which provided ample space for the white-headed langur to coexist with human communities. However, since 1956, within 10 years, intense human activities have led to the insularization of the white-headed langur's habitats. Two interspersed but relatively discrete ecosystems gradually formed: an agro-ecosystem established on the peak cluster depression and peak forest valley, and the karst tropical monsoon rainforest ecosystem that remained in primeval state on the talus and karst hills. Rich humus and rainwater storage make the soil in the talus the most fertile in the Nongguan Mountain range, which allows seeds to germinate and prosper into forests. It is the karst hills and talus that have provided a natural refuge for the white-headed langurs.

Keywords

Nongguan mountains · Natural refuge · Peak cluster depression and peak forest valley · Habitat insularization · Talus

Since the day we unearthed the fossil of the ancestor species of white-headed langurs living 1.4 mya, the natural habitat sustaining the species has undergone many changes. When the tropical monsoon rainforests on the karst wetlands and valleys were chopped down, many of the larger animals there were also eliminated. Only vegetation on the karst hilltops and talus around the karst hills remained. Fifteen years ago, poachers had exterminated 4–6 of the white-headed langur populations, which were distributed on six fragmented habitats at the south bank of Zuo Jiang River. It was by sheer luck that two populations survived with small numbers, one of them was the population in the Nongguan Mountains.

5.1 The Nongguan Mountains Natural Refuge

The concept of "natural refuge" emphasizes the leading role of nature in the conservation of biodiversity. A natural refuge specifically refers to a distinct geographical region and the biology therein where natural forces alone (physical, chemical, and biological) could effectively prevent the usual agricultural reclamation by humans and thus protect the biodiversity of that region. The region could be big or small but with no exceptions must be entirely composed of wilderness uninhabited by humans. Humans may enter the natural refuge at particular times, but were never successful in settling there for long. When humans were still in prehistoric years, the sort of regions were everywhere on earth, and they were described as crude wildernesses in legends. But such regions are shrinking nowadays with over seven billion humans. Nevertheless, natural refuges are of vital importance to the safekeeping and continuance of all living beings.

On a "small land mass" surrounded by Zuo Jiang River and Ming Jiang River, the Nongguan Mountains is among the six karst hill white-headed langur habitats. The Nongguan Mountains has a total area of 24 km^2, which is only 5% of the previously intact white-headed langur distribution. However, that size is appropriate for scientists to conduct field research, since the researcher could travel anywhere in the Nongguan Mountains in one day by walking, and check the dynamics of the white-headed langur population in a timely manner.

W. Pan, *The Population Ecology of White-Headed Langur*, https://doi.org/10.1007/978-981-33-4118-0_5

Additionally, it is a relatively isolated geographical location, a "lone island" surrounded by farmlands with no "green bridge" linking it up with other karst hill regions. Thus, the white-headed langurs there could be taken as a relatively discrete regional population. The third reason is that the natural habitat and the white-headed langur population were being ruined at that time. Both human and white-headed langurs were under the same environmental strain. We wondered whether we could put a stop to the ruination of the natural habitat and whether it is possible to improve the living standards of the people. For these reasons, we decided to station there and take up the challenge of examining and revitalizing the area.

We began by looking into the natural history of the Nongguan Mountains to see whether it could function as a natural refuge.

5.1.1 Geologic Background of the Nongguan Mountains

The geologic development of the Nongguan Mountains can be traced back to the late Paleozoic Era (from the late Devonian Period 3700 mya to the Permian Period 2900 mya) when the region where Chongzuo is situated today was still a part of the vast Paleo Tethys Ocean. In that period, the geological history here was still in the early stage of the development of the sea basin quasi platform. The mantle and the crust on the sea floor were split. At the same time, with the eruption of basic volcanic lava, the Paleozoic seafloor of Guangxi formed a state of alternating concave and convex, which resulted in the difference of marine sedimentary environment and geological formation type. Substances of earthen and silicon properties (SiO_2) settled on the troughs deep under water, while carbonates ($CaCO_3$, $MgCO_3$, etc.) are deposited on the raised shallow platform. Those deposits became the alternately distributed "karst hills" (carbonate sediments) and "earthen hills" (earthen and silica sediments) that can now be seen everywhere in southwest Guangxi. It is also the reason why the existing habitat of white-headed langurs is distributed in discontinuous "patches."

The Nongguan Mountains is one of the "patches" with white-headed langur distributions. The fundamental properties in the geologic composition in Nongguan Mountains include:

- Rock layers in the karst hills contain high proportion of soluble carbonate substances and little acidic insoluble substances.
- The fold angle of rock strata on karst hills is small, it is wide and flat, and it has water permeability.
- The karst hills have sturdy profiles, immense thickness (sediments could be as thick as hundreds to thousands of meters), with few and thin interspersed layers that are hard to dissolve.
- The carbonate layer exposed to the earth's surface is generally hundreds of meters high at present.

These properties gave the Nongguan Mountains its hilly wetland topography and the many limestone caves wherein white-headed langurs take shelter.

5.1.2 Particularities of the Nongguan Mountains Topography

In analyzing the topography specific to the Nongguan Mountains, we put our attention on the following three characteristics:

- The ratio of the composing chemical and physical properties in the carbonated rocks in the Nongguan Mountains (CaO/MgO relative ratio >22, solubility >1, calcite content >65%) have resulted in strong soluble strata. In the vertical direction, carbonate layer is eroded into karst peak cluster and peak forest landform; in the horizontal direction, many caves and platforms are eroded.
- Through the long period of time from the late Cretaceous Period (about 70 mya) to the middle of the Pleistocene Epoch (about 1 mya), areas in Chongzuo were for a long time under the regular influence of moist and hot tropical monsoon winds. The abundant rainfall and continuous high temperature triggered the CO_2 (gas)–H_2O (liquid)–Ca (solid) cycle, which results in the large-scale dissolving of the pure carbonate rocks and carved out the terrain we see today.
- Because of the third phase of the Himalayan Orogeny (2.6 mya to 700,000 years ago), the earth's crust keeps pushing upward slowly and gradually. We estimate the caves that become the white-headed langurs' night shelters in the Nongguan Mountains were formed around 1.8 mya, and by the time the Ancient Javan Langurs reached the Chongzuo Basin, there were already caves 30–40 m aboveground level providing them safe night shelters.

From the above, we may conclude that the peak cluster depression and peak forest valley in the Nongguan Mountains (Fig. 5.1), which befits the white-headed langur inhabitants, were the gradual outcome of the coeffects of tectonics movements and the climatic forces since the Early Pleistocene Epoch. In addition, the Ancient Javan Langurs had arrived at the south bank of Zuo Jiang just in time, when caves had been formed on the karst hills in the Nongguan Mountains.

Fig. 5.1 The peak cluster depression and peak forest valley terrain of the Nongguan Mountains (Photo by Feng Chunguang)

Starting from the Nongguan Mountains, we investigated six existing and historical areas where white-headed langurs once lived (Table 5.1). We found that the common topographical characters of the six regions are a series of steeply inclined karst hill columns, extremely low-lying wetlands, and in particular, the many caves that once were subterranean drainage channels before the mid-Pleistocene Epoch but have by now appeared high up in the karst hills, providing hiding places for white-headed langurs. The field research records show that the columns of karst hills and the talus at their foothills have become the best habitats and only natural refuges sustaining the survival of white-headed langurs and many other medium and small sized mammalians.

5.1.3 The Particular Climate of the Nongguan Mountains

Nongguan Mountains are located at the south of the Tropic of Cancer, at the junction of tropical and subtropical monsoon belts where the overall climate is characteristically arid with intense sunlight, heavy rainfall, and long frost-free seasons. The annual precipitation rate, however, is below its evaporation rate, whereas the simultaneity of warmth and water benefits the growth of vegetation.

5.1.3.1 Sunshine

The hours of sunshine[1] in Chongzuo City between 1971 and 2000 ranged from 1312.2 h/year to 1901.3 h/year, and averaged to 1632.6 h/year. Sunshine hours vary with seasons. Figure 5.2 is a line graph of monthly sunshine hours curve of Chongzuo City (1971–2000). From the value of the annual average sunshine hours and the geographical location, the Nongguan Mountain area belongs to the area with little sunshine.[2]

We can also use the percentage of sunshine[3] to explain the influence of climate conditions in different seasons (mainly cloud, rain, fog, dust, sand, etc.) on the duration of sunshine. The percentage of sunshine can exclude the influence of different days in each month and different days in each season on the absolute sunshine hours. Collected data has shown that the percentage of sunshine hours in Chongzuo City averages between 31% and 43%, with slight differences in different years. On a monthly basis, the percentage of sunshine from January to April is generally below 30%, between 42% and 57% in the months from May to

[1] The hours of sunshine when the radiation from the sun hitting perpendicularly on the earth surface is ≥ 120 W/m^2.

[2] Distribution of yearly sunshine hours in China: https://wenku.baidu.com/view/ea64e43943323968011c929e.html.

[3] Percentage of sunshine hours refers to the proportion of actual sunshine hours against theoretic sunshine hours (the supposed sunshine hours on a cloudless day).

Table 5.1 Topography of the karst hill regions where white-headed langurs are distributed, and of the natural refuge of wild animals including white-headed langurs

Karst hill region/natural refuge	Natural habitat topography	The highest karst hill in the region and its elevation	States of the white-headed langurs	
			Before 2000 AD	After 2000 AD
Nonglin Mountains, Fusui County	Peak cluster depression–peak forest valley[a]	Nongguanshan 380 m	Existing	Existing
Nongdou Mountains, Fusui County	Peak cluster depression–peak forest valley	Rongmaobaishan 361.3 m Nongdoushan 398.2 m	Once existed	Disappeared (exterminated by hunting)
Bazhiqi Mountains, Jiangzhou District	Peak cluster depression–peak forest valley	Balonglou 412.4 m	Once existed	Disappeared (exterminated by hunting)
Longfeng Mountains, Jiangzhou District	Peak cluster depression–peak forest valley	344.1 m	Once existed	Disappeared (exterminated by hunting)
Nongguan Mountains, Jiangzhou District	Peak cluster depression–peak forest valley	Nongguanshan 431 m	Existing	Existing
Longrui Mountains, Ningming County	Peak cluster depression–peak forest valley	510 m	Once existed	Disappeared (exterminated by hunting)

[a]Peak cluster depression–peak forest valley is a terrain of combination forms. Peak cluster depression refers to low-lying wetland encircled by one or more high-rising karst hills, and in places where the wetland is completely enclosed, a deep sinkhole would be developed. Peak forest valleys generally develop on the periphery of peak cluster depressions where the hill is gradually descending and consists of a series of long stripe-shaped open-ended swamps

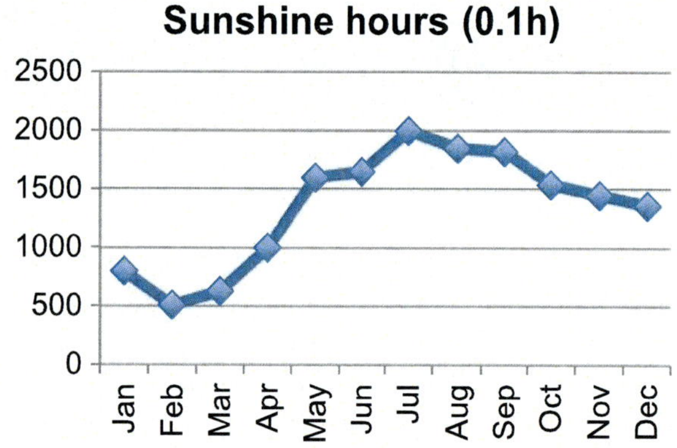

Fig. 5.2 Line graph of sunshine hours in varying months in Chongzuo (1971–2000)

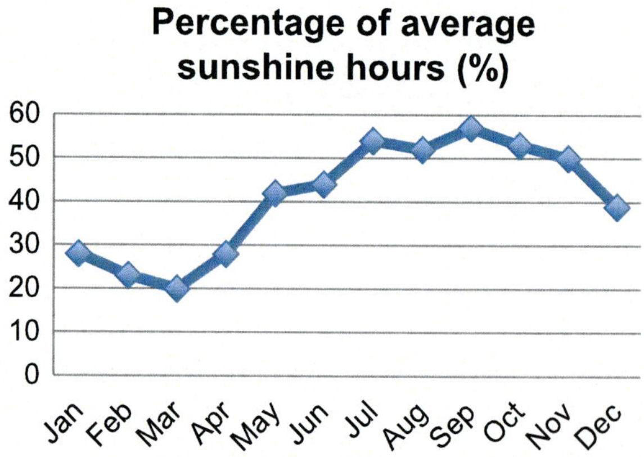

Fig. 5.3 Line graph of percentage of average sunshine hours in varying months in Nanning City (1951–1980) (Nongguan Mountains in those years was under Nanning City administration)

November, and in December it is 39%. A maximum of 57% may be reached in September, whereas the minimum can be as low as 20% in March. Figure 5.3 is a line graph of the monthly sunshine percentages in Nanning City (1951–1980).[4]

5.1.3.2 Temperature

According to records by Chongzuo Weather Station, from 1957 to 2012 the mean temperature over the years was 22.5 °C (range from 20.8 to 23.0 °C), the mean January temperature was 13.8 °C and the July average was 28.1 °C, as shown in Fig. 5.4.[5]The highest annual average temperature in history was in 1943, when Chongshan County hit 23.6 °C, Zuo County hit 23.9 °C, and Luobai County where the Nongguan Mountains are situated recorded 22.0 °C. That was consistent with northern tropical climates, and the mentioned regions were among the hottest in Guangxi. On top of directly determining the rate of growth and the intensity of biochemical reactions in plants, temperature can also cause changes in other factors (such as humidity, precipitation, wind, and oxygen solubility in water) in the environment, thus indirectly affecting the growth and development of plants.

5.1.3.3 Precipitation and Evaporation

The Nongguan Mountains lies in the tropical-subtropical climatic belt, which subjects it to atmospheric circulation, topographic effects, and stark contrasts in precipitation through the year. The average annual precipitation is about 1200 mm (1150–1450 mm), and the uneven distribution

[4]"Statistics of sunshine percentage in China." Source: https://wenku.baidu.com/view/289bb54c336c1eb91a375df3.html.

[5]Source : http://www.gxqxj.com/gzfw/gz_qhbj_uave.asp.

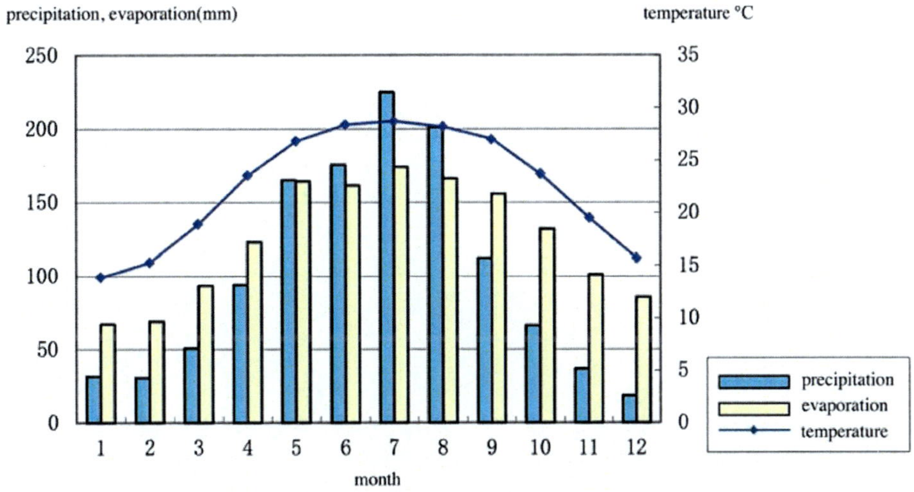

Fig. 5.4 Line and bar graph of yearly average temperature, precipitation, and evaporation in varying months in Chongzuo region (1971–2000)

within the year gives rise to obvious dry and wet seasons. From Fig. 5.4, we could see that the average monthly rainfall from May to September reaches above 100 mm with evaporation at either the same rate or slightly below. This period is decidedly the rainy season. With the higher temperature and ample rain in this period, the vegetation biomass is at the most profuse phase, and it is the season when white-headed langurs have the most abundant food supply. From November to February in the following year, the average monthly rainfall is well below 50 mm and evaporation far exceeds precipitation. This is the dry season. During this period, the temperature and rainfall are the lowest, some plants are deciduous, and the plant biomass is the lowest in a year. In between the dry and rainy seasons are two short transitional periods. Hence we can divide a year in the Nongguan Mountains into a rainy season (May–September), a dry season (November–February in the following year), a dry to wet transiting season (March–April), and a wet to dry transiting season (October). In general, the average annual evaporation of this area is 1492 mm (1971–2000), 292 mm more than the average annual precipitation, so it belongs to the arid area.

In addition, the annual frost-free period in this area is 346 days, with 15 days of frost, and little snow in winter. The average annual humidity stands at 78%. August has the highest relative humidity reaching 82%, whereas the lowest is in January at 75%. In normal years, there are mainly easterly or easterly winds, less frequently are southeast easterly and northeast wind, even rarely is southwest wind. Due to the influence of oceanic monsoon climate, the winter summer wind alternates obviously. The maximum wind speed in July and August can reach 17–22 m/s.

In conclusion, according to the comprehensive climate data, Nonguan Mountain district has the nature of tropical subtropical transition, with obvious dry season and rainy season. Although it belongs to the short sunshine area in terms of the average annual sunshine hours, it has good water and heat synchronization, which is conducive to the growth and recovery of plants.

Throughout the history of spatial distributions of the white-headed langurs, before the 1950s, the distribution range of the whole populations of the white-headed langurs was more extensive than now, and there were more langurs in each regional population. Although the mountain areas are separated, the primitive tropical monsoon rainforest between them is connected with each other; thus, all the white-headed langurs can live in the same composite population until the 1950s when their habitat was fragmented into "patches." This geographical–ecological disconnection had its topographic and latitudinal factors due to the heterogeneity in the region's geologic foundation from ancient times, yet the more decisive factor was the agricultural exploitation by humans on the peak cluster depression and the peak forest valley since the twentieth century. The introduced agricultural ecosystem on the mountain tracts had compelled the white-headed langurs to withdraw to precipitous precipices beyond human reach. The power of nature had heaped up rocks of various sizes on the foothills and formed talus around the precipices (Fig. 5.5), preserving bases for the tropical rainforests, which provided natural refuge for white-headed langurs and other wild animals.

5.2 The White-Headed Langur's Habitat Insularization in the Nongguan Mountains

When we came to the Nongguan Mountains in December 1996, we happened to witness the final episodes of the pristine habitat being turned rapidly from "fragments" into

Fig. 5.5 Talus (the yellow dotted-line-circled foothill) in the insularized habitat (Photo by Liang Zuhong)

"islands." Farmers living in villages and cantonments around the Nongguan Mountains went into the mountain one after another, burning the original plants to exploit land for farming. In just a few years, the white-headed langur habitat became pieces of "isolated islands" on a vast sea of farming fields, and the white-headed langurs were thus confined to live in the karst hills on isolated "ecologic islands." We termed this phenomenon "habitat insularization." In the following sections, we will discuss the reasons for this phenomenon.

5.2.1 Human Population Expansion and Social Development in the Nongguan Mountains

In a long historical period, the aborigines in Nongguan Mountain area all depended on the natural resources there for their livelihood, kept the primitive hunting and picking habits, and gradually transferred to primitive agricultural activities. They lived on riverbanks and cultivated very small pieces of paddy fields by slash and burn techniques, which supported only a very small population. As the population was sparse, human activities in the Nongguan Mountains had for a long time posed no harm to the biodiversity of the region. It was during the population expansion that came in two birth peaks in the 1950s–1960s and in the 1990s when the need for agricultural land and the demand for firewood as daily fuel became pressing. In spite of this, the Nongguan Mountains had still for a long time remained a backward agricultural community, and there is still not a proper industrial enterprise.

5.2.1.1 Human Population Expansion in the Nongguan Mountains

There were no systematic historical records on the human population in the Nongguan Mountains in its long history. We can only use the data collected of Guangxi and of Chongzuo County to estimate the population growth in the Nongguan Mountains. The early day chronicles of Guangxi Province and Chongzuo County showed that there was only a sparse population in the entire Guangxi Province and similarly in Chongzuo County. It can be inferred that the population of Nongguan Mountain area, which was in a very remote and inconvenient traffic condition, must be smaller. According to historical records, Chinese culture spread to Lingnan (southern China) only after Emperor Hanwu succeeded the throne in 140 BC. It was also recorded that in the year 2 AD (West Han Dynasty) that the population was about 200,000 in the entire Guangxi Province, and the average density was below 1 person/km^2. It could be safely estimated that no human lived in the Nongguan Mountains at that time. It was not until 1662–1722 (Qing Dynasty) that Chongshan county had a population of 7860, with a population density of less than 5 people/km^2. Most of the people lived in or around the county town. In 1910, a population of 35,029 was registered, and the population density became 22 people/km^2. In 1947, the aggregate population of Chongshan County and Zuo County was 94,945 with an average density of 60 people/km^2. In 1982, the population in Chongzuo County (county name after the amalgamation of Chongshan County and Zuo County) was 281,187 with an average density of 97 people/km^2. In 1985, Chongzuo County had a population of 297,314 with the average density reaching 102.5 people/km^2. In 1997, Luobai Township had a population of 24,591 with the average density being 155.6 people/km^2.[6]

From the above data, we can see that in the past 60 years, the population has increased by 3–5 times (Fig. 5.5). More resource was needed to support the increased population. Since 1949, due to emptiness of property rights, random falling of trees, arbitrary cultivation of arable land, mountain blasting, and so on, the white-headed langur's habitat was broken up into blocks of standalone "ecologic islands."

5.2.1.2 Social Development in the Nongguan Mountains

First, we have to define the geographic boundaries of the Nongguan Mountains. There is a map in the *Chongshan County Archive* (2011) of Luobai Tribal County in the years before the implementation of the "Replacing Tribal Governor with Dynasty Delegated Governor" policy (Fig. 5.6). The map is not a modern standard map, but it shows a general overview of the area. Many names of places and villages had since been changed, so it is difficult to calibrate key cartographic elements such as direction, altitude, and distances between the mountains, valleys, villages, roads, etc. Still, we can see clearly on the map that Luobai Tribal County was surrounded by Xinning, Jiangzhou, Shangsi, and Zhongzhou in those days, and villages under the County administration were all scattered on the outskirts of the tribal county. The tribal county bureau was situated on the present-day Luobai Street of Luobai Township. It was recorded in history that "Luobai Tribal County has an east-west span of 45 *li* and a north-south span of 45 *li* with 32 villages under its administration." Converting the Qing Dynasty measurement unit *li* into the metric system we are using today, we get the length and width of Luobai Tribal County to be both 21.6 km, making up a total area of 466.56 km^2. The area was about one sixth of the initial Chongzuo County and roughly equaled to the combined areas of Luobai Township and Banli Township today. Luobai Tribal County was not only small in area, but was also mountainous with little flatland. The soil was infertile, and the social unrests in those years led to a constant shortage of labor and government officers. Social development was thus hampered and the economy of the County has always remained weak and impoverished.

We can divide the developments of human society in the Nongguan Mountains into three periods:

The first is the Luobai Tribal County period, which had lasted 803 years (1053 AD to 1856 AD).

"Tribal County" happened at a time of specific social conditions and ethnic relations in history. It was to be distinguished from "dynastic county" where the governor was delegated by the dynasty. Ever since Zhao Lun, denominated as King Wu of Nanyue (name for Southern China in early days), claimed kingship at the end of the Qin Dynasty and practiced the ethnic policy of "peaceful grouping with all *Yue* tribes" (*Yue* is a collective name for aborigines in southern China), the subsequent dynasties through the long Chinese history from the Han Dynasty to the Qing Dynasty had adhered to the policy of "*Yue* autonomous rule" in ethnic minority tribal settlements in southern China. It was proven to be an apt policy as it demonstrated the Dynasty's respect to the folk customs of the *Yues* and maintained intact its upper class and economy, which helped keep society unified and stable.

The formation of "Luobai Tribal County" was granted by the Song Dynasty in the 35th year of the rule of Emperor Renzong (1053 AD). Up until the sixth year of Xianfeng in the Qing Dynasty (1856 AD), the tribal county bureau had remained situated at the center of the most forbidding zone in the Nongguan Mountains where nowadays is the Nongguan Mountains wetland (location indicated in Fig. 5.7). While the forbidding terrain there offered good protection in tumultuous years, the transportation in days of peace when the society was in good order proved to be inconvenient. Furthermore, the terrain was disadvantageous for further development, so there were deliberations of relocating the tribal county bureau elsewhere. All through the 803 years, the chiefs of Luobai Tribal County had stuck to the tradition laid down by their *Zhuang* forefathers, "rely on the forbidding mountains," "make a living in the karst mountains," "reside in the caves in wilderness," and "live in a tribal group." The population was small back then and the very primitive production methods they employed had no harmful effects on the ecology of the Nongguan Mountains. In *Taiping Prefecture Archive—landscape passages*, Luobai Tribal County was portrayed as "surrounded by beautiful mountains and streams." We have to note that the description could not have meant the entire Luobai Tribal County, but refers only to the Nongguan Mountain wetland where the tribal county bureau was located. For it was also recorded in the passage that "the distance from Luobai to the prefecture

[6] Source: Administration registers offices of Luobai Township and of Chongzuo County.

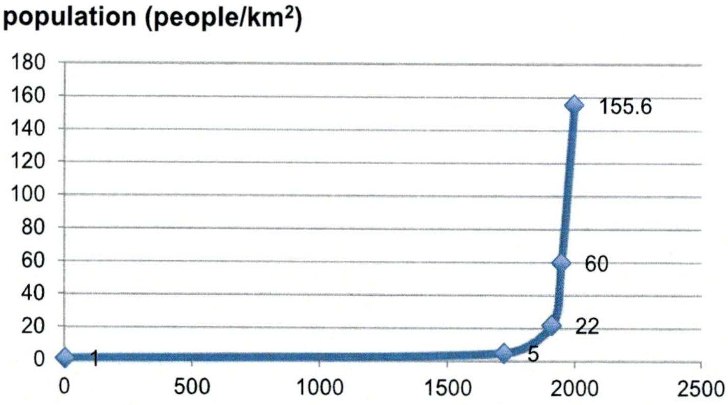

Fig. 5.6 The time-scaled population growth of Luobai Township, Chongzuo County, Guangxi. Remark: The curve has revealed that the population in Chongzuo County started to increase only after 1772, and a typical exponential increase pattern began to appear after 1997

seat (present day Taiping Town in Jiangzhou District of Chongzuo City) is 50 *li*" (24 km). It also needs to be noted in particular that at the time of the writing of *Taiping Prefecture Archive*, the Tribal County Bureau had not yet been relocated, and there was not a Luobai Street but instead a Longtou Village. When we were doing our field surveys, we often heard aged villagers say there was once a high-ranking officer living in Nongguanshan, and he was the top officer of the district. We think the high-ranking officer must be the Luobai Tribal County Chief. In addition, many villagers had recalled that in the past there were houses, residents, and farmlands in the region from Sangengnonghua to the entry of Nongguanshan. The region has nowadays been turned into sugarcane fields, and there are still many ruined huge stonewalls, and relic of the enigmatic "stone phalanx" is still remaining in the wetland in Nongguanshan. We tracked with effort with the clues we got and found there were still many more relics of the Tribal County in the Nongguan Mountains. The villagers' words were verified, and we also found that the *Zhuang* ancestral customs of "relying on the forbidding mountains, each make one's living" as recorded in chronicle are true. Our discoveries include:

1. Relics of a huge stone rampart in a valley (Fig. 5.8). Sangengnonghua is a northeasterly–southwesterly oriented valley in Nongguan Mountains. The valley is broad with a gentle slope, but the karst hills on the flanks are steeply inclined. There was once a stone rampart (in two sections) built across the valley separating

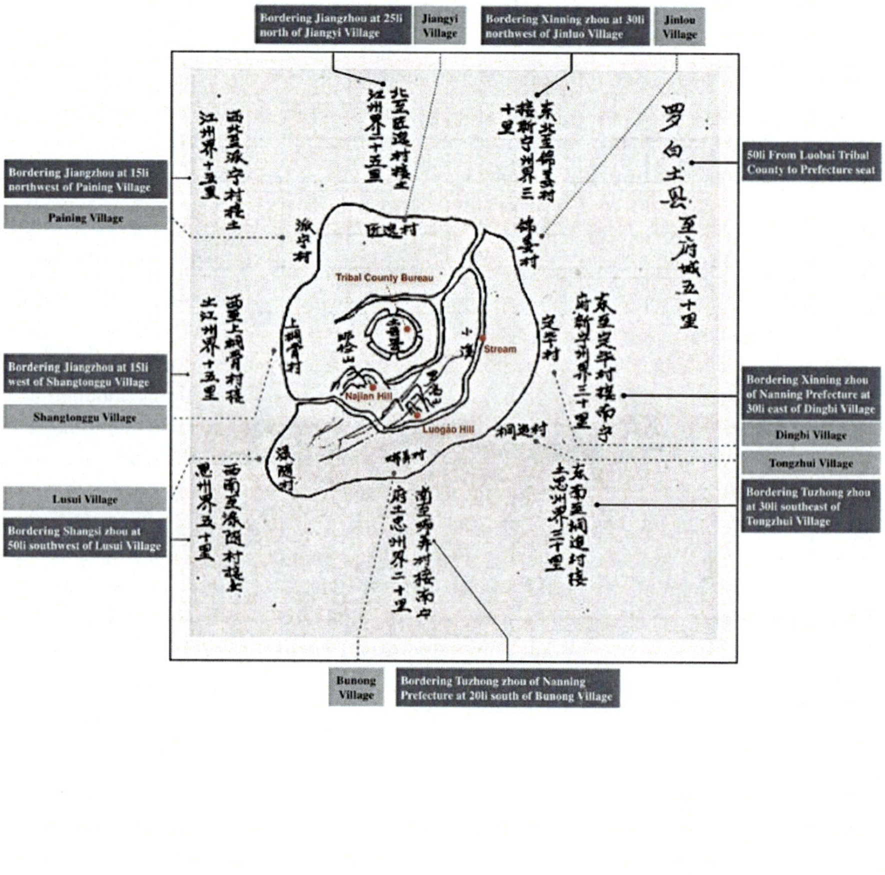

Fig. 5.7 Luobai Tribal County map

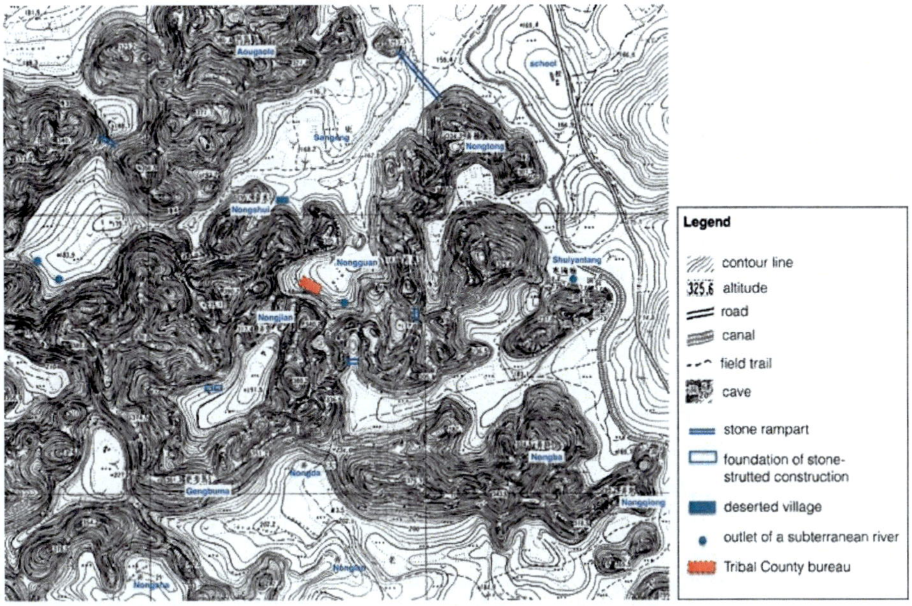

Fig. 5.8 Relics of Luobai Tribal County

Muzishan from Nongtongshan. The rampart was built with uncrafted rocks at natural shapes and sizes and structured in way of "non-plastered rock stacking," meaning not applying lime paste or plasters of any kind when fixing the rocks together. It had a total length of more than 200 m, 4 m in height with a base width of 3 m, and a top width of 2 m. In the mid-section (where a country road nowadays runs through), an entry of 1.9 m in width was built, and the two stones serving as sides of the entry remain buried on the ground until today. The country road runs in a straight line from the entry in the southeast toward Muzishan in the northwest. The rampart bedrocks underneath the country road are nowadays still clearly visible, and the thickness of the rampart has come to be the breadth of the country road. A ficus was planted each on the left and right sides of the entry, but nowadays the rampart and the ficus trees are all gone. Only in the arable land near the Nongtongshan side, there are still a pile of stones that have been demolished but not removed. Most of these stones have not been processed, and only a few traces of sharp tools have been chiseled. We were led to discover this relic, and thus able to verify all this, by a 67-year-old man surnamed Chen who had extensive knowledge of the above because he was employed in the deconstruction of the rampart in 1958.

2. Relics of huge stone ramparts on mountain passes. Apart from the rampart that sprawled across the valley plain at Sangengnonghua. There are also three stone ramparts in the other three mountain passes:
   - The stone rampart at Nongtong. It was built in the mountain pass between Nongtong Valley and Nongguanshan. It was laid out in a north–south position, barring people's passages in and out of the mountain pass, and was built by way of "non-pasted stacking of stones of natural shape and sizes." It had a length of roughly 50 m, and the height and width at the top were measured to be 4 and 2 m, respectively. The base width was not measured as the rampart foundation had been buried by thorny thickets and weeds. There were embrasures 2.5 m above the base of the rampart looking toward Nongtong Valley. The embrasures were rectangular in shape, 35 cm in width, and 15 cm in height, with wider openings on the outside that got narrower on the inside. The southern section of the rampart was extremely ruined while the northern section was preserved.
   - The stone rampart at Nongbaga. This rampart was built at the mountain pass on the northerly tip of the northwestern end of Nongbaga Valley (Nongbaga Valley has a northwestern–southeastern layout) at a strategic position that held entry to the Nongguanshan wetland. The rampart spanned from east to west across the mountain pass with a length of more than 40 m. It was built in the same way as the rampart at Nongtong was, by stacking up non-pasted un-crafted rocks. Today the west portion of the rampart has collapsed.
   - The stone rampart at Nongxiang. Between Nongxiang Wetland and Nongniu Wetland, there is a mountain pass, and a stone rampart was once built across it. A pity that it was demolished completely with only traces left.

   On a large map, it can easily be noted that in those days as long as the tribal county controlled all those ramparts, there was almost no other place for the invaders to enter Nongguanshan. For the rest of the boundaries are circled by craggy peaks of more than

Fig. 5.9 The "Stone Phalanx" relics in the Nongguan Mountain (Photo by Lu Hongyi)

100 m in height with steeply inclined exteriors that are hard to climb. Inside the enclosure settled the Liang clan of Luobai Tribal County, and it was upon those forbidding mountains that the tribal county government bureau relied on for security.

3. Relics of the Tribal County Bureau. Walking from Sangengnonghua along the flank of Nongtongshan deeper into the valley, one will get to the entry to Nongguanshan. On the southwestern side of the entry is an extraordinary piece of flatland in front of a mountain cliff. It was on that flatland that the Luobai Tribal County Bureau once stood, but today the place has been turned into a sugarcane field. The flatland was surrounded by mountains on three sides with a subterranean stream outlet on its west. Water flows steadily on the front exterior of the flatland. On the whole, the landscape does match the "surrounded by beautiful mountains and streams" portrayal. The piece of flatland had been measured, and it has an area of about 320 m^2. It has a very flat and neat terrain that reveals traces of human activity long ago. Such a feature distinguishes this piece of land from the gently sloping farmlands neighboring it. It is a pity that no remnants of any sort of the Tribal County Bureau had been found as the place has been transformed by planting since more than a hundred years ago.
4. Stone Phalanx (Fig. 5.9). At the west side of the Nongguanshan entry, there is a rugged mountain track leading up to Nongguanshan. There are two pieces of wetland along the way, one bigger than the other. The big wetland is circled on three sides by mountains, and the "Stone Phalanx" is found at the bottom of a precipice in the mountain on the north side. The "Stone Phalanx" appears to be the foundation of two blocks of "pillar-strutted" constructions, a style that was known as "thousand legs on the ground." It constituted four arrays of stone struts.[7] The stone struts stood 40 cm aboveground with indentations on the top of each one for resting wood beams. Beams were then interlaced atop the pillars forming the floor, and then the floor was circled up with walls, covered with a ceiling, and the "pillar strutted" construction was finished. Building the floor aboveground could prevent sogginess and flooding. Today the area is still the most obscure place in Nongguan Mountains, and without doubt it was the most secure shelter in those early years when trees were so lush and dense that even sunlight was blocked.

To sum up, if we draw the distribution position of the ramparts from the topographic map, we could see clearly that it matches with the cliff formed by the karst landform on the outside of the surrounding area, forming a very safe and

[7] The survey diary that I used to record the number of stone struts and the distance between the struts had been soaked with rain, and thus, I was unable to work out the occupying area of the construction.

closed internal environment. The forbidding topography had made it like a fort amidst mountains, which could be defended easily from the inside while difficult to attack from the outside. At the same time, there were extensive farmlands and forests inside the fort, as well as water resources that could sustain quite a big population. In later years, people relocated out of the Nongguanshan wetland and built Qiandong Village at the west of the Nongguanshan entry. The village was since ruined, but in the ruined village we found remnants of broken bricks, broken tiles, and a Guanyin temple in a karst cave. The treacherous but excellent natural geography made the *Zhuang* people's ancestral custom of "relying on forbidding mountains" very fit for the stage of economic development when "each making one's living." The relics have provided bits of evidence of Luobai Tribal County's past, which were out of the records of written history.

The second period began when Loubai Tribal County was relocated away from the Nongguan Mountains, and it lasted 102 years (1856–1958).

It was a time when China had already entered into modern history. There were two relocations. The first relocation took place in *Xianfeng* year 6 (1856 AD) when the tribal county government set its eyes on Qulu Village, which was just recently founded during *Jiaqing* years of the Qing Dynasty. In *Zhuang* language, "qu" means stream, "lu" means mountain, and the village was situated by a stream in front of a mountain, hence the name. The mountain at the back of Qulu Village is Muwanshan, which has an altitude of 244 m above sea level, and in front of the village is Luobai Creek that flows from the west to the east. That place was the first abode of the tribal county bureau since leaving the Nongguan Mountains. The earthen structure of the new bureau was very crudely constructed, and after just 6 years, the bureau was destroyed during the wars between the Qing Dynasty and the Taiping rebels led by Hong Xiuquan. That prompted the second relocation. It was recorded in *Archive of Name of Places in Chongzuo* that "Luobai Tribal County was previously located in the place called Jiuxian Village (Qulu Village was by then renamed Jiuxian Village and nowadays is named Jiuxian Encampment). Because of a former plague, people thought the place had bad *fengshui* (supernatural influence of landscape) and for that reason the bureau was relocated to the present address at Cheng Street (the street is today named Luobai Street)." Hence it was recorded in *Chongzuo County Archive* that "County Bureau is located in Luobai Street." That was the second and last abode of the tribal county bureau since moving away from the Nongguan Mountains, and it took place in the *Tongzhi* year 6 (1862 AD). Since then, the tribal county bureau remained at Luobai Street until the "Replacing Tribal Governors with Dynasty Delegated Governors" policy put the Tribal County out of history. Until 1956, there were no more inhabitants regularly living in the Nongguan Mountains. Villages with inhabitants were all scattered on the outskirts of karst hills. People lived on river banks and made a living on small plots of paddy fields. They would go into the mountain only when they had to chop trees for firewood, hunt a wild animal, or gather wild plants for food. Up until the 1980s, most mountain tracts within the administration of Luobai Township and Banli Township remained as tropical monsoon rainforests with animals like the muntjac, which is exclusive to Asia, Reeves's muntjac, wild boars, hog badgers, masked palm civets, hares, Asian golden cats, as well as medium- and big-sized mammals like spotted leopards, Sumatran serows, tigers, etc. Shrub thickets were growing profusely at the lower stratum of the tropical forests, and herbaceous plants spread all the way from the foothills of karst hills to the open stretches of mountain tracts. Big pythons, world's biggest king cobras, and tortoises of many peculiar kinds could be found anywhere. And there were still troops of white-headed langurs living in the karst hills.

The third period began when agricultural exploitation took place in the Nongguan Mountains after "Replacing Tribal Governors with Dynasty Delegated Governors" (1956 AD to present).

The destruction of the biodiversity in the Nongguan Mountains really started in this period. Locally, the continual population expansions were inducing heavy living pressures on the farmers in the Nongguan Mountains and increased their demand for farmland, so they had to exploit the comparatively flat land in the river valleys and slopes for dry farming. On a governmental scale, in 1956, the State decision-makers, due to lack of ecological knowledge, mistakenly propounded the "More people, higher spirit, more vigorous power" slogan, encouraging unrestrained population growth which today has become an enormous burden in social development. In addition, against the backdrop of the "Big Leap Forward," thousands of trees in the Nongguan Mountains were chopped down. Subsequently, the historically unprecedented "Cultural Revolution" that brought huge calamities to the Chinese people took place. During that 14 years (1966–1980), the destruction of the natural environment intensified, and wild animals were hunted at anyone's discretion. Since the demand for firewood outpaced plant growth, the vegetation cover did not have a chance to regenerate, and the habitat for the white-headed langur was drastically reduced. The karst hills that were once covered with tropical monsoon rainforests gradually started the stony desertification process. The once relatively intact northern tropical natural habitats in history was thus shattered into pieces of isolated "ecologic islands" on a sea of agricultural farmlands.

5.2.2 Consequences of Agricultural Developments

With the expansion of population, farmers around the Nongguan Mountains could not find sufficient arable land, so they had to expand their agricultural exploitation deep into the Nongguan Mountains. Up to the 1980s, people were cutting trees in the Nongguan Mountains at a rate that outpaced the growth rate of the trees. In the 1990s, all pristine tropical-subtropical monsoon rainforests on karst peak cluster depression and peak forest valley at the lower altitudes had been completely chopped clear. Only second-generation small trees and shrub thickets remained in small areas, where people found it difficult to penetrate. Since the trees on the plains and level slopes had been completely eradicated, people turned to the talus where small trees and shrubs were growing in clusters with new seedlings coming out every year.

We had surveyed a total of eight villages and encampments closely neighboring the northwest of the Nongguan Mountains, and found there was an agricultural population of 3131 people (year 2000). Their living resources were obtained from the Nongguan Mountains. We supposed if one person needed in 1 day, 1 kg of dried wood (that is the lowest estimation), the annual demand for firewood would be 1142.82 tons. This figure shows that the basic yearly demand for wood by the population in the region had far exceeded the annual growth of the forests. In Leizhai for instance, in 1951, there were 1500 *mu*[8] of forest-covered areas, but there were only 40 *mu* remaining in 1999. The forested area had been reduced 64-fold in 48 years. When we moved into the Nongguan Mountains in mid-November 1996, what we saw was even more stunning. Many people from outside of the above-mentioned eight villages and encampments also came to cut the trees in Nongguan Mountains. There were also several wood factories, which obtained their resource materials by chopping the trees in the Nongguan Mountains. People watched passively, while the trees in the Nongguan Mountains were almost entirely obliterated. In the winter of that year, a primary school teacher from Leizhai led a group of kids driving an ox-cart. They trudged 6 km to the area near our research base and cut the trees on the talus for firewood. When they loaded the wood onto the ox-cart, we noticed that 72% of the trees they cut were white-headed langur's diet species. We further noted that most of them were saplings that sprouted just earlier in the same year. Confronted by our hesitating stares, the teacher said helplessly, "I feel no less pain in my heart than you. I feel like culling a piglet with each chop, and I feel sad without allowing it to grow up, but there's no other way to sustain a family..." In the many years afterward, we were still staggered to find almost no blossoming trees in the valley every time we passed through, and that was all because the trees were chopped before they could reach maturity. Similar incidents also happened in other regions. In a report by Lu Liren and Huang Chengming on their survey in Nonglin Mountains of Fusui County in 1991, it was written, "Thousands of people are cutting the trees in the mountain tracts, obliterating all the medium and large trees in the habitat" (Fig. 5.10). Alongside the destruction of the forests, people kept exploiting the wetlands and valleys year after year, turning them into framing plots. When the farming plots were all joined together, the white-headed langurs, out of innate instinct, were reluctant and did not dare cross such a vast expanse of open land. They could only confine their life to the isolated pieces of stand-alone "ecologic islands."

[8] Chinese area measurement unit, 1 mu roughly equals 667 m^2.

5.3 Formation of Two Relatively Discrete Ecosystems in the Nongguan Mountains

Through 2.6 million years of evolution, the various ecologic factors had coordinated to make the Nongguan Mountains one relatively distinct ecologic region with self-adapting and restorative capabilities. But human blunders in the mid and late years of the twentieth century and the excessive demand for firewood had tipped the balance of the ecosystem in the Nongguan Mountains and seriously damaged its ability to restore itself. In the last decade of the twentieth century, human activities have powerfully altered the ecology in the Nongguan Mountains and made it into two discrete ecosystems. An agro-ecosystem had been established on the peak cluster depressions and peak forest valleys while the karst tropical monsoon rainforest ecosystem remains on the talus and karst hills.

5.3.1 The Agro-ecosystem of Farmers and Sugarcane Fields

From the day we stationed at the Nongguan Mountains in the winter of 1996 until 2004, we witnessed all the wetlands, valleys, and slopes that could be adapted to planting sugarcane had their pristine northern-tropical monsoon rainforests completely wiped out and were turned into sugarcane fields by the local farmers (Fig. 5.11), since they need to raise their living standards. The original habitat had become "sterile agricultural lands" that rely on chemical fertilizers and poisons to grow sugarcane, and at the same time, biodiversity was reduced to the lowest level. The rapid decline of natural fertility in the soil, the increasing farm pests, and the rising labor costs have made the farmer increase their revenue input. Once there is an economic downturn (e.g., sugar price drop) or political turmoil, it would make farmers unable to break

(a) (b)

Fig. 5.10 (**a**) The rampant slash and burn accelerated the devastation to the ecosystem in the Nongguan Mountains. (**b**) As people's demand for firewood increases, the white-headed langurs in the Nongguan Mountains will suffer more (Photo by Pan Wenshi)

Fig. 5.11 Reclamation in the Nongguan Mountains have turned all the level lands and valleys into sugarcane fields. (Photo by Feng Chunguang)

their costs and find growing sugarcane uneconomic, and the sugarcane field ecosystem would not be able to endure. Hence, the present agro-ecosystem in the Nongguan Mountains is an unstable system.

5.3.2 The Monsoon Rainforest Ecosystem of White-Headed Langurs and Karst Hills

The present monsoon rainforest ecosystem of white-headed langurs and karst hills in the Nongguan Mountains is a consequence of the fragmentation of the natural habitat. They were divided by the sugarcane fields and appear as stand-alone "ecological islands." If we were to examine the natural habitat fragmentation effect in the Nongguan Mountains in the general view of the "island biogeography" theory in oceanography, we would come to conclude that as the fragmented habitat pieces become isolated, species loss is bound to occur, since the population would be reduced. However, such an outcome is not pronounced in the Nongguan Mountains for the time being, since there is a time lag from the fragmentation of habitat to the decline of species abundance to the level of equilibrium point, which is called the relaxation time.

We hold the view that the "island biogeography" theory in oceanography may not be aptly applicable to the ecology of stand-alone "terrestrial ecologic islands" surrounded by farm fields. For 20 years, we have continued tracking the progress of biodiversity restoration on the stand-alone "monsoon rainforest karst hill ecologic island" in the Nongguan Mountains, and we have registered rapid restoration of the tropical monsoon rainforests after farmers stopped or reduced cutting down firewood.

Some species from the Pleistocene years had disappeared. For instance, the tiger, which is the top carnivorous predator in the tropical monsoon rainforest food chain, had disappeared from Nongguan Mountains for 30 years. Its disappearance could well be due to the decline in numbers of its prey, such as deer, antelope, and wild boar, which would have compelled the tigers to move to faraway places and rapidly disappear from the area. The disappearing process for some other species could be very slow, like porcupines, ferret badgers, wild boars, masked palm civets, leopard cats, tree-shrews, etc. They can still be found inhabiting in those "karst hill islands" with monsoon rainforests. The spotted-leopard (the top predator in the fauna since the tigers disappeared) had not disappeared quickly either. In every winter in the last decade of the twentieth century, in sugarcane harvest seasons, a spotted-leopard would climb the walls of Chongzuo Ecologic Park into the relatively undisturbed place within the bounds of our research district.

The white-headed langurs in the Nongguan Mountains, the flagship species of the karst hill monsoon forest ecosystem, are restoring their population and multiplying at a fast rate after poaching activities had been stopped. At the same time, with the restoration of vegetation on the karst hills, we found that the biodiversity was recovering rapidly (Fig. 5.12). For example, the bird varieties did not declined, but there were even newly joined species. If the succession of vegetation within the bounds of our research base could be a hint, the white-headed langurs and karst hill monsoon rainforests' ecosystem is in a state of 'progressive succession.'

5.4 The Eco-line Between the Two Ecosystems in the Nongguan Mountains

Agricultural developments by humans in the Nongguan Mountains have brought three changes to the area's natural habitat:

- In order to sustain the ever-increasing population, inhabitants in the Nongguan Mountains had nearly developed all arable lands into farm fields, and in the beginning of the twenty century had planted all the fields with sugarcane.
- Native vegetation on the karst peak cluster depression and peak forest valley had been drastically reduced, compelling the white-headed langurs to live only on the plants growing on the talus or on the karst hills.
- The talus below the karst hills are critical zones in maintaining the biodiversity and biomass of the Nongguan Mountains.

From an ecological perspective, the three changes above were inevitable. The question is whether we could prepare ourselves to handle the possible coming crisis and draw up contingent plans to save what was doomed to disappear. From the perspective of maintaining the livelihood of the farmers in the Nongguan Mountains, the new agro-ecosystem on the mountain tracts that came into shape in the early twenty-first century should not be discarded. However, on the perspective of conserving the biodiversity, the gradual restoration of the tropical monsoon forests and karst hill ecosystem is also critical. In this case, it is very important to ensure the eco-line between the two ecosystems and keep it in a relatively stable state.

In the past 20 years, the research and practice of conservation biology in Nonguan Mountains shows that the eco-line between the agro-ecosystem and the karst hills wild life ecosystem lies on the talus.

The Muzishan example:

Fig. 5.12 The recovering biodiversity in the Nongguan Mountains (Photo by Liang Zuhong, Feng Chunguang). (**a**) *Sus bucculentus*; (**b**) *Prionailurus bengalensis*; (**c**) *Atherurus macrourus*; (**d**) *Strix leptogrammica*; (**e**) *Nyctyornis athertoni*; (**f**) *Cacomantis merulinus*; (**g**) *Gorsachius melanolophus* (birdling); (**h**) *Cinnyris jugularis*; (**i**) *Zosterops japonicus* (breeding); (**j**) *Ophiophagus hannah*; (**k**) *Hypsipetes leucocephalus*; (**l**) *Watasenia scintillans*

(i) (j) (k) (l)

Fig. 5.12 (continued)

Muzishan, situated at the heartland of the Nongguan Mountains, is a relatively stand-alone karst hill. Everywhere surrounding it had been reclaimed into a large area of agricultural land for sugarcane planting, which is archetypal of the new agro-ecosystem in karst hill regions (Fig. 5.13). Nonetheless, in Muzishan, there is still a troop of white-headed langurs, which are dependent on the vegetation on the karst hill and the talus. The talus around the Muzishan represents the states of talus in the entire Nongguan Mountains: they have become the interface zones isolating the two ecosystems. On the karst hill side of the zone, an ecosystem of pristine northern tropical wilderness maintains itself, whereas on the other side of the zone, an agro-ecosystem of sugarcane planting has been put in place (Fig. 5.14a, b).

Figure 5.14a shows rocks of various sizes buried underground or piling aboveground circling around the karst hill. They are the fundamental elements of the talus. Figure 5.14b shows how the talus around the foothills of Muzishan is separating the karst hill from the farm fields. From this, we can get an idea of how the talus is dividing the entire Nongguan Mountains into two ecosystems.

The talus cannot be easily destroyed by humans. We had witnessed in the years between the mid and late twentieth century when farmers around the Nongguan Mountains, in order to feed their enlarged families, had to go to the hills to exploit the area for planting. They torched the wilderness during winter, sometimes it burned all the way to the hilltop, but they soon discovered that the talus could not be utilized. There were rocks everywhere, underneath, and aboveground. The rocks were of various size, some were big as houses or heavy-duty trucks and could not be removed. The stones have been falling from the mountain's top since 2.6 mya, piling up around the mountain and forming layers. Every year, between the cracks, they are filled with tropical forest leaves and branches. Over time, they become rich humus, mixed with the weathered sand and debris on the mountain. In every rainy season, rainwater would percolate and be stored underground, turning the soil on the talus into the most fertile in the Nongguan Mountains. Seeds fallen there would germinate and prosper into forests. The fertile soil made of water and humus also nourished an abundance of vines creeping on the trees all over the karst hill, filling the hilltops with dashes of vitality. The trees there would blossom and bear fruit and their seeds would also fall on the talus and germinate. It was the bless of nature; she decided the talus to be the eco-line, encircling within it the slopes of rocky terrain and karst hills that are reserved to be a repository for biodiversity, while leaving on the exterior mountain tracts for humans to cultivate on (Fig. 5.15).

Fig. 5.13 Photographic view of Muzishan (Photo by Gu Tieliu)

For 2.6 million years, the talus have provided the material basis for the survival and growth of the tropical monsoon rainforest in the Nongguan Mountains. We calculated roughly the total area of the whole Nongguan Mountains to be about 24 km^2 (including the mountain tracts, wetlands, valley plains, and karst hills with their talus), wherein the karst hills and the talus together take up about 12 km^2. Among the 12 km^2, talus take up about 57%, which means

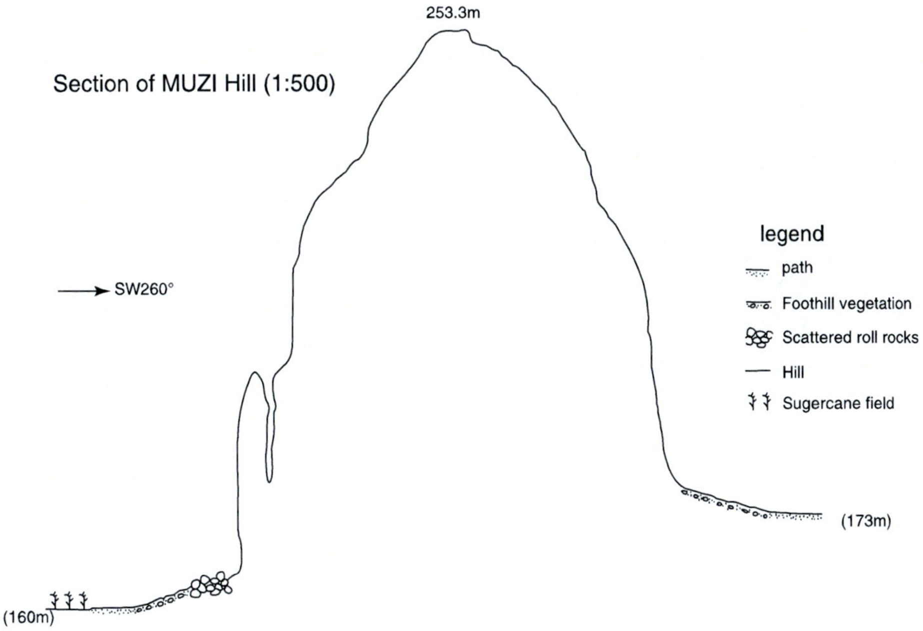

Fig. 5.14 (**a**) Sectional drawing of Muzishan (Scale 1:500). (**b**) Vegetation diet distributions on Muzishan and talus for the Muzishan white-headed langur troop

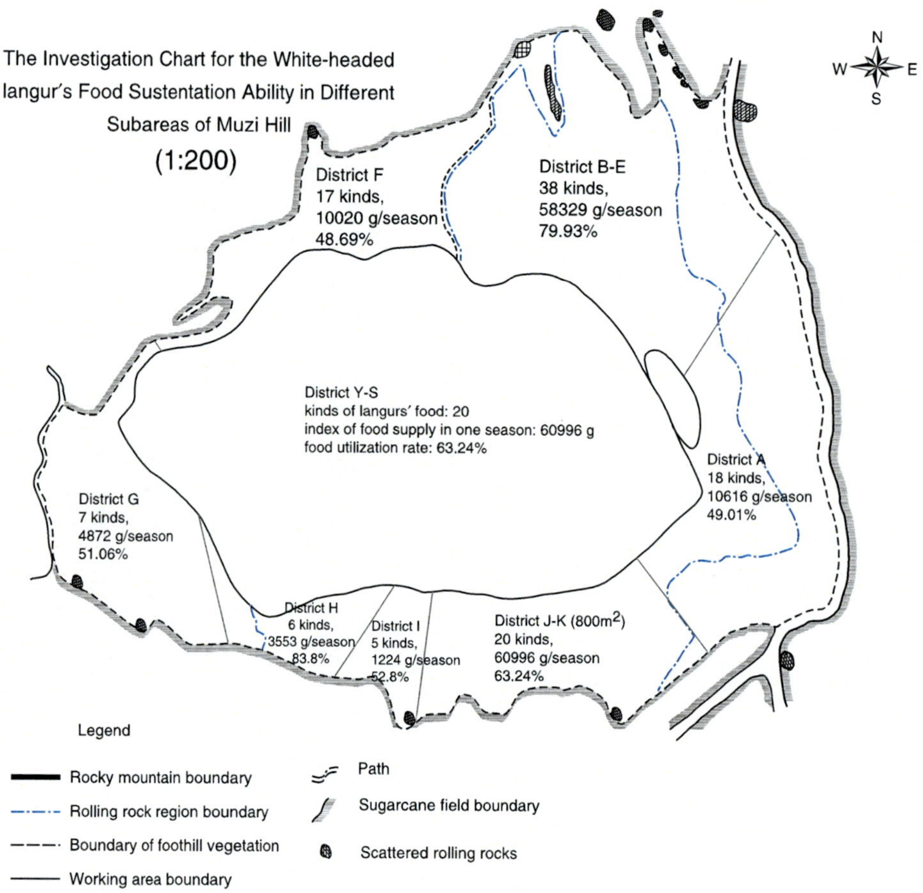

Fig. 5.14 (continued)

Fig. 5.15 The recovering vegetation in the Nonguan Mountains (Photo by Gu Tieliu, Liang Zuhong). (**a**) Vegetation on the top of the karst hills; (**b**) Vegetation on the talus

there are about 6.8 km^2 of talus. In 2002, when the vegetation on the talus were heavily devastated, there were still 339 plant species. Not only did they provide food for more than 200 white-headed langurs, but they also produced annual firewood for more than 10,000 people around the Nongguan Mountains.

In the last years of the twentieth century, since the basic demand for fuel by inhabitants in the Nongguan Mountains

Fig. 5.16 Professor Pan Wenshi helped villagers build biogas digesters with his $100,000 prize from Ford Conservation and Environmental Grants (Photo by Liu Hongji)

was exceedingly huge, farmers went into the Nongguan Mountains to chop the trees for firewood at an unprecedented level, the volume of vegetation that were cut exceeded the volume of growth. The drastic diminishment of the forests was critically threatening the sustainability of white-headed langurs as well as the inhabitants in the Nongguan Mountains. In the face of a doomed future foreboded by the day-to-day dwindling of the variety and volume of vegetation on the talus, how were we going to resolve the situation? After thorough considerations, we decided it had to be tackled by joint efforts from the government and the community. With support from the Ministry of Agriculture, Ford China Co ltd., Royal Dutch Embassy to China, philanthropists from overseas, and domestic communities, we had successfully helped build each household a biogas digester in the 14 villages on the outskirts of the Nongguan Mountains (Fig. 5.16). Biogas produced from cow dung and weed replaced stoke fuel, and almost all the villagers had since stopped going to the hills to cut the trees there. As a result, seeds from different tree species, preserved in the clefts among the rocks in the talus, germinated and had chances to grow, blossom, and produce more seeds. After more than 10 years of restoration, by 2015 there were 409 plant species in the Nongguan Mountains. The region is thriving vigorously, verdant with lush vegetation, and is providing white-headed langurs with sufficient food resources as well as a sustainable natural refuge. The talus is also providing uncountable benefits to the inhabitants of the Nongguan Mountains, which include herb medicines, firewood, conserved water resources, edible wild plants, as well as materials for scientific, cultural, and ethical education.

Reference

Chongshan County Archive. Chongshan County Archive, Minguo year 26 edition (Photocopy edition) (崇善县志(民国二十六年辑) (影印本)). Nanning: Guangxi Renmin Publishing House; 2011.

6 The White-Headed Langur's Homeland in the Nongguan Mountains

Abstract

From November 1996 to May 2015, we registered in detail the total range of eight resident family troops and two transitory troops and discovered that their home ranges became smaller as vegetation biomass, and thus food, was more abundant. We estimated that 0.26 km^2 is the optimum home range size for white-headed langur families. The langurs would pick different kinds of food in different forage quarters according to seasonal changes. The langurs were highly territorial, maintaining both fixed and drifting territories. The fixed territory of each family amounts to about 12% of their home range. Females of the family are restricted to that area, which is inherited from earlier generations. The role of protection of the fixed territory befalls the resident alpha male. Typically the alpha male's territorial behavior includes monosyllabic "Ga" barks, jumping, shaking branches, and successive braying to threaten intruders. When the intruder goes beyond what the alpha male perceives as a safe distance, male offspring older than 3 years would help the alpha male to protect their territory. When a white-headed langur troop is confronted with competitors on its way to forage, they would show temporary territory defending behaviors. This kind of temporary "drifting territory" is about half the size of the core area.

Keywords

Home range · Foraging · Fixed territory · Drifting territory

6.1 Delineation of the Living Space of White-Headed Langurs

Before we proceed to discuss the living space of white-headed langurs, we have to lay out the definitions of the following terms.

The home area of an animal is professionally termed "home range." In our case, it defines the area where a white-headed langur or a troop of white-headed langurs regularly move and forage, usually including food and water resources. In some cases, when the distribution density is high, the "home range" may be the same as the "total range" of their activities, which means they never leave a certain area. When the distribution density reaches its saturation state, the home range is usually identical to its territory. In other words, when the distribution density is high, some individuals (or troops) would prevent others of its kind from entering its home range. Nevertheless, we find in our observations that until now, in Nongguan Mountains, the total area in which a white-headed langur or troop conducts activities is still larger than its home range, while its home range is more extensive than its territory. It shows that the white-headed langur population in the Nongguan Mountains has not yet reached its saturation state.

Territory refers to the area occupied by an individual or a troop where it would, with overt defenses, prevent others of its kind from entering. The most common type and most distinguishable territory is a fixed geographical location occupied by an individual or a troop where it must defend. This type is known as "fix territory." However, sometimes territory is spatiotemporal, shifting as the animal moves. This type of territory is known as "drifting territory" where an animal engages in defense only at particular instances.

The core area is the area most frequently used by the animal for its home activities. In a white-headed langur society, we found that their main core areas are often their night shelters, and their daytime core areas only occupy a small portion of their home ranges. Some examples are the small areas with lush trees, where under its shades, a white-headed langur troop fixedly takes afternoon rests and prevents other troops from entering. In a sense, the core area is the nucleus of the animal's territory. The white-headed

W. Pan, *The Population Ecology of White-Headed Langur*, https://doi.org/10.1007/978-981-33-4118-0_6

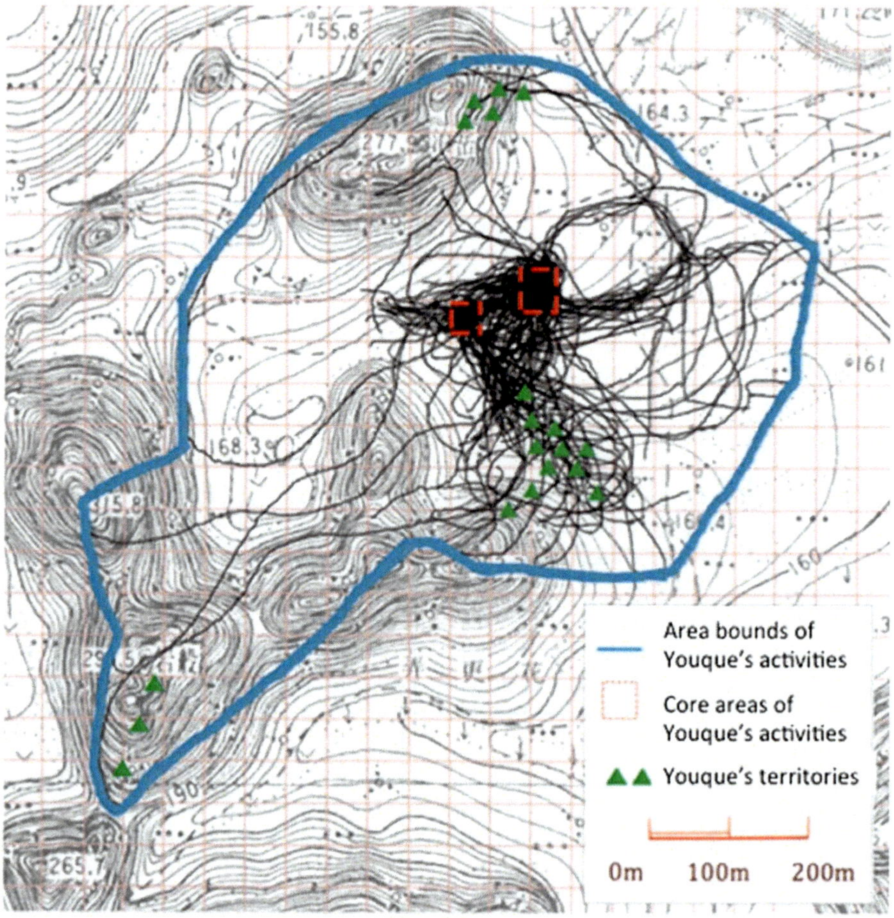

Fig. 6.1 Movement tracks, home range, and core territories of Youque

langurs' home ranges often overlap with each other, but their respective core areas never overlap.

The total range of a troop's (or an individual langur's) living space refers to all the areas where it makes movements and conducts activities, either in its lifetime or during a specific period. Regarding all white-headed langurs in FJC, the total range of their living space covers all their night shelters, forage regions, and all the territories where they would defend.

Take the example in Fig. 6.1 in which the black lines depict the movement tracks of a white-headed langur named Youque, the third son of Yintangxiaotu at the age of a sub-adult, in Eco-Park from September 2013 to May 2014. Its home range thus is the area framed by the blue line, and the red squares highlight its two core areas (night shelter area or core area of activities), which were both very small (Table 6.1). Its home range had an east–west span of about 900 m and a north–south span of about 700 m. The green triangles indicate areas where Youque had one time or another repelled outlier individuals from entering, in other words, places where spatiotemporal defense behaviors had taken place.

Table 6.1 Youque's home range and core areas sizes

Item	Area, km^2
Home range	0.436250
First core area	0.001875
Second core area	0.002500

6.2 The Home Range of FJC White-Headed Langurs

It was on November 19, 1996, that we first went into FJC. We saw a troop of white-headed langurs sitting on the small ledges at FJC Small Cave. In the evening of the next day, we saw the same troop of white-headed langurs climbing up the precipitous precipice on the west of FJC to spend their nights on the crannies there. On the sixth day, we waited at FJC until dark but did not find the langurs on the precipices on the east, south, or west of FJC. At that time, we did not know where the white-headed langurs lived and whether they had their territories. We also had no idea of their movement patterns. Hence we began our research by tracking them to document their living space.

In those days, only the above-mentioned troop was living in FJC and the surrounding habitat. On January 27, 1997, we had the luck to observe them at a closer distance and noted that there were two notches on the ear of the resident alpha male, so we gave the troop the name "Queque troop." Since then, "Queque troop" had become the first white-headed langur troop of our research in the FJC region of the Nongguan Mountains. In fact, it was the founding troop in that area. For 20 years, we have kept researching and tracking the troop and the other eight family troops and two transitory troops that derived from it.

From November 1996 to May 2015, we had registered in detail every encounter with white-headed langurs in FJC region. In this report, we randomly picked ten encountering locations from each month in our field registers and flagged the locations on a 1:10,000 scale map in order to analyze the home ranges of the respective white-headed langur families:

310 episodes from "Queque family troop" (including Queque transitory troop) (1996.11–2000.07)
410 episodes from "Archeng family troop" (1998.04–2002.06)
410 episodes from "α-Gonghou family troop" (2002.06–2006.09)
103 episodes from "β-Gonghou family troop" (2002.06–2006.06)
600 episodes from "Yintangxiaotu family troop" (2006.09–2012.07)
271 episodes from "Yuweng family troop" (2013.07–2015.02)
80 episodes from "Laoxie family troop" (2015.03–2015.12)
93 episodes from "Xishan family troop" (2015.03–2015.12)
120 episodes from "Yintangxiaotu transitory troop" (2012.07–2015.03)
85 episodes from "Bokyueshan transitory troop" (from outlier) (2015.03–2015.12)

We then put coordinate grids of the 1-ha unit on top of the location-flagged map to get the home ranges of the eight family troops and two transitory troops that inhabited FJC from 1996 to 2015. As shown in Fig. 6.2, home ranges are framed in red and night shelter positions are denoted by a red dot. It can be noted from Fig. 6.2 and Table 6.2 that the home ranges of the eight family troops got smaller each year.

We can see in Table 6.2 that the average area of home range of the seven family troops founded since 2002.06 (sequence b–h) is 0.259 km^2, which is only 25% of the home range of the founding troop in 1998. It also reveals that with the restoration of vegetation in FJC in 2006, the white-headed langurs were able to find enough food within smaller areas. In any case, 0.26 km^2 is for the time being the optimum and most befitting home range size for a white-headed langur family.

Every dawn, different troops of white-headed langurs in FJC would come out from their night shelters at almost the same time (before daybreak). They would first take a rest around their night shelters for about half an hour to 2 h (the time taken differs between winter and summer, as well as between rainy days and sunny days), and then begin to move to their foraging areas on their regular routes. We got the impression that, in each troop, an experienced adult female would decide the route, the pace of their movement, and the length of time they spent at feeding places. The route of any particular troop differs between days. But if we consolidate the movements and the times allotted for different activities in each day by various langur troops, the rhythm of their movements could be revealed.

Over the 20 years, regarding the five family troops which had FJC Big Cave as their core area of activities, some had identical movement routes, and some had different routes:

First, from November 1996 to May 2002, the "Queque family troop" and "Archeng family troop" had taken more or less the same three movement routes. Every day, after they came out from their night shelter, they would decide on which direction to move. It could be eastwards, crossing the Tianxian Pass into forage place at Taohuagu, or northwestwards, crossing the Guoyuan Pass into forage place at Guoyuan, or northwards, climbing over the hilltop to other forage places. Regardless of winter or summer, but particularly in hot summers, they must take a long rest under the shades of trees (usually from 11:00 to 15:00) at noon. On some days before evening, they would go to the puddle by the subterranean river inlet below the precipice on the south side of Guoyuan Pass to quench their thirsts. Only after sunset would they return to sleep in Big Cave on the precipitous precipice, in Small Cave, or the Xishan night shelter. These troops had 3–5 night shelters (see Fig. 6.2a, b, and Table 6.2).

Second, from June 2002 to December 2015, the core areas of the "α-Gonghou family troop," "Yintangxiaotu family troop," and "Yuweng family troop" all withdrew to the eastern side of FJC. Their forage routes were similar to the "Queque family troop" and "Archeng family troop," but since the vegetation in FJC had been restored as time goes by, they did not have to move as far as the "Queque family troop" and "Archeng family troop" did to forage. The subterranean river inlet at Guoyuan was clogged by new vegetation, so the three family troops had to find new water resource. We noticed that they often moved to drink in the small pool of water at the foothill of Xiaoshan at the southernmost point of FJC. But that small pool had to be shared with other troops from Fangniushan, clashes occurred quite often between them. The daytime movement and activity rhythm of all FJC groups are very similar.

Third, from 2012.07 to 2015.12, one after another, the "Yintangxiaotu transitory troop" and "Laoxie family troop" had their core areas and night shelters at FJC Small Cave. Besides moving northward crossing Tianxin Pass into Taohuagu, more frequently, they went south, crossing the rugged hill ridges behind Small Cave and then moving along the hill ridge at the southern margin of Taohuagu to

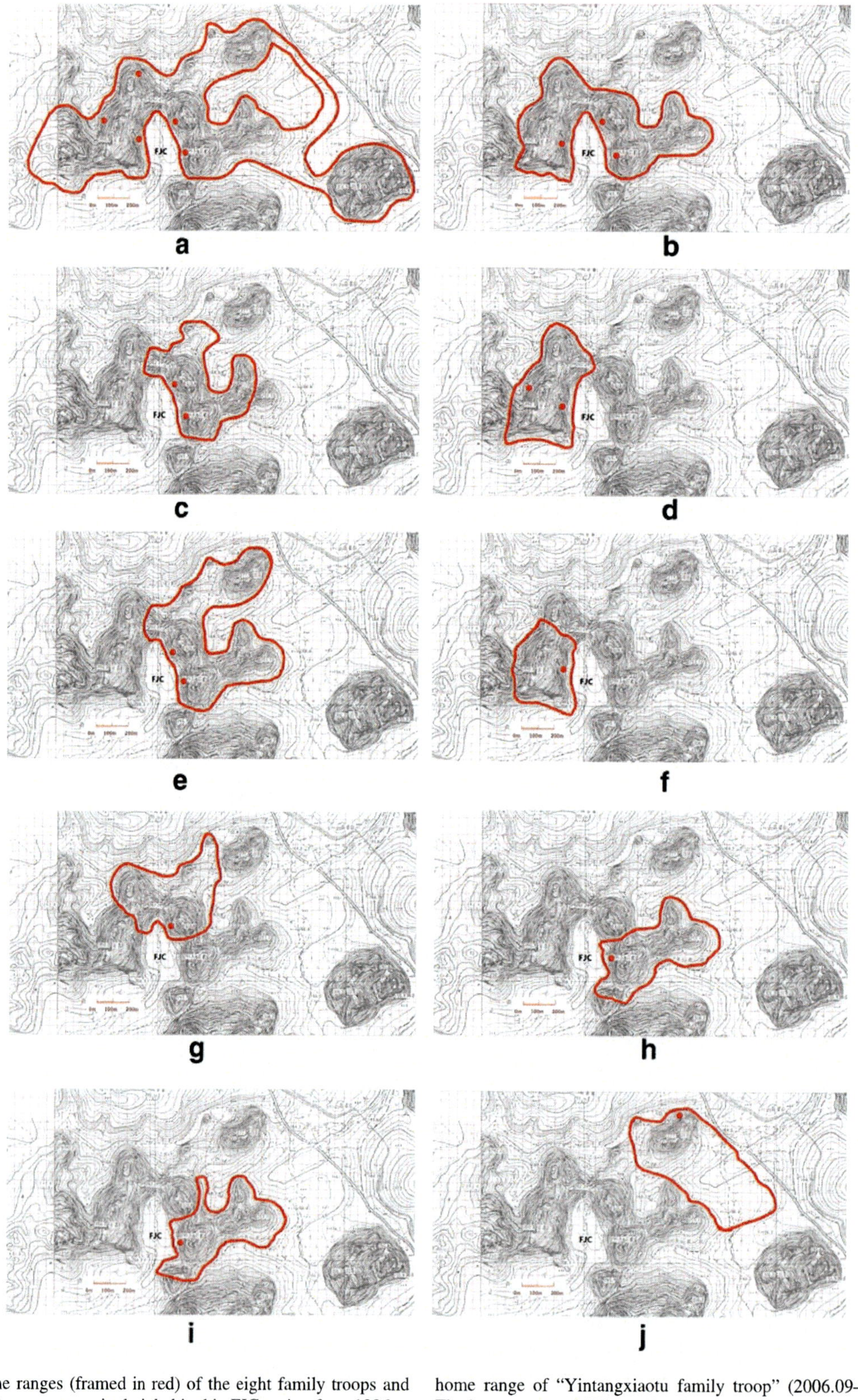

Fig. 6.2 Home ranges (framed in red) of the eight family troops and two transitory troops successively inhabited in FJC region from 1996 to 2015 (red dots denoting night shelter positions). (**a**) The home range of "Queque family troop" (and transitory troop) (1998.04–2002.06). (**b**) The home range of "Archeng family troop" (1998.04–2002.06). (**c**) The home range of "α-Gonghou family troop" (2002.06–2006.09). (**d**) The home range of "β-Gonghou family troop". (2002.06–2006.06). (**e**) The home range of "Yintangxiaotu family troop" (2006.09–2012.07). (**f**) The home range of "Xishan family troop" (2015.06 to). (**g**) The home range of "Yuweng family troop" (2013.07–2015.12). (**h**) The home range of "Laoxie family troop" (2015.03–2015.12). (**i**) The home range of "Yintangxiaotu transitory troop"(2006.09–2012.07). (**j**) The home range of "Bokyueshan transitory troop" (2015.03–2015.12)

Table 6.2 The sizes of home ranges and core areas of the successive white-headed langur family troops and transitory troops in FJC from 1996 to 2016

Sequence	Troop	Duration	Home range size			Core area size			No. of core areas and night shelters
			km^2	ha	*mu*	km^2	ha	*mu*	
a	Queque family troop	1994–1998.4	1.026	102.6	1539.0	0.0988	9.88	148.2	5
b	Archeng family troop	1998.4–2002.6	0.386	38.6	579.0	0.0613	6.13	91.95	3
c	α-Gonghou family troop	2002.6–2006.9	0.250	25.0	375.0	0.0613	6.13	91.95	2
d	*β*-Gonghou family troop	2002.10–2006.6	0.231	23.1	346.5	0.0375	3.75	56.25	2
e	Yintangxiaotu family troop	2006.9–2012.7	0.383	38.3	574.5	0.0613	6.13	91.95	2
f	Xishan family troop	2012.6–(still existed by 2016.12)	0.231	23.1	346.5	0.0375	3.75	56.25	1
g	Yuweng family troop	2013.7–(still existed by 2016.12)	0.238	23.8	357.0	0.0350	3.50	52.50	1
h	Laoxie family troop	2015.3–(still existed by 2016.12)	0.241	24.1	361.5	0.0263	2.63	39.45	1
i	Yintangxiaotu transitory troop	2012.7–2015.3	0.241	24.1	361.5	0.0263	2.63	39.45	1

arrive at Fanghoushan where our Research Base is situated. They also preferred to drink at the small reservoir in the Research Base. Their daytime movements and activity rhythm were the same as that of the other troops.

6.3 White-Headed Langurs' Use of Their Home Range

We will take the daily movements of "Yintangxiaotu family troop" from September 2006 to July 2012 as an example to see how the white-headed langurs were making use of their home range.

Every morning, after leaving the night place, they enter the forest within the home range from a certain direction and move while searching for food, until late evening. On the next day, they would usually leave their night shelter and move to another foraging area within the troop's home range, and then at sunset, they would return to their night shelter. This repeats day after day. We noticed that they would not forage at the same spot as the day before, which means they alternated between the foraging quarters.

6.3.1 Forage Quarters

We divide the home range of "Yintangxiaotu family troop" into four forage quarters as shown in Fig. 6.3:

First forage quarter: Taohuagu region, including from FJC Big Cave to the surroundings of Taohuagu.

Second forage quarter: Big Cave–Small Cave talus region, including from Big Cave on the east side of FJC to the talus below the Small Cave on the south.

Third forage quarter: Fanghoushan region, including from the back of the Small Cave along the ridge east to Fanghoushan where the Research Base is.

Fourth forage quarter: Baizhiya–Guoyuan region, including from the hill ridges at the northwestern flank of Big Cave crossing Baizhiya Pass at the northwestern corner of FJC to the mid-hills at Guoyuan (Table 6.3).

6.3.2 Forage Probability in Each Forage Quarter in Relation to Weather Conditions

Concerning the four ecological weather conditions (dry season, dry to wet season, wet season, wet to dry season) in the Nongguan Mountains, we calculated the movement probabilities of the "Yintangxiaotu family troop" in each of the four forage quarters in the year 2011.

From the tracking videos in the monitor in 2011, we registered that there were a total of 410 movements in each day and night by the family troop. Out of these movements, 210 were leaving the night shelter in the early morning to move to a forage quarter, and 200 were leaving the forage quarter at sunset to return to the night shelter. To find out how they made use of their home range, we sorted out in Table 6.4 and 6.5 registers of 112 days in which the troop's daytime movements are clearly discerned.

Tables 6.4 and 6.5, and Fig. 6.4 have revealed that:

- In 2011, the "Yintangxiaotu family troop" foraged in the four forage quarters inside its home range in turns. They rarely foraged in the same forage quarter for more than

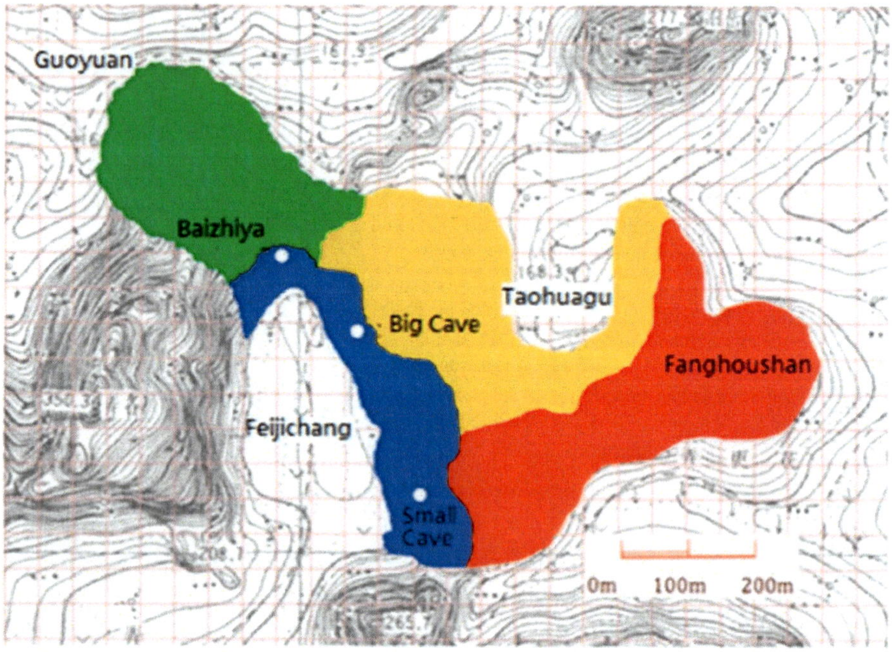

Fig. 6.3 The four forage quarters of "Yintangxiaotu family troop"

Table 6.3 Areas of the various forage quarters of "Yintangxiaotu family troop"

Forage quarter		Area		
Quarter	Designation	km^2	ha	*mu*
First forage quarter	Taohuagu Quarter	0.0750	7.50	112.50
Second forage quarter	Big Cave–Small Cave talus Quarter	0.0500	5.00	75.00
Third forage quarter	Fanghoushan Quarter	0.0975	9.75	146.25
Fourth forage quarter	Baizhiya–Guoyuan Quarter	0.0700	7.00	105.00
Total		0.2925	29.25	438.75
Mean		0.0731	7.31	109.69

several days consecutively. In general, they would forage in a given quarter for 1–3 days then switch to another quarter, and it would be another 3–15 days until they returned to forage again in the given quarter. We found that their regular change of forage quarters had provided recovery time for the diet vegetation. In our opinion, it was the white-headed langurs' way of preserving food for themselves.

- The bottom of the precipitous precipices around Big Cave, Small Cave, and the talus were the most foraged areas among the four forage quarters by "Yintangxiaotu family troop," taking up 38.4% of the year's register. During the dry and cloudy days in the months in Jan–Feb and Oct–Dec of 2011, the troop often foraged in this quarter. The quarter is shaded from strong sunlight by higher grounds, and the humidity over the talus is relatively higher. It has the most stable food resources in all weather conditions throughout the year.
- Taohuagu region was the primary forage quarter in the dry turning wet season for "Yintangxiaotu family troop." In the 112 days of foraging registered in 2011, the troop had foraged in this quarter for 38 days, which was 33.9% of the year's record. Throughout the year, this quarter receives the strongest sunshine and has the most number of sunny days. Particularly during spring, in March through May, the vegetation in this region is the earliest to sprout.
- From the back (east side) of Small Cave to Feilaishi and Fanghoushan, there is a row of low hill ridges, which was also one of the routes taken by "Yintangxiaotu family troop" on their way from their night shelter to Taohuagu. They regularly foraged as they moved, and on average, they would forage there once every 9 days. The probability of foraging in this area accounted for 10.7% of the whole year. It was not their primary forage quarter; they routinely foraged there because it was a way to provide the vegetation in the chief forage quarters time to regenerate.
- In the dry turning wet season of 2011, the "Yintangxiaotu family troop" was often seen in the Baizhiya–Guoyuan Quarter, but they rarely foraged there in the wet turning

Table 6.4 The forage quarter choice by "Yintangxiaotu family troop" in different ecological weather conditions

Sequence	Dry season			Dry season to Wet season			Wet season			Wet season to Dry season		
	Date	Weather condition	Forage quarter	Date	Weather condition	Forage quarter	Date	weather	Forage location	Date	Weather condition	Forage quarter
1	Nov.3,	sunny	■	Apr.9,	sunny	●	Jun.12,	raining	■	Aug.22,	sunny	⬟
2	Nov.9,	cloudy	■	Apr.12,	sunny	■	Jun.16,	raining	⬟	Aug.23,	sunny	■
3	Nov.11,	sunny	■	Apr.13,	sunny	●	Jun.19,	raining	▲	Aug.24,	cloudy	▲
4	Nov.11,	cloudy	▲	Apr.14,	sunny	■	Jun.20,	raining	▲	Aug.25,	cloudy	■
5	Nov.18,	cloudy	●	Apr.15,	cloudy	▲	Jun.21,	raining	▲	Aug.28,	sunny	■
6	Nov.19,	sunny	▲	Apr.16,	sunny	▲	Jun.24,	raining	■	Sep.15	cloudy	■
7	Nov.20,	sunny	■	Apr.17,	cloudy	●	Jun.25,	raining	■	Sep.16,	sunny	■
8	Dec.1,	sunny	■	Apr.18,	cloudy	⬟	Jul.7,	sunny	▲	Sep.17,	cloudy	▲
9	Dec.2,	sunny	■	Apr.19,	sunny	●	Jul.8,	cloudy	●	Sep.22,	sunny	▲
10	Dec.13,	sunny	■	Apr.20,	cloudy	■	Jul.9,	sunny	■	Sep.23,	sunny	▲
11	Jan.3,	cloudy	▲	Apr.21,	cloudy	▲	Jul.11,	cloudy	●	Sep.24,	sunny	●
12	Jan.4,	cloudy	⬟	Apr.23,	sunny	▲	Jul.16,	sunny	⬟	Sep.26,	raining	▲
13	Jan.6,	cloudy	■	Apr.24,	raining	▲	Jul.17,	sunny	●	Sep.27,	raining	■
14	Jan.9,	raining	●	Apr.25,	cloudy	■	Jul.18,	sunny	▲	Sep.29,	raining	▲
15	Jan.10,	cloudy	▲	Apr.26,	sunny	▲	Jul.22,	raining	●	Sep.30,	raining	■
16	Jan.13,	cloudy	▲	Apr.27,	raining	▲	Jul.23,	sunny	⬟	Oct.10,	sunny	■
17	Jan.17,	cloudy	■	Apr.28,	raining	▲	Jul.25,	sunny	■	Oct.12,	sunny	▲
18	Jan.18,	cloudy	●	Apr.29	raining	▲	Jul.26,	sunny	⬟	Oct.13,	sunny	■
19	Jan.20,	cloudy	■	Apr.30,	cloudy	▲	Jul.27,	sunny	⬟	Oct.14,	cloudy	■
20	Jan.21,	cloudy	▲	May 1,	sunny	●	Jul.28,	sunny	●	Oct.18,	sunny	■
21	Jan.22,	cloudy	■	May 2,	sunny	▲	Jul.30,	raining	■	Oct.24,	cloudy	■
22	Jan.24,	cloudy	■	May 3,	raining	■	Jul.31,	raining	▲	Oct.25,	cloudy	⬟
23	Jan.25,	cloudy	⬟	May 4,	cloudy	■	Aug.1,	sunny	●	Oct.28,	raining	●
24	Jan.28,	cloudy	●	May 5,	raining	▲	Aug. 2,	cloudy	■	Oct.29,	cloudy	▲
25	Jan.29,	cloudy	■	May 7,	sunny	▲	Aug.5,	cloudy	■			
26	Feb.10,	cloudy	⬟	May 8,	sunny	▲	Aug.6,	cloudy	■			
27	Feb.11,	cloudy	●	May 9,	sunny	▲	Aug.8,	raining	▲			
28	Feb.12,	cloudy	■	May 10,	sunny	▲	Aug.9,	raining	●			
29				May 11	cloudy	⬟	Aug.10,	raining	⬟			
30							Aug.18,	cloudy	■			
31							Aug.19,	raining	■			

Note: Different forage quarters are denoted by different colored symbols.

▲ Taohuagu Quarter (1st forage quarter)

■ Big Cave - Small Cave skirting slopes Quarter (2nd forage quarter).

⬟ Fanghaoshan Quarter (3rd forage quarter)

● Baizhiya – Guoyuan Quarter (4th forage quarter)

Table 6.5 Forage probability of each forage quarter by "Yintangxiaotu family troop" in different ecological weather conditions in the year 2011

Forage quarter	Dry season (Nov. Dec. Jan. Feb.)		Dry to Wet season (Mar. Apr. May)		West season (Jun. Jul. Aug.)		Wet to Dry season (Aug. Sep. Oct. Nov.)		Total of the forage quarter	
	days	Forage probability (%)	days	Forage probability (%)	days	Forage probability (%)	days	Forage probability (%)	days	Forage probability (%)
▲	6	21.4	16	55.2	8	25.8	8	33.3	38	33.9
■	14	50.0	6	20.7	11	35.5	12	50.0	43	38.4
⬟	3	10.7	2	6.9	5	16.1	2	8.4	12	10.7
●	5	17.9	5	17.2	7	22.6	2	8.4	19	17.0
total	28	100.0	29	100.0	31	100.0	24	100.0	112	100.0

*Different forage quarters are denoted by different colored symbols.

▲ Taohuagu Quarter (1st forage quarter)

■ Big Cave - Small Cave skirting slopes Quarter (2nd forage quarter).

⬟ Fanghaoshan Quarter (3rd forage quarter)

● Baizhiya – Guoyuan Quarter (4th forage quarter)

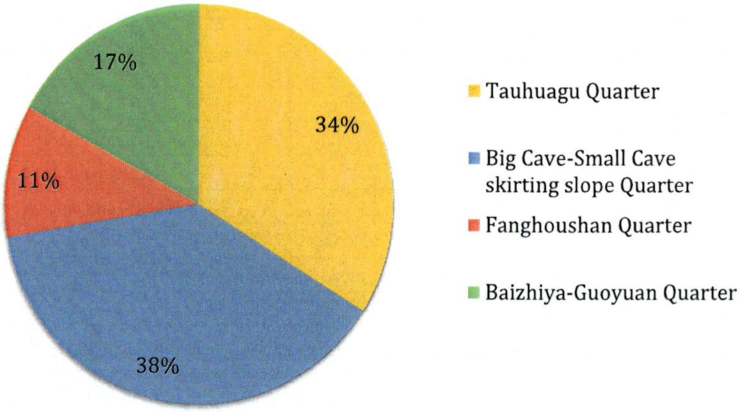

Fig. 6.4 The forage probabilities of the forage quarters by "Yintangxiaotu family troop" in 2011

dry season. The annual forage probability in this quarter was 17%. The dry turning wet season is the coldest season, and any foraging in this quarter might be associated with using the sun to get warm.

- The talus around and below Big Cave and Small Cave was the chief forage quarter of the "Yintangxiaotu family troop," and the second primary forage quarter was the Taohuagu Quarter. Together they occupied an annual forage probability of 72.3%.

6.3.3 The Appropriate Forage Strategies

Every day, the white-headed langurs have to obtain more energy in their forage movements than it had to expend. Thus, they must adopt appropriate strategies in foraging in its home range; in other words, decide on what to eat and where to feed.

The white-headed langur is a species with a varied diet (generalist), capable of digesting various kinds of plants. We will describe the plant species of the white-headed langur's diet in detail in Chap. 7. The optimal forage theory states that in an environment of abundant food resources, an animal would forage relatively fewer kinds of food, whereas if there are insufficient food resources in the environment, an animal will feed on a larger variety of food (Shang 2005). Since 2011, we have been looking for an opportunity to observe whether the forage approach of the "Yintangxiaotu family troop" agrees with the optimal forage theory in zoology.

In the middle of February in 2015 when the moist southeastern wind was blowing over Shiwanda Mountain, where clusters of small white flowers of tiexianlian (*Clematis florida*) started blossoming and were soon to spread all over

(a) (b) (c)

Fig. 6.5 (**a**) Flower cluster of *Clematis florida*; (**b**) Common tailorbird singing on the branch; (**c**) A flock of scaly-breasted munias pecking the seeds of *ban-mao* (*Saccharum arundinaceum*) (Photo by Liang Zuhong)

like wildfire, we saw the first common tailorbird that signaled the end of winter by singing on a gou tree (*Broussonetia papyrifera*). It warbled loudly, but not one female bird was responding.

We knew that the singing of one bird did not signify the end of winter, yet after several days when a flock of scaly breasted munias came breaking through the hazy mist and landed on the seeded ears of ban-mao (*Saccharum arundinaceum*), we knew for sure spring had come (Fig. 6.5c).

We waited for the warm and moist air to come and the flowers to bud on the bare branches of gou tree (*Broussonetia papyrifera*). Every year when warm winds blow into the valleys in FJC, the troop had allowed us opportunities to observe their forages at near distances.

The following are the observation registers of the "Yintangxiaotu family troop" forage movements in 2015 (Fig. 6.6):

2015.02.25 We measured the leaf buds on the gou tree growing at the bottom of FJC Big Cave to be only 1.09 cm (n = 50). There were not any white-headed langurs.

2015.02.26 The 'Yintangxiaotu family troop' came to the talus below FJC Big Cave to forage for leaves on the kudzu (Pueraria thunbergiana), qingtan (Pteroceltis tatarinowii), Chinese banyan (Ficus microcarpa), Hainanese ventilago (Ventilago inaequilateralis), Chinese chestnut (Sterculia nobilis), yinbeiteng (Argyreia capitiformis) and huimaojiangguolian (Cipadessa cinerascens). They left after they had filled their stomachs. The Gou tree leaf buds by then grew to 1.6 cm in length (n = 10), but not one white-headed langur was interested.

2015.02.28 The 'Yintangxiaotu family troop' went up the leafless gou tree branches and consumed the 3 newly sprouted male inflorescences buds which by now had grown to 4.2 cm in length (n = 50). Very soon, they turned to forage the leaf sprouts of qingtan, Chinese banyan, and shanganteng (Cansjera rheedei) and left only when their stomachs were full.

2015.03.01–03 For three days we did not find the 'Yintangxiaotu family troop' foraging at the talus in FJC. We guess they had gone to forage in other forage quarters.

2015.03.04 The 'Yintangxiaotu family troop' came to the FJC talus to feed on qingtan, kudzu, and Chinese banyan until they had enough. Still, they paid no attention to the gou tree though the leaf buds on its stems by now had reached 5.5 cm in length (n = 10).

2015.03.05 The 'Yintangxiaotu family troop' came again to feed on the qingtan and kudzu at FJC talus until they were full.

Fig. 6.6 Forage behaviors of "Yintangxiaotu family troop" on *Broussonetia papyrifera* (Photo by Liang Zuhong). (**a**) Yintangxiaotu checking up the *Broussonetia papyrifera* male inflorescences (2015.02.28). (**b**) Sample tasting of the male inflorescences of *Broussonetia papyrifera* (2015.03.06). (**c**) Feeding on swollen *Broussonetia papyrifera* male inflorescences (2015.03.10). (**d**) In the middle March, male inflorescence of *Broussonetia papyrifera* is the major diet of the white-headed langur's diet

Still, not one of them fed on gou tree leaf buds even though they had grown to 6.0 cm (n = 10).

2015.03.06 The 'Yintangxiaotu family troop' came to forage at the FJC talus. We noticed the father and 2 elder brothers of 007 fed on the now 8.2 cm (n = 10) gou tree leaf buds but only stayed for 5 minutes before turning to join the rest of the family in foraging qingtan, kudzu, and Hainanese ventilago. They left when they had enough.

2015.03.07–09 For 3 consecutive days, the 'Yintangxiaotu family troop' fed only on qingtan and Hainanese ventilago without paying attention to the gou tree leaf buds though they had by now grown more than 10 cm (n = 10).

2015.03.10 At last, we saw the 'Yintangxiaotu family troop' coming to the FJC talus and feeding on the male inflorescences of gou tree. They fed until their stomachs were full, but they only consumed the male inflorescences. They were not observed to have fed on other plants. We noted on that day the gou tree buds had sprouted into nascent leaves and the male inflorescences were overall bigger than 15 cm in size.

2015.03.13 We came to notice that by this time in whichever forage quarter the 'Yintangxiaotu family troop' went, male inflorescences and nascent leaves of the gou tree became their chief diets. At times we found them picking and eating other plants along their way to the forage quarters but only in insignificant amounts.

The presence and absence of the "Yintangxiaotu family troop" to forage at the FJC Big Cave–Small Cave talus Quarter in the days from 2015.2.15 to 2015. 3.10 revealed their forage strategies and that they had to make precisely correct choices on what to eat.

- We found that only when experienced individuals in the "Yintangxiaotu family troop" were able to make correct judgments on the sizes and weights of the gou tree inflorescences, leaf buds, and leaf sprouts, could every individual in the troop obtain maximum energy gains from their forages. On February 26, when the "Yintangxiaotu family troop" went to forage at the FJC talus, they did not feed on the still very short gou tree leaf buds. Instead, they fed on the other seven plant species until their stomachs were full. The choice they made was in accord with food availability and conforms to the economic principle of "time input and net energy gain." For if they were to pluck the gou tree leaf buds which at the time had a mean length of only 1.6 cm, they would have expended more time in feeding with fewer energy gains.
- The field registers have revealed that the presence and absence of the "Yintangxiaotu family troop" in FJC between the end of February and the beginning of March was in fact dictated by choices on food. The research from 2015.2.25 to 3.10 provided a case example to show that the white-headed langur has an ability of food sampling and trade-off consideration. On February 28, when the gou tree leaf buds had grown to 4.2 cm, the "Yintangxiaotu family troop" carried out "sampling feeding," but they did not have enough, and they had to rely on another three kinds of plants to feed. In the following days, they came to the FJC talus every 2–3 days to check on the gou tree inflorescence buds. The white-headed langurs came to check on the tree three times, and despite the gou tree shoots, sprouts, leaf buds, and inflorescences buds growing day by day from 5.5 cm to

6.0 cm to 8.2 cm, they did not touch them but fed on the other three plant species to fill their hunger. When 10 more days passed, on March 10, when most of the gou tree buds had grown longer than 15 cm, they fed on only the gou tree inflorescence buds until they were full. The white-headed langurs might have been able to make correct discernments about the gou tree by distinct senses of smell and sight inherent in them, or by the fact that there was an experienced adult langur (Yintangxiaotu) that led them. Apparently, this adaptive behavior from evolution had been instilled in their genetics.

6.4 The White-Headed Langur Is a Territorial Animal

The white-headed langur is a typical territory-occupying animal. To all readers who are not cognizant of the natural history of white-headed langurs, the defending territory behavior typical of male langurs can be seen in Fig. 6.7.

Our observations and video recordings through the entire year of 2012 had documented 198 counts of territory-related behaviors by the resident alpha male langur Yintangxiaotu, while female langurs in the family had performed 45 counts of territory-related behaviors. This shows that males had 4.4 times more territory-related behaviors than females. While behaviors of protection, vigilance, alertness, intimidation, and driving away other male langurs make up the total of the males' territory-related behaviors, the females had only 6.7% of their territory-related behaviors involved in protection and vigilance, there is a 15-fold difference between the two. While driving away outlier females took up the other 93.3% of territory-related behaviors of females, male langurs do not drive away outlier female langurs. There we can see the intrinsic difference in territory-related behaviors by different sexes in the family (Table 6.6).

Fig. 6.7 Barking of the adult resident alpha male is the most common characteristic territory-related behavior of white-headed langur (Photo by Li Xinyang)

6.4.1 Alpha Male and Male Offspring Defend Their Territory

We had registered the following territory-related behaviors by alpha male Yintangxiaotu and its male offspring:

- Only Yintangxiaotu and its male offspring over 3 years old carried out territory-related behaviors.
- There were distinct territory boundaries between the "Yintangxiaotu family troop" and other troops. Yintangxiaotu and its male offspring over 3 years old would show themselves imposingly to male intruders of its same species or drive them right away (Fig. 6.8).
- Yintangxiaotu and its male offspring over 3 years old would jointly defend their land from their rivals and usually win in the struggle.
- When protecting their territory, Yintangxiaotu and its adolescent male offspring would take up different positions at the territory boundary, sit high up on outcropping tree branches, and spread their legs revealing their genitals to show their male characteristics, and the white hair on their heads and shoulder would appear particularly striking under the sunlight.

Table 6.6 Comparison of the territory-related behaviors by resident alpha male langur ♂ and adult female langur ♀ in "Yintangxiaotu family troop" in the year 2012

		Territory-related behavior statistics from monitoring system				
		Resident alpha ♂		Adult ♀		
Item	Behavior*	Counts	% of total	Counts	% of total	Remark
1	Protection	48	24.24	2	4.44	*All territory-related behaviors (100%) by resident alpha ♂ were targeted at intruding ♂ of its same species **93.3% of territory-related behaviors by adult ♀ were targeted at driving away outlier ♀ of its same species
2	Vigilance	85	42.93	1	2.22	
3	In alert	57	28.79			
4	Intimidation	2	1.01			
5	Driving away	6	3.03	42	93.33**	
Total		198	100.00	45	100.00	

(a) (b) (c)

Fig. 6.8 Male langurs' territory defending behaviors (Photo by Liang Zuhong). (**a**) "Yintangxiaotu" (point A) and "Zuoque" (point B), "Wuque" and "Youque" (point C) jointly defend "outlier male monkeys" (point D). (**b**) The male monkey "Yintangxiaotu" roared and rushed over his wife and children to drive away foreign invaders. (**c**) The male langur seriously injured in the fights is still guarding his family and territory

- In most cases, the territory-related behavior by Yintangxiaotu was limited to monosyllabic "Ga" barks, jumping, shaking branches, and successive braying to threaten the intruder while the male offspring would sit silently and motionlessly on the treetops, waiting for the alpha male's command. Only when the intruder goes beyond what the alpha male thinks to be a safe distance, would male langurs engage themselves in fighting without making a sound. In such cases, fatalities and body injuries may happen.
- All the fights to drive away intruders in its territory by Yintangxiaotu were for protecting its wives and pups, and the sons helped their father in protecting the mother, aunts, and siblings. The fights were usually brutal and might bring serious wounds such as bitten-off tails, broken limbs, or lacerations on the face and body. Females and pups below 2 years old would not engage in fights and would hide inside groves or behind rocks to wait for the fight to end.

6.4.2 The Initiating of Territory Boundaries

In the various family troops in FJC Big Cave, the female white-headed langurs were mostly closely blood-related. In every family, the females by heredity occupied the core areas in the families' territories, but we almost never see them act imposingly in their territory-related behaviors, though there were registers of them driving off outlier females. In January 2015, a sub-adult female from the "FNS transitory troop" attempted to get into the "Yintangxiaotu transitory troop" and was driven away by the adult females in the latter troop. It then turned to join the "Wuque brothers troop" and was accepted. In March 2015, when Laoxie replaced Yintangxiaotu as the resident alpha male in FJC Small Cave, it adopted that sub-adult female (we named her Wailaimei).

With every white-headed langur family, every 4 years there would come an outlier adult male that invaded and took over the family. It would kill all suckling pups of the previous male langur and reproduce new pups of its own. At the same time, this new resident alpha male would take up the responsibility of defending the core territories against other outlier male langurs and protecting the females from the former family and its own pups. When did such biological cycles begin? Even today, our theory is still hypothetical.

In April 1998, we observed the resident alpha male langur Queque of the "Queque family troop" defeated by an outlier male langur. It led its three daughters and three sons away from FJC Big Cave and began roving in the region of Guoyuan and Bokyueshan, showing its offspring ways to look for food and water. In June 2000, Queque, leading its offspring, roved along the southeast wall of the Research Base and reached Fangshaoshan (FSS) to explore new territory (Fig. 6.9). There at the time was an isolated "hill-island"

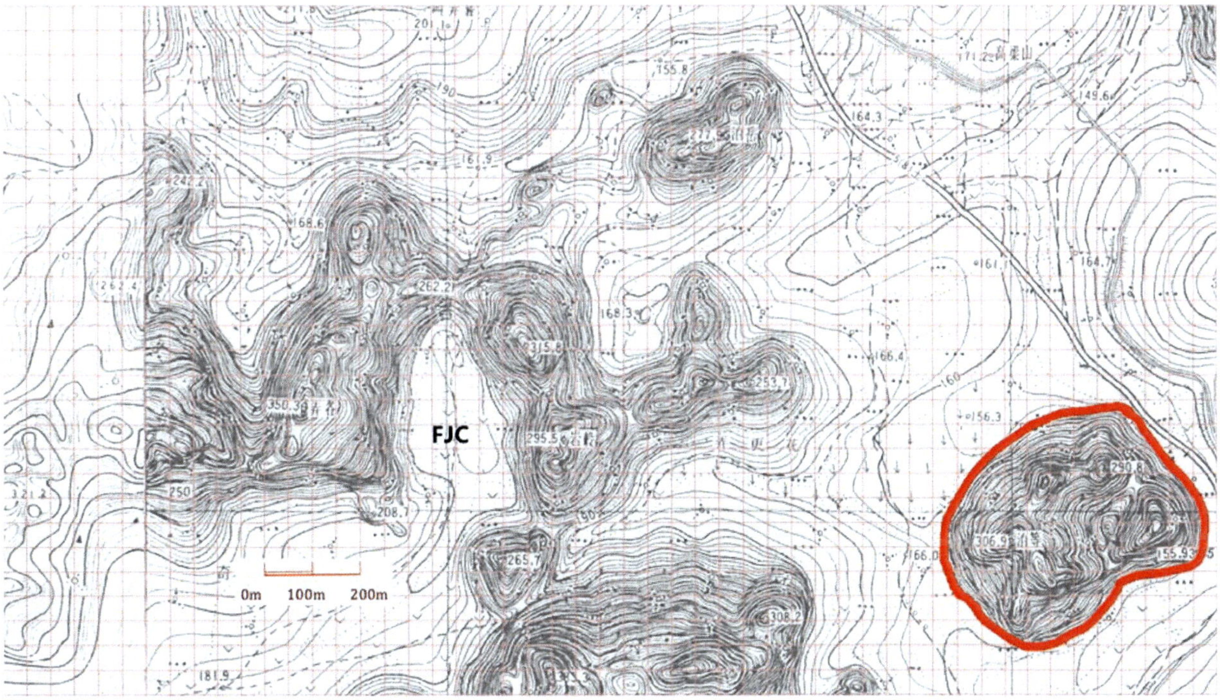

Fig. 6.9 Home range (circled in red) of "FSS troop," the troop grafted from FJC Big Cave (2001.08)

without any white-headed langurs. The three daughters, Dagege, Ergege, and Sangege by then had reached puberty. In July of the same year, adult male langurs roaming in the vicinity of Fangshaoshan one after another kept attempting to intrude Fangshaoshan. Fighting was as fierce as ever, and Queque was killed in the fighting in Fangshaoshan. Its three sons, Daarge, Erarge, and Sanarge, immediately left Fangshaoshan for Bokyueshan and had ever since never returned to Fangshaoshan.

The three daughters of Queque stayed in Fangshaoshan and had become the contesting targets for many outlier male langurs. In February 2001, at the bottom of a precipice on the western side of Fangshaoshan, we had in 3 days picked up two yellow-haired pups infanticided by invading male langurs, which shows the fierceness of the fights for reproduction rights. Later in August, we found that a robust male langur of prime age with a scarred face and a tail with the tip bitten off had mastered Fangshaoshan; we named it FSS Duanwei. It had been seriously wounded but also had defeated all other male langurs and eventually brought stability to the family that centered on three young female langurs. Between January and March 2002, three yellow-haired pups were born to this "FSS troop." In the past 15 years, the "FSS troop" has become a big family troop in which Dagege, Ergege, and Sangege are living not only with their daughters but also their granddaughters and great-granddaughters. Its home range area is 0.185 km^2. We term the way that the three young females left their original family to initiate a new family in a new territory (FSS) as "grafting".

There were clear territory boundaries between the various FJC family troops. The boundaries were results of family splits, and they were passed on from generation to generation. How did it get started? So far, we have in our registers only the founding of the "FSS family troop" to base our discussions on. In the beginning, it was the home range of several females; territories were afterward founded and defended by the resident alpha male. The behavior is beneficial in maintaining the stability of white-headed langur society in that:

First, several adolescent females explored a piece of land to sustain their livelihood, and it became their home range. Males were attracted to come and start a family with them, and the home range became fix territory in which a lineage was created and passed on from generation to generation.

Second, with the arrival of a resident alpha male, a solid family is founded. The resident alpha male would by instinct protect all its reproductive resources, which means its wives and the offspring it breeds. Hence, a territory that needed to be defended against other invading males came into place, and it is carried on by turns of resident alpha males that replace one after another.

Third, the fix territory is passed down to the next generation of females, while the responsibility to defend the territory lies on the shoulders of successive resident alpha males. However, a resident alpha male only prevents males of its own species from invading, and after four years, it will be replaced by another outlier adult male. Females in the family sometimes also take up the responsibility of defending the territory, but it is limited to driving away outlier females. We had observed on four occasions adult females from two family troops driving away outlier females. However, the females that were driven away, in the end, joined the two families.

6.5 The Territory of FJC White-Headed Langurs

Upon our research on the home ranges and territories of the eight white-headed langur family troops and transitory troops that once inhabited FJC Big Cave and FJC Small Cave, we discovered that their territory-related behaviors were not limited to a single territory. The entire region was a place in which they displayed their vigorous vitality.

In 1996, when we just came to station ourselves at FJC, we often heard monosyllabic barks of resident alpha male langurs. They could be sitting alone high up on treetops, or atop a big rock, or maybe hiding inside a cave. For instance:

- In the morning of December 28, 1996, male langur Queque was sitting alone on the branch of a shrub atop Big Cave, where its night shelter was, and barked successively 559 monosyllabic "Ga" barks in 38 mins. During this period, there appeared no male langurs invading.
- On January 23, 1997, the valley was drenched by cold drizzles and heavily covered under the misty fog. The "FNS family troop" stayed in the night shelter cave to stay away from the cold weather and would not come out. The new resident male Tutu, while staying inside the cave at Wojian, kept barking monosyllabic cries from 9:00 till 11:30. We noticed that there were no invader on the hill slope or the precipice outside the cave during the time it was barking. Even if there were invaders, Tutu, being inside the cave, would not have seen it.

Hence we got the opinion that those monosyllabic barks by white-headed langurs were not directed at invaders but were flaunting broadcasts by a territory occupier. This is a labor-saving and effective behavior. The function of the territory-related behaviors by resident alpha males is to protect its reproductive resources and its next generation. These resources enhanced the individual fitness of white-headed langurs as well as the genetic fitness of the white-headed langur population in the Nongguan Mountains. As a result, such behavior is maintained in its evolution.

We found that in each of the white-headed langur families in the Nongguan Mountains, there was a piece of fix territory that allowed no breaching by others. This territory was reserved for breeding and raising offspring. Apart from that, during their forages, white-headed langurs would also drive away other white-headed langurs they accidentally encountered that do not belong to its troop, which is a spatiotemporal territory behavior. Hence, white-headed langur territories can be divided into fix types and drifting types.

6.5.1 Fix Territory

We take the three family troops which in December 2015 inhabited the northern, southeastern, and western FJC as an example, since they all had to defend their territories and their core areas. The core area of each family was the little piece of area that generations of the family fixedly live on and is inherited. In whichever family, the core area was just a tiny portion of its home range. As shown in Fig. 6.10 and Table 6.7, the core area of "Xishan family troop" was 16.2% of its home range, that of "Yuweng family troop" was 14.7% of its home range, that of "Laoxie family troop" was 10.9% of its home range, and that of "Bokyueshan transitory troop" was only 7.3% of its home range.

The common feature of these core areas is that they were closely around the family's night shelters, and their terrain is characterized by their back against the mountains and facing an open space (mountains and plains), as shown in Fig. 6.11. The resident alpha male langurs could see the frontline of the core area with their excellent eyesight and can quickly rush to drive the invaders away with their agile skills.

Figure 6.11 shows the core area of the "Yuweng family troop" (red-lined area), the "Laoxie family troop" (yellow-lined area), the "Xishan family troop" (blue-lined area), and part of the core area of the "Bokyueshan transitory troop" (white-lined area).

The fix territories (core areas) of the above four white-headed langur troops took up only from 7.3% to 16.2% of the area of their respective home ranges, indicating that a white-headed langur family does not need a big territory or core area. As there are rich food resources in FJC at present, a fix territory averaging 12% of its home range for a resident alpha male to protect its wives and offspring would be sufficient. A big core area would be impractical for the resident alpha male langur. Since if the core area is too big, the resident alpha male langur would not be able to take sufficient care of it. The columns of the karst hills and the lush northern tropical forests would impede its view, making it unable to defend it.

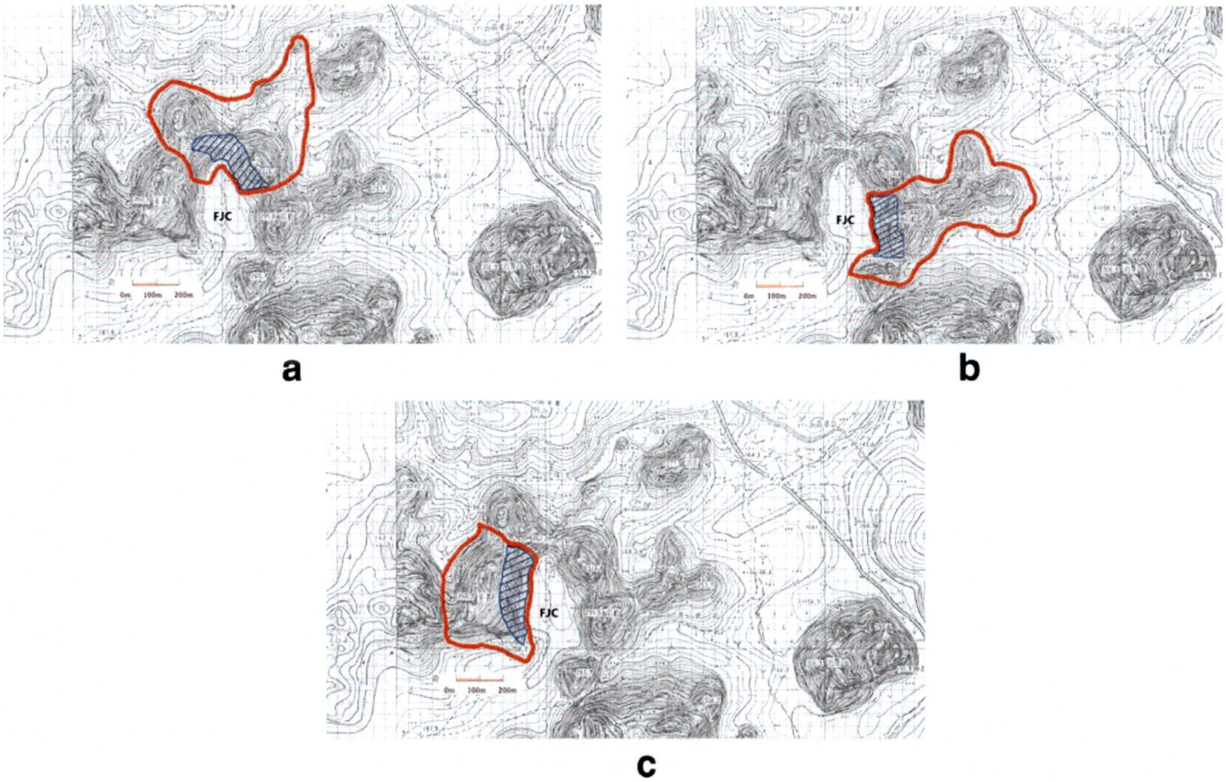

Fig. 6.10 The home ranges (red-lined) and core areas (blue-shaded area) of the three FJC white-headed langur family troops in the year 2015. (**a**) The home range and core area of "Yuweng family troop." (**b**) The home range and core area of "Laoxie family troop." (**c**) The home range and core area of "Xishan family troop"

Table 6.7 The home ranges and core areas of various white-headed langur troops in FJC

	Home range			Core area		
Troop	km^2	ha	*mu*	km^2	ha	*mu*
Yuweng family troop	0.238	23.8	357.0	0.0350	3.50	52.50
Laoxie family troop	0.241	24.1	361.5	0.0263	2.63	39.45
Xishan family troop	0.231	23.1	346.5	0.0375	3.75	56.25
Bokyueshan transitory troop	0.310	31.0	465.0	0.0225	2.25	33.75
Sum area	1.020	102.0	1530.0	0.1213	12.13	181.95
Average area	0.255	25.5	382.5	0.0303	3.03	45.49

6.5.2 Drifting Territory

Some animals have their territories only spatiotemporally or drifting. A typical example can be found with bitterlings (*Rhodeus amarus*) (Breder 1933), which would stick around river mussels in their reproductive seasons, for their fertilized eggs could only develop inside the valve cavity of river mussels. As the river mussels move, the bitterling follows, in order to defend its drifting territory.

The following are our registers of drifting territories seen in the several white-headed langur families in FJC.

6.5.2.1 Drifting Territory on the Forage Route

We shall illustrate the drifting territories observed of the "Bokyueshan transitory troop" on its regular forage route as an example.

The "Bokyueshan transitory troop" spent their nights in the night shelter cave at the northern hilltop of Bokyueshan (BYS). In the morning, they came down from the night shelter and always moved up the same tree-shaded route along the talus at BYS foothill and on Eco-Park, foraging as they moved. When they would encounter competitors of its same species or signs of insecurity, they would show

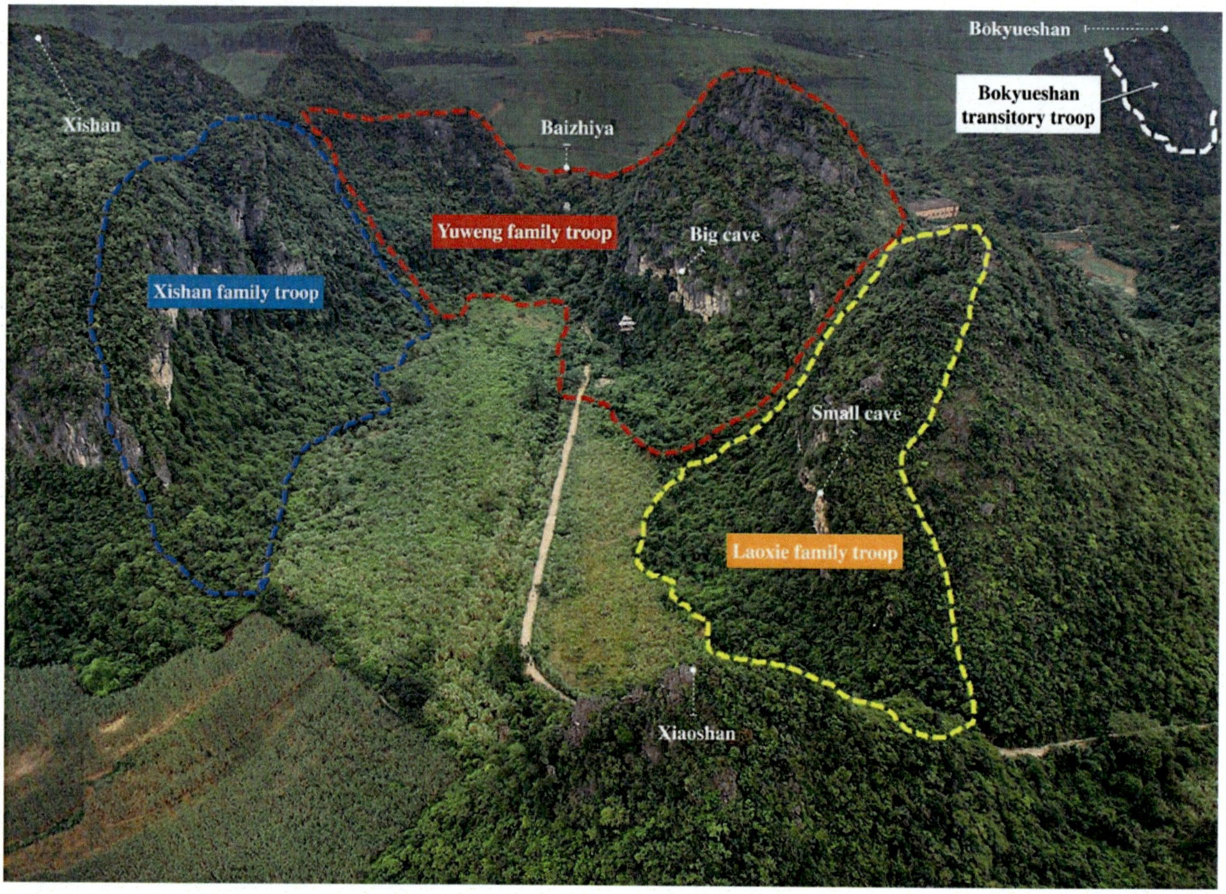

Fig. 6.11 Photo view of the core areas of the four white-headed langur family troops in FJC

territory-related behaviors. Since the position of the "Bokyueshan transitory troop" on the tree-shaded route kept changing, territory clashes with competitors could happen at any time and anywhere on its route to forage. When our observers got too near, the resident alpha male would show territory-defending behaviors as well. We termed the temporary location where territory-related behaviors happened as "drifting territory" (Fig. 6.12).

On December 29, 2015, we trailed the "Bokyueshan transitory troop" to observe and register their behaviors on their forage route by every hour. If we got too close when they were picking leaves, the resident alpha male would immediately consider us as a potential danger and take defensive measures. Its territory-related behavior was jumping to the front of the trailing observer and put itself between the observer and the troop to protect its family members. We registered and flagged on a map the drifting territories of the "Bokyueshan transitory troop" every hour and noted the area size of each drifting territory (the green patch on Fig. 6.13). All the drifting territories had very small areas (0.002–0.007 km^2) with an average of 0.012775 km^2 (equivalent to 1.2775 ha or 19.1625 *mu*), which was about half the size of the core area of the "Bokyueshan transitory troop" (blue-shaded on Fig. 6.12). Table 6.8 lists the sizes of various registered drifting territories.

6.5.2.2 Drifting Territories of Invading Males

Drifting territories could also arise in the contest for reproduction by male white-headed langurs. All the cases we are citing below are concerning Yintangxiaotu. They showed how the all-male troop of Yintangxiaotu and its sons made use of drifting territories as a base to invade the family troops of the other male langurs and the outcome.

Yintangxiaotu was an amazing male white-headed langur. It has achieved the most among all the resident alpha male langurs we observed in FJC region. It successfully invaded FJC Big Cave and founded the "Yintangxiaotu family troop" in September 2006. In July 2012, it yielded and withdrew to FJC Small Cave to form the "Yintangxiaotu transitory troop," and in March 2015 further yielded and formed the

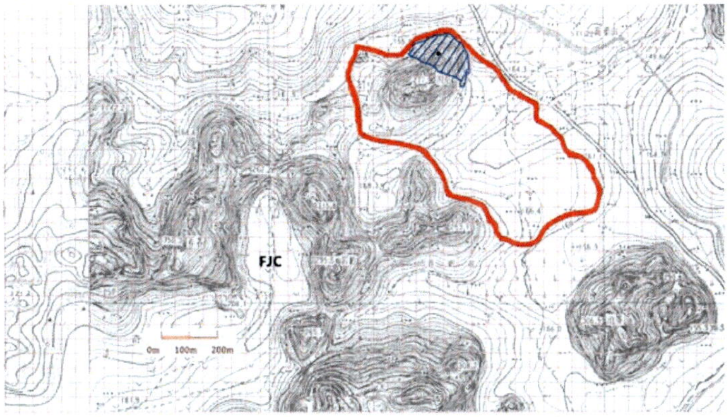

Fig. 6.12 Home range (red-lined area) and core area (blue-shaded area) of "Bokyueshan transitory troop"

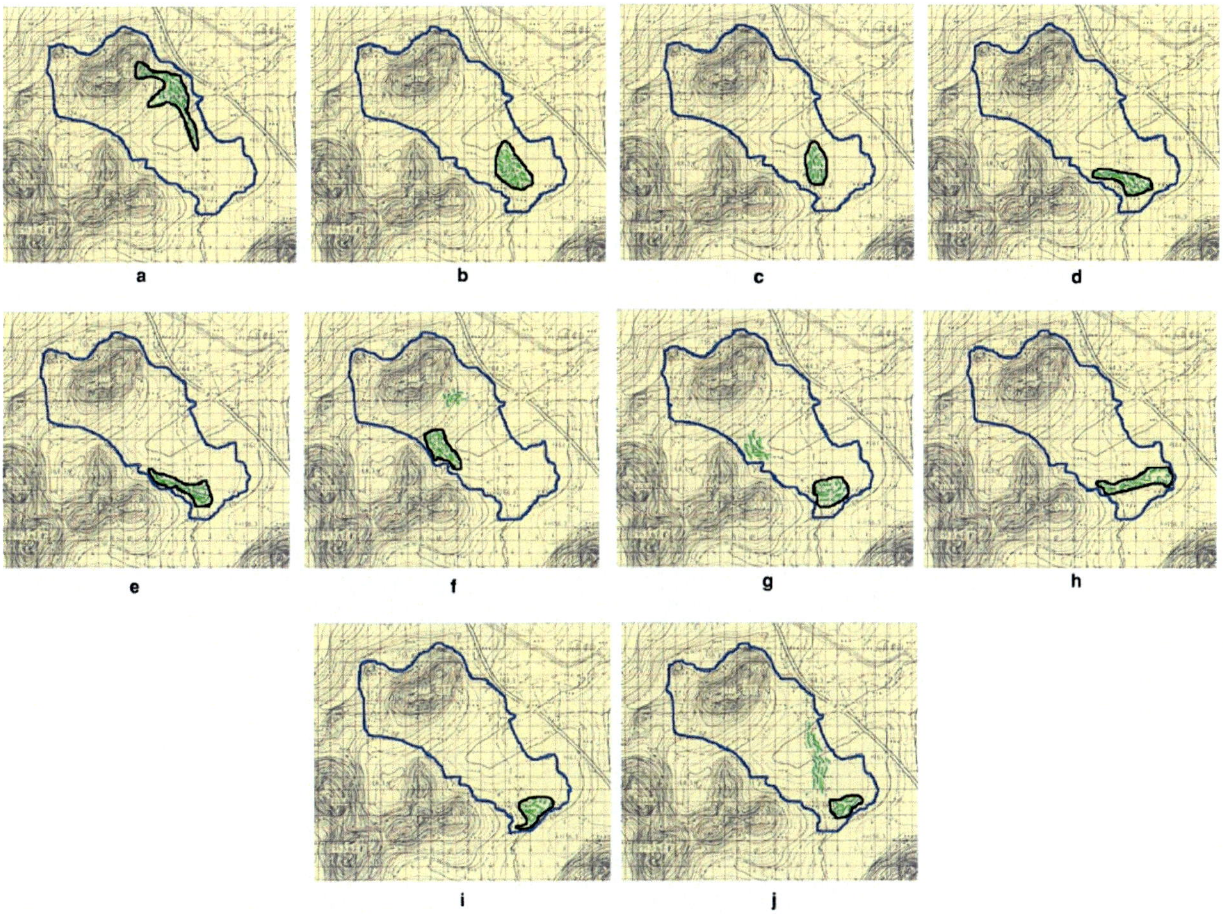

Fig. 6.13 2005.12.29 The ten drifting territories of "Bokyueshan transitory troop" on its day movements

Table 6.8 Area size of the drifting territory in every hour of daytime movements by 'Bokyueshan transitory troop'

Sequence	Time slot	Area of the drifting territory		
		km^2	ha	*mu*
a	08:00–09:00	0.02000	2.000	30.000
b	09:00–10:00	0.01625	1.625	24.375
c	10:00–11:00	0.01000	1.000	15.000
d	11:00–12:00	0.01125	1.125	16.875
e	12:00–13:00	0.01375	1.375	20.625
f	13:00–14:00	0.01250	1.250	18.750
g	14:00–15:00	0.01200	1.200	18.000
h	15:00–16:00	0.01500	1.500	22.500
i	16:00–17:00	0.01000	1.000	15.000
j	17:00–18:00	0.00700	0.700	10.500
Sum area		0.12775	12.775	191.625
Average area		0.012775	1.2775	19.1625

"Yintangxiaotu all-male troop." Since then, it started on a new journey with its sons, meandering like a stream through the Nongguan Mountains, sowing its DNA.

The following three cases exemplify what drifting territory means to the lives of langurs in an all-male white-headed langur troop.

Case 1: Drifting Territory on the Rooftop of Wudong

2013.10–2015.03

In this period the eldest son of Yintangxiaotu, Wuque, leading its brothers, had formed an all-male troop, which we called "Wuque all-male troop." The troop often broke away from their parent and sisters in the "Yintangxiaotu transitory troop" and established themselves a territory of drifting nature on the rooftop of Wudong, which is situated at about 500 m away from FJC Small Cave (Fig. 6.14). Wudong was the new three-story building of our Research Base with a roof area of about 300 m^2at the center of Eco-Park. Wuque and its brothers took the rooftop as their primary night shelter, and sometimes they slept on the narrow window ledges. Occasionally, they would return to spend a night with their parent at FJC Small Cave. During the one and a half year period, "Wuque brothers troop" had over a hundred clashes with "FNS transitory troop" coming from the southwest and "BYS transitory troop" from the northeast. At the peak of the clashes, there were more than three fights in a week.

Since after March 2015

Yintangxiaotu, taking along two of its 1-year-old pups, left FJC Small Cave and came and joined the "Wuque brothers troop" (the female pup Liaoshaotou was wounded during the outlier male invasion and later died), turning the troop to the "Yintangxiaotu all-male troop." Yet the territory of the "Yintangxiaotu all-male troop" was smaller than that of the "Yintangxiaotu transitory troop" as well as of the "Wuque brothers troop" (Fig. 6.15). Yintangxiaotu always offered help to its sons by taking up vigilance and guard positions. Out of their closeness to us, they took the rooftop on Research Base as their primary resting place at night (Fig. 6.16).

2015.04.15–2015.05.18

Yintangxiaotu was still leading the all-male troop roaming in FJC region. During this time, it had helped its eldest son Wuque seize hold of 12 females inhabiting the Nongxiang

Fig. 6.14 The home range and territory of "Wuque brothers troop"

Fig. 6.15 2015.3–2015.5 Photo view of the home range and territory of "Yintangxiaotu all-male troop"

Fig. 6.16 "Yintangxiaotu all-male troop" often rested on the rooftop of the research base (Photo by Liang Zuhong)

wetland and initiated the "Nongxiang family troop," also called the "Wuque family troop."

2016.02

Eight golden yellow–haired pups appeared in the "Wuque family troop," which signified that Wuque with the help of its father and brothers had successfully invaded another troop.

Case 2: Drifting Territory on the Mid-hill at the South Flank of Nongbaga to Support the Sons' Failed Invasion Attempt (Fig. 6.17)

2015.5.19.

Yintangxiaotu abandoned the territory on the rooftop of Wudong, taking along the rest of his 11 sons, departing its familiar FJC region for the south. We tracked and witnessed

(a) (b) (c)

Fig. 6.17 "Drifting territories" of "Yintangxiaotu all-male troop" on their way of migration (Photo by Yang Yanping). (**a**) Night shelter in Nongbaga; (**b**) Night shelter at the north entrance of Nongbaga; (**c**) Night shelter in Beiqishan

that they traversed the FSS foothill, passing through Muzishan (MZS) while moving south, reaching the 1500-m-long Nongbaga valley (NBG). NBG is at a straight-line distance of about 2.5 km away from FJC but the troop had to go twists and turns to bypass many obstacles, making their journey take 7–8 km. Nongbaga is in the heartland of the Nongguan Mountains where hilly wetlands and swampy valleys are interspersed, with its entirety covered by complex terrain and vegetation.

2015.06.06.

Yintangxiaotu and its sons appeared on the entry of a mountain track at east NBG, and the next day, they clashed with an unnamed male langur. The fighting was intense, and in the following days, 2–3 sons of Yintangxiaotu were seriously wounded.

Since 2015.06.16.

"Yintangxiaotu all-male troop" pushed to the mid-section of the NBG mountain tract, where on the far western end is an abrupt precipice with 7–8 female langurs and a number of yellow-haired pups. North of the precipice is another narrow mountain track extending for about 200 m with an entry at its northern end, which we named the Nongbaga North Entrance. Crossing the Nongbaga North Entrance, one could reach the highest craggy peak (440 m above sea level) in the Nongguan Mountains and the enigmatic Nongguan wetland, where several white-headed langur family troops lived. The resident alpha male of the troop inhabiting Nongbaga North Entrance was very strong and powerful, and it had taken the initiative to strike the "Yintangxiaotu all-male troop" more than ten times, injuring Yintangxiaotu and its sons one after another. Besides, "Yintangxiaotu all-male troop" was also attacked by other male langurs coming from Nongtung at the northeast of Nongbaga.

By the end of June 2015.

"Yintangxiaotu all-male troop" had intended to invade the "Nongbaga North Entrance family troop" but failed. Since then the whole troop gave up the drifting territory on the mid-hill at the mid-section of Nongbaga, climbed over the mountain ridge, and moved to Nonglao (NL).

Case 3: Occupying Drifting Territory at Beiqishan Targeting to Invade the "Nongfeng Female Troop"

Having climbed over the mountain ridge at the mid-section of Nongbaga and moving south into Nonglao, "Yintangxiaotu all-male troop" found a night shelter on a precipice at the northern stretch. We called it North Nonglao night shelter, which was in effect the drifting territory at the time being, and the all-male troop continued their contest for reproduction.

2015.07.05–2015.10

One day there appeared a troop of 13 + 1 female white-headed langurs (13 adult female and 1 juvenile female) on the precipice at Nongda (ND), about 500 m from North Nonglao night shelter of the "Yintangxiaotu all-male troop." We named the troop the "Nongguan West female troop." Bordering on the north of Nongda precipice is Nongguashan wetland, and we guessed that the female troop had their night shelter there.

On that same day, the "Yintangxiaotu all-male troop" left north Nonglao and moved west across the Nongda precipice to establish a new territory 1200 m away on Beiqishan (BQS) (Fig. 6.17c). The "Yintangxiaotu all-male troop" made Beiqishan their core activities area. First, they went to the southwest corner of Nongda to drive the all-male troop there away, which had five male langurs in it, and then they came back to their BQS night shelter. Until October 2015, the "Yintangxiaotu all-male troop" kept in on and off friendly relationships with the 13 + 1 female langurs troop. The two troops maintained friendliness, and langurs of both troops could come and go between each other.

Since after 2015.10

The "Yintangxiaotu all-male troop" drove away several male langurs with steep features from the Beiqishan region to prevent them from getting near the 13 + 1 female langurs. Among the ones driven away was one we named Duanzhang, for its palm was fractured, and one we named Duaner, who had a small bit of tail broken away.

2016.02–2016.03

The "Nongguan West female langur troop" had come to stay more often in BQS night shelter.

2015.08–2016.05

The three sons of Yintangxiaotu, Gualai, Xiaoliu, and Chongsheng had mingled with "Nongguan west female langur troop" for more than ten times, hoping to be the chosen one by the female langurs.

The way it goes, the "Nongguan (west) female troop" would likely accept a son of Yintangxiaotu, since the two troops have the same drifting territory and are ready to fight alongside each other. However, contests for reproduction would also happen between brothers. Our tracking and research are still in progress, and we shall wait and see what happens.

We have discovered that the grown-up sons of Yintangxiaotu would, on their way in their migration, attempt to invade the family troops they came across, leading to frequent fights for territory and females. Since there are abundant food resources in the Nongguan Mountains, fights between the males were generally not food-related. Hence the territory-related behaviors of "Yintangxiaotu all-male troop" were part of their reproductive behaviors in an attempt to get

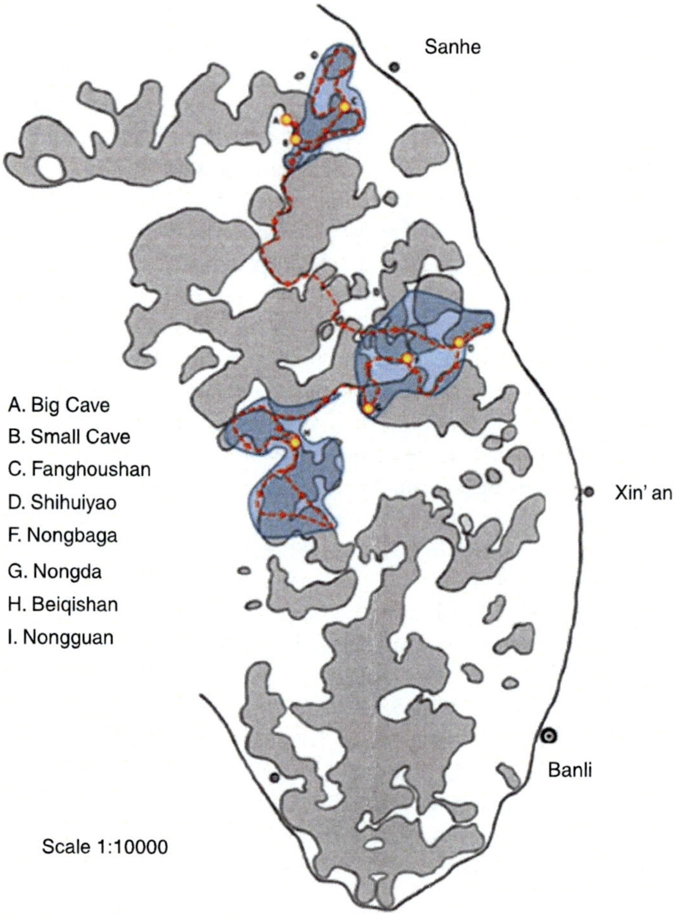

Fig. 6.18 The migration route of "Yintangxiaotu all-male troop" from June 2016 to the present

hold of female langurs. The migration of all-male troop was to disperse its genes, as well as propagate the population (Fig. 6.18).

References

Breder CM. Rhodeus amarus spawning in American mussels. Copeia. 1933;3:147–8.

Shang Y. Animal ethology. Beijing: Peking University Press; 2005.

7 The White-Headed Langur's Diets and Forage Behaviors

Abstract

From 1996 to 2015, we surveyed the composition of diets of white-headed langurs during three sampling periods. According to plant species verified in their diets, we found that when the biomass of plants in the environment was low (1996–2001), white-headed langurs would make up the quality and quantity deficiencies in diets by extending their choices. When vegetation biomass started to restore, langurs showed a preference for particular species (2002–2006). During the period (2008–2015) when the plant community was fully restored, we found an increased number of plant species in their diets, indicating their choice of food had increased. Nonetheless, in all three periods, 21 species could be reckoned as the main diets of white-headed langurs. Among them were *Iodes vitiginea*, *Pueraria Montana*, *Broussonetia papyrifera*, *Ficus tinctoria subgibbosa*, *Pteroceltis tatarinowii*, *Sterculia nobilis*, and *Radermachera sinica*, which were vital to the survival of white-headed langurs. According to our detailed direct observations of daily diets of white-headed langurs and the chime found in the stomach of a 4.5-year-old male that died from electrocution, the mass of the food adults consume every day is estimated at around 800–900 g.

Keywords

Vegetation biomass · Adaptive foraging · Daily food consumption

Foraging takes up 16.62% of the white-headed langurs' activities. This chapter studies what they eat, how they eat, and particularly if they could obtain enough food in dry seasons when plants had withered or at the times when their habitat was heavily devastated.

7.1 The Characteristics and Composition of Vegetation in the Nongguan Mountains

In the early nineteenth century, southwestern Guangxi was as tranquil as it was hundreds of millions of years ago. In the tropical limestone evergreen monsoon rainforest on the northern slope of Shiwanda Mountain, tall and erect broadleaf trees like *Horsfiedia hainanensis* and *Saraca chinensis* were still silently towering toward the sky. White-headed langurs climbed and leaped about on the precipices on karst hilly wetlands and swampy valleys, enjoying their lives in the forests.

From 1950 to 1990, as the human population grew incessantly, the once widely distributed tropical limestone evergreen monsoon rainforest suffered through repeated chopping and devastation. By the late 1990s, almost all the river valleys, mountain tracts, and hill slopes that were relatively level had been exploited for farmlands and cleared for villages or small plantations, while most karst hills had been stripped bare after long periods of cutting for firewood. People planted crops like corn, sugarcane, and paddies to sustain the large populace. Eucalyptus plantations, pine plantations, and tung tree plantations were introduced and developed in small lots from place to place on hills and wetlands. The village margins and open spaces next to the houses were planted with fruit trees like longan, prunnus, wampi, pomelo, etc., or dicotyledons trees like banyan, camphor, kulian (*Melia azedarach)*, xiangchun (*Toona sinensis*), etc. Vast stretches of the original tropical monsoon rainforest ecosystem with rich biodiversity had been rapidly transformed into one unitary agro-ecosystem in the short span of tens of years. The original native species suffered unprecedented damage. Until the last year in the twentieth century, did the situation begin to alleviate when people in the region started using biogas to replace firewood and

W. Pan, *The Population Ecology of White-Headed Langur*, https://doi.org/10.1007/978-981-33-4118-0_7

gradually stopped devastating the karst hills and the surrounding areas. The seeds buried in the talus around the karst hills once more had a chance to germinate and grow, and the Nongguan Mountains were gradually restored.

Succession is one of the most significant characteristics in flora dynamics. A region's flora that had been repeatedly harvested and devastated, in our case tropical evergreen monsoon limestone rainforests, would gradually self-restore after the destructive external forces were eliminated. Nevertheless, the restoring process and the time it takes varies in different areas, since they suffered through different levels of devastation. Since 1996, we have been studying white-headed langurs every day on the mountain tracts and hill slopes in the Nongguan Mountains and have witnessed the succession of floras that brought restoration to the tropical monsoon rainforests in the region. On the desolate mountain tracts and talus with relatively thick soil, herbaceous plants like *Themeda triandra* and *Miscanthus floridulus* noiselessly germinated from bare ground. They dispersed everywhere and are nurtured by sunshine and rain. In time, the tall *Saccharum arundinaceum* appeared and outgrew the other plants, with long and broad leaves swaying in the wind during all the seasons. Small woody plants like *Berchemia polyphylla* and *Croton euryphyllus* started to appear among the flora and lianas like *Pueraria lobata*, *Erythropalum scandens*, and *Broussonetia kazinoki* started climbing up the precipices. On relatively level tracts, *Cipadessa cinerascens* and *Sterculia nobilis* gradually took precedence and became the dominant species among other trees. On the steeply inclined karst hills, *Pteroceltis tatarinowii*, *Ficus*, and *Pistacia chinensis* were steadfastly growing. In 20 years, the bare red earth and mottled cliffs become invigorated with colors that changed with the seasons, from soft yellow to fresh green to thick green. Grasses, lofty trees, and lianas are vigorously growing and spreading to wherever they can. Different plant species, in turn, keep altering and remaking the surrounding environment. They can turn bare ground with the most miserable conditions into grass bushes, then into thorny shrub thickets, then into forests dominated by lofty second-growth trees.

Since the beginning of our research, we paid close attention to the flora composition in the Nongguan Mountains. We had kept track of the restoration progress after conservation policies were put in place. However, it is more critical for us to know which plants are included in the diets of the white-headed langur.

7.1.1 Research Method

In the years from 2000 to 2006, we had, on many occasions, collected plants from the Nongguan Mountains, pressed them into specimens and invited specialists to verify them.

In 2006, 2009–2010, and 2015, Prof. Li Xiangdong and his research team by way of combining the line transect method and the belt transect method conducted meticulous surveys on the main white-headed langur activity areas, which are on karst hills and neighboring flatlands, wetlands, and slopes.

By the line transect method, we registered data like plant species, their heights, coverage, etc. along a straight measured (meters) distance in a fixed direction.

General surveys and comprehensive surveys were two approaches of the belt transect method. In general surveys, researchers register dominant plant species and common plant species of the flora, their heights, coverage, and appearance of the flora in different seasons, etc. along a certain route. In comprehensive surveys, researchers square out quadrants of 10 m × 10 m sample plots to conduct a stratified survey in each plot. They stratify the flora into trees, shrubs, and herbs, find the height and coverage of each stratum, and then do the survey stratum by stratum.

7.1.2 Flora Composition in the Nongguan Mountains

Over the 20 years of research, we have found and verified a total of 420 plant species in our plant surveys, which comprise 282 genera and 102 families (sub-families) in the Nongguan Mountains. Among them are 18 fern species belonging to 12 genera in eight families, three gymnosperm species belonging to two genera in two families, 358 angiosperm dicotyledon flowering plant species belonging to 232 genera in 82 families and 41 angiosperm monocotyledon flowering plant species belonging to 36 genera in 10 families (sub-families). They are listed in the "Nongguan Mountains Plants Directory" in the Appendix. The statistical data shows that angiosperms are the most abundant in the Nongguan Mountains. The families, genera, and species they belong to take up 80.4%, 82.3%, and 85.2%, respectively, of the flora in the territory.

7.1.3 Essential Plant Life Forms in the Nongguan Mountains

We had briefly analyzed the chief life forms of the plants that were verified and found. There were 21 families of trees, 57 families of shrubs, 24 families of woody vines, two families of herbaceous vines, and 32 families of herbs. We further categorized the life forms, i.e., trees, shrubs, woody vines, herbaceous vines, and herbs, concerning their genera and taxonomic ranks to find out the relative proportion of each life form to get an overview of the Nongguan Mountains vegetation cover. Table 7.1 shows the results.

Table 7.1 Statistics of the chief plant life forms in the Nongguan Mountains

	Trees	shrubs	Woody vines	Herbaceous vines	herbs	Total
No. of families	21	57	24	2	32	102
Proportion[a] (%)	20.6	55.9	23.5	2.0	31.4	
No. of genera	47	150	45	7	68	282
Proportion[a] (%)	16.7	53.2	16.0	2.5	24.4	
No. of species	74	205	74	11	83	420
Proportion[a] (%)	17.6	48.8	17.6	2.6	19.8	

[a]As relatively significant variances may exist in the characteristics of plants within the same family, genus, or even species, overlaps occur in the statistics from identifying by characteristics, making the sum of the proportions bigger than 100%. For instance, there could be trees, shrubs, and herbs in the *Euphorbiaceae* family

Table 7.2 Phytogeography of the families of plants in Nongguan Mountains (adapted from Liu Hongying 2010 with revisions)

Distribution	Family	No. of families (proportion)
Worldwide	*Papilionaceae, Rhamnaceae, Asclepiadaceae, Balsaminaceae, Liliaceae, Compositae, Araceae, Euphorbiaceae, Rosaceae, Dioscoreaceae, Pteridaceae, Lygopodiaceae, Adiantaceae, Aspleniaceae, Cyperaceae, Bambusoideae, Lamiaceae*	17 (16.7%)
Tropic	*Myrtaceae, Bignoniaceae, Icacinaceae, Sterculiaceae, Sapotaceae, Burseraceae, Lauraceae, Cucurbitaceae, Bombacaceae, Hernandiaceae, Anacardiaceae, Capparaceae, Myristicaceae, Ebenaceae, Hypericaceae, Combretaceae, Acanthaceae, Passifloraceae, Connaraceae, Palmaceae, Polypodiaceae, Nyctaginaceae, Commelinaceae, Guttiferae Juss, Loranthaceae, Caricaceae, Blechnaceae*	27 (26.5%)
Tropic-subtropic	*Annonaceae, Rubiaceae, Vernenaceae, Mysinaceae, Magnoliaaceae, Menispermaceae, Aquifoliaceae, Theaceae, Convolulaceae, Hipocrateaceae, Olacaceae, Zingiberaceae, Agavaceae, Melastomataceae, Meliaceae, Moraceae, Sapindaceae, Alangiaceae, Apocynaceae, Escalloniaceae, Sargentodoxaceae, Smilacaceae, Cycadaceae, Lythraceae, Gesneriaceae, Davaliaceae, Gnetaceae, Thelyteriaceae, Hydrangeaceae, Flacourtiaceae, Umbelliferae, Capparaceae, Hamamelidaceae*	33 (32.4%)
Tropic-temperate	*Celastraceae, Urticaceae, Vitaceae, Santalaceae, Aristolochiaceae, Caesalpiniaceae, Mimosaceae, Araliaceae, Symplocaceae, Oleaceae, Solanaceae, Ulmaceae, Rutaceae, Loganiaceae, Malvaceae, Agrostidodeae, Caprifoliaceae, Amaranthaceae, Tiliaceae, Boraginaceae, Actinidiaceae, Juglandaceae*	22 (21.6%)
Temperate	*Ranunculaceae, Polygonaceae, Chenopodiaceae*	3 (2.9%)

From Table 7.1, we can see that shrubs make up more than half, or almost half, of the total number of families, genera, or species, and the results reflect that the shrub is the chief plant life form in the flora of the Nongguan Mountains. Life forms are more steady and identifiable in species than in genus and family. We can see in the species tier that, apart from shrubs, the proportions of trees, woody vines, and herbs are more or less the same at 17.6%, 17.6%, and 19.8%, respectively, while herbaceous vines are comparatively less distributed among the flora. According to the above statistics, we could know that by the combined effects of climate and soil, abundant woody plants are growing in the flora of the Nongguan Mountains. They are mostly in shrub life forms, giving rise to clusters of small trees and clusters of shrubs, which are typical of evergreen monsoon rainforests in karst landscapes.

7.1.4 Phytogeography Composition in the Nongguan Mountains

The phytogeography of the above-mentioned 102 plant families includes worldwide distribution species, tropic distribution species, tropic-subtropic distribution species, tropic-temperate climate distribution species, and temperate climate distribution species. Among them, 60 families are tropic and subtropic distribution species, which accounts for 58.8% of the total families, whereas there are only 17 families that are worldwide distributing, which represents 16.7% of the total (Hou Kuanchao 1982) (see Table 7.2).

The Nongguan Mountains, situated at the northern border of the tropic, have a tropic-subtropic transitional climate, and the transitional character is well reflected in the phytogeography of its flora.

7.2 The White-Headed Langur's Diet Varieties

7.2.1 Specialized Diet Species and Generalized Diet Species

Animals can be sorted into specialized diet species (specialists) and generalized diet species (generalists) based on the variety of food in their diets. The term "specialists" refers to those animals which only have a small variety in their diets, or those feeding on just one specific variety of plant or animal, or even just a specific part of that one plant or animal. Generally, specialists live in habitats with abundant food resources, whose supply is relatively steady through all seasons. Generalists are opportunists; they could feed on whatever edible and available in abundance. Generalist species usually distribute in temperate climate regions, where climate and the availability of food resources vary markedly with the seasons (Shang Yuchang 2005). We may regard the divergence in diet habits as adaptations to different environmental conditions. Although the white-headed langur lives on the northern margin of the tropic, it is a generalized diet species. Since the localized climatic conditions and the vegetation composition in the Nongguan Mountains bring about seasonal variances in food varieties and quantities, and disturbances from human activities had reduced food resources to scarcity, feeding on multiple varieties of plants was doubtlessly a wise choice.

7.2.2 A Review of the Research of White-Headed Langur Diets

The white-headed langur's regular diet includes a large variety of plants, including ones it tried as a sample. The details have been registered in many researchers' reports.

The first document was by Shen Lantian (Lantian and Hanhua 1982), who recorded that white-headed langurs dieted on a variety of leaves, twigs, sprouts, flowers, and fruits, and their favorites were the leaves and figs of banyans (*Ficus* genus), sprouting leaves of *Millettia* genus plants, *Oroxylum indicum*, *Pleomele cambodiana*, and the *Ulmaceae* family and *Mimosaceae* family plants, and they also often fed on flowers of mumian-shu (*Bombax ceiba*), mulberries, shanlongyan (*Helicia grandis Hemsl*), bananas, shajimu leaves, and bamboo leaves.

Lai Yuemei (1987), after studying white-headed langurs in the wild and in caged conditions, summed up 21 plant species that made up their chief diet.

From 1991 to 1992, Qiang et al. (1994) surveyed wild white-headed langurs in Bapen Township in Fusui County of Guangxi and inspected the traces of their forages. They then verified their findings by feeding the same thing they observed to caged white-headed langurs. Afterward, they came up with the report that white-headed langur diets included no less than 30 plant species (27 genera in 18 families).

From 1997 to 1998, Li Zhaoyuan et al. (2003) researched the foraging ecology of white-headed langurs in the Fusui Nature Reserve. He registered 50 plant species from 1381 forage surveys and found among them ten species that were in white-headed langur's chief diets. They were the *Broussonetia kazinoki*, *Capparis viminea*, *Celtis austrosinensis*, *Celtis sinensis*, *Cuscuta chinensis*, *Ficus microcarpa*, *Millettia pachycarpa*, *Pteroceltis tatarinowii*, *Pueraria thunbergiana*, and *Tetracera asiatica*; together they made up 61.8% of the total diet.

In 2000, Huang Chengming et al. (2000) reported that white-headed langurs inhabiting the Nonglin habitat in Bapen Township in Fusui had 42 plant species included in their diets.

In 2002 and 2003, Zhou Qihai et al. (2013) researched the Fusui Nature Reserve and found that white-headed langurs fed on 109 plant species, with 11 species among them taking up 84.9% of the total diet. The 11 species were the *Pithecellobium clypearia*, *Celtis sinensis*, banyan (*Ficus*), *Broussonetia papyrifera*, *Ficus subulata*, Japanese dodder (*Cuscuta japonica*), *Oroxylum indicum*, *Maclura cochinchinensis*, *Ficus tinctoria*, *Ficus hispida*, and *Securidaca inappendiculata*.

Most of the research described above was conducted in Bapen Township in Fusui, Guangxi, where the variety of plant species included in the white-headed langurs diets may be different from those in the Nongguan Mountains due to differences caused by localized climates, landscapes, and human disturbances. Hence, we are not able to make particular remarks about their research with what we found in the Nongguan Mountains. However, from the registers above, we can at least conclude that:

- The white-headed langur lives on a diet of multiple varieties of plants.
- Among the variety of plant diets, some species are in its chief diet.
- Diet composition could be different for different troops of white-headed langurs in the same area or for the same troop at different times.

7.2.3 Chief Research Method

In situ observations, feces analysis, cage-feed, and stomach content analysis are regularly used research methods in studying primate diets, but every one of them has its limitations. A majority of white-headed langur diet and nutrition research had adopted in situ observation methods (Lantian and Hanhua 1982; Lai Yuemei 1987; Qiang et al. 1994; Li et al. 2003; Huang Chengming et al. 2000; Zhou et al. 2013) while some adopted cage feed methods (Lai Yuemei 1987; Qiang et al. 1994). We adopted the in situ observation method in our research. In 20 years of research, we eventually established our unique research method.

These semi-arboreal semi-saxicolous primates are shy and hard to approach. It is hard to obtain in fullness and with accuracy, what plant species were eaten and in what quantity.

From 2002 to 2006, we had begun using HD video recording equipment to record white-headed langur forage behaviors. The equipment included the CANON XLHI 3CCD HD camcorder, the SONY HVR-M15C HD video cassette recorder, the AMOI LC-27HWT2A LCD TV, and the ASUS video data analysis and storage system, among others. We tracked the langur troops to observe and record them from daybreak when the langurs left the night shelters to forage until they returned to rest at sunset. We recorded all their behaviors (forage, moves, and recess) during the entire day. Then, in laboratory, we used the video data analyzing system to watch replays of the footage on forages. We enlisted specialists to help verify the plant species the white-headed langurs gathered and which parts of the plants they had eaten. We also recorded the length of the feeding time. In cases when we could not determine the plant species from the recorded image, we returned to the foraging site to collect specimens and verify them.

Since 2008, our reorganized research team had again upgraded our photo shooting, filming, and image analyzing equipment. Camera models and other related equipment includes the SONY PDW-700 professional HD digital blue-ray disc camcorder fitted with the FUJINON HA42x9.7BERD lens, the SONY NEX-FS700 HD camcorder fitted with the SONY 70–300 mm (or 18–200 mm) lens, the SONY PMW-EX1R HD handheld camcorder, the SONY HDC-310 HD video camera with the CANON 4F 19x VCL-719BXS lens, and the CANON EOS 1D MARK III digital camera fitted with the CANON EF 600 mm f/4 L lens for shooting in the wilderness. Video and photo analyses were processed with the Mac Pro Mid 2010 and other computer systems.

The high-quality cameras and camcorders had made it easier to accurately identify the type of plants the white-headed langurs had fed on. However, the many years of tracking and observing were indispensable. They had provided researchers with clues on the daily movement routines of the white-headed langurs and a general idea on what plant species they feed on. These factors had significantly been useful in planning for and making successful video recordings. After many years, accustomed to the presence of researchers, the langurs became more relaxed and could be filmed at much closer distances, increasing the success rate in identifying the plant species they were feeding.

Many researchers had contributed significantly to the field research. Prof. Li Xiangdong from the Forestry College of Guangxi University and his work team had provided plenteous expert advice on verifying the plants which the langurs fed on. It is worth mentioning that Prof. Li, despite his elderly age and feebleness, presented himself to survey the plants on the dangerously inclined karst hills where the white-headed langurs inhabit (Fig. 7.1). His dedicating spirit had since been a source of great encouragement for us. Since 2015, Prof. Li Hua and Deng Rongyan, Ph.D. from Guangxi University, as well as Prof. Liu Yan from the Guangxi Institute of Botany, Chinese Academy of Science, had provided us tremendous help in verifying the species in langur diets.

7.2.4 Diet Composition and Alterations

Even today, there are still lots of primordial tropical plant species in the Nongguan Mountains. They had persisted throughout the devastating calamities in the twentieth century. Many parts of the flora also regenerated on bare grounds through phases of herb bushes and thorny shrub thickets, and then gradually grew to be lofty trees that dominated second-growth forests. Through that long process, the white-headed langurs in the Nongguan Mountains had demonstrated their rapid adaptability of generalized diet species. They lived on different plant diets in different phases. In respect of different research methods, we categorize our 20 years of research findings into three periods in the following pages. The three periods reflect the phases from the devastations by human activities to the gradual restoration of the flora.

7.2.4.1 Diet Composition in the Period of Intense Human Activity Disturbances (1990–2001)

From 1996 to 2001, we verified the plant species in langur diets primarily by observing through binoculars and collecting specimens from the plants that langurs had foraged on. In that period, we had verified 138 plant species belonging to 108 genera in 57 families that were consumed by white-headed langurs, and most of them were dicotyledon angiosperms. There were only six monocotyledon species belonging to five genera in five families in the white-headed langurs' diet, and no gymnosperm plants and fern plants were consumed (Xiangdong and Shiyang 2005). Specifics are listed in the "Nongguan Mountains Plants Directory" in Appendix denoted by ○1, and they constitute the white-headed langur diet composition in the period of intense disturbance caused by human activity (1990–2001).

We can see from the above-mentioned statistics of the langur diet that 30 families had each only one species in its composition, 20 families each had two species, and seven families had five species each or more. The seven families were *Moraceae* with 11 species, *Euphorbiaceae* with 11 species, *Papilionaceae* with nine species, *Rhamnaceae* with seven species, *Caesalpiniaceae* with six species, *Vitaceae* with six species, and *Asclepiadaceae* with five species. The total 55 species in the seven families took up 40% of the composition of 138 species. Under the disturbances of intense human activities, most of the flora in the habitat was composed of thorny shrub thickets, and the white-headed langur's diet composition was complex, with no notably dominant plant families. Moreover, the intense chopping for firewood had reduced the biomass of trees and shrubs to a

Fig. 7.1 Vegetation in the Nongguan Mountains (Photo by Feng Chunguang)

low level. Under those conditions, the white-headed langurs had to make up the quality and quantity deficiencies in their diets by extending their choice to include more plant species.

By analyzing the 108 genera in the diet composition concerning the 15 phytogeographic zones put forward by Mr. Wu Zhengyi, we found that the white-headed langur diet consisted chiefly of tropic distributing types. There were 32 genera of pan-tropic distributing types, 20 genera of old-world tropic distributing types, 15 genera of tropical Asia distributing types, 13 genera of tropical Asia–Oceania distributing types, five genera of tropical Asia–America distributing types, and four genera tropical Asia–Africa distributing types, totaling 89 genera that make up 82.4% of the diet composition. The proportions resembled the overall plant phytogeography in the Nongguan Mountains very much. The diet composition had reflected the tropical attribute of white-headed langurs and provided evidence that they adapted to the ecosystem by way of generalizing their diet.

7.2.4.2 Diet Composition in the Early Period of Flora Restoration (2002–2006)

From 2002 to 2006, we had introduced high-definition video recording gear into our field observations and equipped image-analyzing systems in our laboratory to conduct systematic sampling and analysis. In that period, we verified 70 plant species (including mutated species) in the white-headed langur diet belonging to 60 genera in 35 families (Li Xiangdong et al. 2008). Specifics are listed in the "Nongguan Mountains Plants Directory" in the Appendix denoted by ○2, and it constitutes the "white-headed langur diet composition in the early period of flora restoration (2002–2006)."

Comparing the findings with those in the previous period, we found that 30 families, 38 genera, and 39 species appeared in both the periods, which represented 85.7%, 63.3%, and 55.6%, respectively, of the diet composition. There were still no gymnosperms and ferns in the composition, and *Pleomele cambodiana* was the only monocotyledon plant in the diet. However, when we analyzed the phytogeographic attributes of the diet composition, we found the proportion of tropic distributing species had increased to 91.5%.

Through watching video recording replays, we registered the specific parts of the 70 species of plants that were eaten by white-headed langurs. Except for *Cuscuta chinensis*, which is leafless, all the other 69 plant species had their leaves eaten, and the white-headed langurs particularly favored sprouting leaves and buds including leaf buds, shoot buds, and flower buds.

Going through our field observation registers, we discovered that white-headed langurs had diet preferences. There were some plant species they particularly favored, some plants they consumed in large quantities, and some plants that they only ate just once in a while. During this period, their favored plant species included the *Radermachera sinica*, *Stereospermum colais*, *Oroxylum indicum*, *Euaraliopsis ciliata*, *Garuga pinnata*, *Pteroceltis tatarinowii*, *Ehretia tsangii*, and *Iodes vitiginea*. Every time the white-headed langurs came across those plants, their sprouting leaves, young shoots, and fruits would all be eaten. However, those plants were not most abundant among the flora, so they made up only limited proportions in the diet of the white-headed langur. There were other plants that made up the chief diet of white-headed langurs, including various species of *ficus* genus, *Broussonetia papyrifera*, *Sterculia nobilis*, *Litsea glutinosa*, *Pueraria lobata*, *Croton euryphyllus*, etc. Those were dominant species in the flora with big biomasses, which was able to provide the white-headed langurs with a steady food supply. There were also a large variety of plants that the white-headed langurs fed on only once in a while and only in small quantities, like the *Impatiens balsamina*, *Thunbergia grandiflora*, *Aristolochia kwangsiensis*, etc. They were beneficial supplements to the main diets of white-headed langurs, in particular at times when food was scarcely available.

7.2.4.3 Diet Composition After 10 Years of Uninterrupted Flora Succession (2008–2015)

With the continual flora successions, thorny shrub thickets were gradually replaced by second-growth mixed-wood forests. Free from human interferences, some trees grew big and became dominant species in the flora. In the years between 2008 and 2015, we had again upgraded our photographic and video recording equipment, which makes precise observations and accurate verifications possible. Up to now, we have verified 88 plant species in langurs' diets belonging to 78 genera in 42 families (listing in Table 7.3). There are still other plants in the diet composition that are pending verification, and they will be included in the table when processed. In the meantime, we are continuing our field surveys on the continual flora successions.

When we compare the 88 langur diet species to those in the two earlier periods, we found that 35 species were among the 138 species in the period of intense human disturbance (1996–2001) and 31 species were present among the 70 species in the early period of flora restoration (2002–2006). Twenty-one species were present in all three periods. They were the *Croton euryphyllus*, *Flueggea virosa*, *Mallotus philippinensis*, *Mallotus repandus*, *Broussonetia papyrifera*, *Ficus tinctoria subgibbosa*, banyan, *Ficus virens*, *Malaisia scandens*, *Dalbergia hancei*, *Pueraria Montana*, *Celtis biondii*, *Pteroceltis tatarinowii*, *Pistacia weinmannifolia*, *Rhus chinensis*, *Sterculia nobilis*, *Oroxylum indicum*, *Radermachera sinica*, *Litsea glutinosa*, *Iodes vitiginea*, and *Lepionurus latisquamus*. They could be reckoned as part of the chief diet of white-headed langurs (Fig. 7.2).

From our field observations, we discovered that out of the above 88 langur diet species, white-headed langurs consumed the *Iodes vitiginea*, *Pueraria Montana*, *Millettia pachycarpa*, *Cayratia japonica*, *Broussonetia papyrifera*, *Ficus tinctoria subgibbosa*, *Ficus microcarpa*, *Erythropalum scandens*, *Pteroceltis tatarinowii*, *Sterculia nobilis*, *Radermachera sinica*, *Argyreia capitiformis*, *Argyreia pierreana*, *Thunbergia grandiflora*, *Cansjera rheedii*, *Paederia scandens*, *Gouania javanica*, *Vitex kwangsiensis*, *Stereospermum colais*, *Pristimera arborea*, *Ventilago inaequilateralis*, *Delonix regia*, and the beans of *Leucaena leucocephala* in large quantities (http://frps.eflora.cn/frps/Leucaenaleucocephala). These plants could be regarded as the chief diet of white-headed langurs in the period of our observations (Fig. 7.3). Among them, *Iodes vitiginea*, *Pueraria Montana*, *Broussonetia papyrifera*, *Ficus tinctoria subgibbosa*, *Pteroceltis tatarinowii*, *Sterculia nobilis*, and *Radermachera sinica* were also among the 31 long-time chief diet species, indicating that they are vital to the survival of white-headed langurs in the Nongguan Mountains.

In terms of seasons, the *Pueraria montana*, *Argyreia capitiformis*, *Thunbergia grandiflora*, *Cansjera rheedii*, *Celtis biondii*, and *Sterculia nobilis* were species consumed by white-headed langurs in large quantities all year round (Fig. 7.4). Particularly in dry seasons when food was scarce, those plants, together with the various evergreen species of *ficus* genus, were the crucial diet species that sustain the white-headed langurs' living. Plant species like the *Broussonetia papyrifera*, *Pteroceltis tatarinowii*, *Millettia pachycarpa*, etc. were food resources richly available in most time of the year but were comparatively less consumed during the dry season from December to February in the following year. All of those were vital and indispensable diet species to white-headed langurs.

Of the langur diet plant species registered in Table 7.3, except the leafless *Cuscuta chinensis*, there were 73 species from which the white-headed langur consumed leaves, especially sprouting leaves and leave buds. Fruits and seeds were consumed from 30 species, which accounted for 34.1% of the total species eaten, and fruits of the various species of *ficus* genus in *Moraceae* family were their particular favorites. They consumed flowers from 19 plant species, which account for 21.6% of the total species eaten. Nonetheless, in terms of available period and volume, fruits, flowers, and seeds were relatively scarce sources of food and could only serve as beneficial supplements to the white-headed langurs' daily diets (Figs. 7.5 and 7.6).

We analyzed the statistics of the 78 plant genera in Table 7.3 concerning the 15 phytogeographic distributions put forward by Mr. Wu Zhengyi (1997). We found that the result was very similar to what we got from our observations. In that period, white-headed langurs' diets were still chiefly made of various tropic distributing species with most of them

Table 7.3 The diet composition of white-headed langurs after ten years of uninterrupted flora succession (2008–2015)

Sequence	Family	Genus	Species (common name in China)	Species (scientific name)	Parts of plant eaten	Consumed in seasons of:	Feeding instances
1	*Alangiaceae*	*Alangium*	Ba jiao feng	*Alangium chinense*	Leaf, fruit	Wet	Fair
2	*Liliaceae*	*Liriope*	Shan mai tong	*Liriope spicata*	Fruit	Dry	Few
3	*Icacinaceae*	*Iodes*	Xia guo wei hua teng	*Iodes vitiginea*	Leaf, fruit, sprouting shoot	Wet, dry	Often
4	*Hipocrateaceae*	*Pristimera*	Er zi bian shuo teng	*Pristimera arborea*	Leaf	Wet	Often
5	*Euphorbiaceae*	*Alchornea*	Hong bei shan ma gan	*Alchornea trewioides*	Leaf	Wet	Few
6	*Euphorbiaceae*	*Croton*	Shi shan b adou	*Croton euryphyllus*	Leaf	Wet, dry	Often
7	*Euphorbiaceae*	*Flueggea*	Bai fan shu	*Flueggea virosa*	Fruit, leaf	Wet, dry	Fair
8	*Euphorbiaceae*	*Claoxylon*	Bai tong shu	*Claoxylon polot*	Leaf	Wet, dry	Few
9	*Euphorbiaceae*	*Mallotus*	Cu kang chai	*Mallotus philippinensis*	Leaf	Wet, dry	Few
10	*Euphorbiaceae*	*Mallotus*	Shi yan feng	*Mallotus repandus*	Leaf	Wet, dry	Fair
11	*Phyllanthaceae*	*Phyllanthus*	Xiao guo ye xia zhu	*Phyllanthus reticulatus*	Leaf, fruit	Wet, dry	Few
12	*Papilionaceae*	*Dalbergia*	Teng huang tan	*Dalbergia hancei*	Leaf, flower	Wet, dry	Often
13	*Papilionaceae*	*Zenia*	Ren dou	*Zenia insignis*	Leaf	Wet	Fair
14	*Papilionaceae*	*Pueraria*	Ge	*Pueraria montana*	Leaf, flower	Wet, dry	Often
15	*Papilionaceae*	*Whitfordiodendron*	Zhu yao dou	*Whitfordiodendron filipes*	Leaf	Wet	Few
16	*Papilionaceae*	*Millettia*	Hou guo ya dou teng	*Millettia pachycarpa*	Leaf	Wet	Often
17	*Papilionaceae*	*Rhynchosia*	Lu huo	*Rhynchosia volubilis*	Fruit (bean)	Dry	Few
18	*Menispermaceae*	*Diploclisia*	Cheng hou feng	*Diploclisia affinis*	Leaf	Wet	Few
19	*Menispermaceae*	*Stephania*	Fen ji du	*Stephania longa*	Leaf, fruit	Wet, dry	Fair
20	*Burseraceae*	*Garuga*	Yu ye bai tou shu	*Garuga pinnata*	Leaf, flower	Wet	Fair
21	*Mimosaceae*	*Adenanthera*	Hai hong dou	*Adenanthera pavonina var. microsperma*	Sprouting leaf	Wet	Fair
22	*Mimosaceae*	*Albizia*	Shan h ehuan	*Albizia kalkora*	Leaf, bean	Wet, dry	Fair
23	*Mimosaceae*	*Acacia*	Taiwan xiangsi	*Acacia confusa*	Leaf	Wet, dry	Fair
24	*Mimosaceae*	*Delonix*	Feng huang mu	*Delonix regia*	Leaf, flower, bean	Wet, dry	Often
25	*Mimosaceae*	*Leucaena*	Yin he huan	*Leucaena leucocephala*	Leaf, bean	Wet, dry	Often
26	*Poaceae*	*Pleioblastus*	Ku zhu	*Pleioblastus amarus*	Sprouting leaf	Wet	Rare
27	*Poaceae*	*Dendrocalamus*	Ma zhu	*Dendrocalamus latiflorus*	Budding shoot	Wet	Rare
28	*Apocynaceae*	*Trachelospermum*	Luo shi	*Trachelospermum jasminoides*	Leaf, fruit	Wet	Fair
29	*Apocynaceae*	*Urceola*	Suan ye jiao teng	*Urceola rosea*	Leaf	Wet, dry	Few
30	*Malvaceae*	*Abelmoschus*	Huang shu kui	*Abelmoschus manihot*	Flower	Wet	Few
31	*Acanthaceae*	*Thunbergia*	Shan qian niu	*Thunbergia grandiflora*	Leaf, flower	Wet, dry	Often
32	*Meliaceae*	*Cipadessa*	Hui mao jiang guo lian	*Cipadessa cinerascens*	Leaf, fruit	Wet, dry	Fair
33	*Meliaceae*	*Melia*	Ku lian	*Melia azedarach*	Fruit	Dry	Few
34	*Polygonaceae*	*Polygonum*	Huo tan mu	*Polygonum chinesis*	Leaf, shoot	Wet	Few
35	*Agavaceae*	*Dracaena*	Jian ye long xue shu	*Dracaena cochinchinensis*	Leaf, flower	Wet, dry	Few
36	*Asclepiadaceae*	*Cryptolepis*	Gu gou teng	*Cryptolepis buchananii*	Leaf	Wet, dry	Few
37	*Vernenaceae*	*Vitex*	Guangxi mu jing	*Vitex kwangsiensis*	Leaf	Wet	Often

(continued)

Table 7.3 (continued)

Sequence	Family	Genus	Species (common name in China)	Species (scientific name)	Parts of plant eaten	Consumed in seasons of:	Feeding instances
38	*Ranunculaceae*	*Clematis*	Zhu guo tie xian lian	*Clematis uncinata*	Leaf	Wet, dry	Fair
39	*Ranunculaceae*	*Clematis*	Shan mu tong	*Clematis finetiana Lévl. et Vant.*	Stem, flower	Wet, dry	Few
40	*Bombacaceae*	*Gossampinus*	Mu mian	*Gossampinus malabarica*	Flower, fruit, leaf	Wet, dry	Few
41	*Oleaceae*	*Jasminum*	Bai e su xin	*Jasminum albicalyx*	Leaf	Wet	Fair
42	*Vitaceae*	*Cayratia*	Wu lian mei	*Cayratia japonica*	Leaf, fruit	Wet, dry	Often
43	*Anacardiaceae*	*Mangifera*	Mango	*Mangifera indica*	Fruit, nascent pip	Wet	Few
44	*Anacardiaceae*	*Rhus*	Yan fu mu	*Rhus chinensis*	Leaf, insect gall	Wet, dry	Often
45	*Anacardiaceae*	*Pistacia*	Qing xiang mu	*Pistacia weinmannifolia*	Leaf, flower	Wet	Often
46	*Rubiaceae*	*Paederia*	Ji shi teng	*Paederia scandens*	Leaf, bud	Wet, dry	Often
47	*Rubiaceae*	*Oxyceros*	Ji zhao le	*Oxyceros sinensis*	Leaf	Wet	Fair
48	*Solanaceae*	*Solanum*	Xiao hua long kui	*Solanum americanum*	Fruit	Wet	Few
49	*Loranthaceae*	*Macrosolen*	Qiao hua	*Macrosolen cochinchinensis*	Flower	Wet	Few
50	*Loranthaceae*	*Taxillus*	Guang ji sheng	*Taxillus chinensis*	Leaf, fruit	Wet, dry	Few
51	*Moraceae*	*Broussonetia*	Gou shu	*Broussonetia papyrifera*	Flower, leaf, fruit	Wet, dry	Often
52	*Moraceae*	*Maclura*	Zhe shu	*Maclura tricuspidata*	Leaf	Wet	Often
53	*Moraceae*	*Ficus*	Xie ye rong	*Ficus gibbosa*	Leaf, fruit	Wet, dry	Often
54	*Moraceae*	*Ficus*	Rong	*Ficus microcarpa*	Leaf, fruit	Wet, dry	Often
55	*Moraceae*	*Ficus*	Zhi mai rong	*Ficus orthoneura*	Leaf, fruit	Wet, dry	Often
56	*Moraceae*	*Ficus*	Dou guo rong	*Ficus pisocarpa*	Leaf, fruit	Wet, dry	Often
57	*Moraceae*	*Ficus*	Lu huang ge shu	*Ficus virens*	Leaf, fruit	Wet, dry	Often
58	*Moraceae*	*Malaisia*	Niu gan teng	*Malaisia scandens*	Leaf	Wet, dry	Fair
59	*Theaceae*	*Camellia*	Fusui Jin hua cha	*Camellia fusuiensis*	Flower	Dry	Rare
60	*Opiliaceae*	*Cansjera*	Shan gan teng	*Cansjera rheedii*	Leaf	Wet, dry	Often
61	*Opiliaceae*	*Lepionurus*	Lin wei mu	*Lepionurus latisquamus*	Leaf	Wet, dry	Fair
62	*Rhamnaceae*	*Gouania*	Mao ju qian	*Gouania javanica*	Leaf, bud, flower	Wet, dry	Often
63	*Rhamnaceae*	*Sageretia*	Que mei teng	*Sageretia theezans*	Leaf	Wet, dry	Fair
64	*Rhamnaceae*	*Ziziphus*	Yin du zao	*Ziziphus incurve*	Leaf, fruit	Wet, dry	Few
65	*Rhamnaceae*	*Ventilago*	Hainan yi he guo	*Ventilago inaequilateralis*	Sprouting leaf	Wet	Often
66	*Rhamnaceae*	*Ventilago*	Yi he guo	*Ventilago leiocarpa*	Sprouting leaf	Wet	Often
67	*Dioscoreaceae*	*Dioscorea*	He bao shuy u	*Dioscorea persimilis*	Leaf	Wet	Fair
68	*Myrtaceae*	*Eucalyptus*	Ning meng an	*Eucalyptus citriodora*	Bark	Wet	Rare
69	*Myrtaceae*	*Rhodomyrtus*	Tao jin niang	*Rhodomyrtus tomentosa*	Fruit	Wet	Few
70	*Olacaceae*	*Erythropalum*	Chi cang teng	*Erythropalum scandens*	Leaf, sprouting shoot	Wet	Often
71	*Sapindaceae*	*Dimocarpus*	Long an	*Dimocarpus longgana*	fruit	Wet	Fair
72	*Sapindaceae*	*Boniodendron*	Huang li mu	*Boniodendron minus*	Leaf	Wet	Fair
73	*Sterculiaceae*	*Sterculia*	Ping po	*Sterculia nobilis*	Leaf, flower	Wet, dry	Often
74	*Convolulaceae*	*Ipomoea*	Fan shu	*Ipomoea batatas*	Leaf	Wet, dry	Few
75	*Convolulaceae*	*Cuscuta*	Tu si zi	*Cuscuta chinensis*	Stem, flower	Wet, dry	Few
76	*Convolulaceae*	*Argyreia*	Tou hua yin bei teng	*Argyreia capitiformis*	Leaf, flower, fruit	Wet, dry	Often

(continued)

Table 7.3 (continued)

Sequence	Family	Genus	Species (common name in China)	Species (scientific name)	Parts of plant eaten	Consumed in seasons of:	Feeding instances
77	*Convolulaceae*	*Argyreia*	Dong jing yin bei teng	*Argyreia pierreana*	Leaf, flower	Wet, dry	Often
78	*Urticaceae*	*Boehmeria*	Qing ye zhu ma	*Boehmeria nive var. tenacissima*	Leaf	Wet	Few
79	*Ulmaceae*	*Celtis*	Zi dan shu	*Celtis biondii*	Leaf	Wet, dry	Often
80	*Ulmaceae*	*Celtis*	Jia yu gui	*Celtis timorensis*	Leaf	Wet, dry	Fair
81	*Ulmaceae*	*Celtis*	Shan wu pu	*Celtis julianea*	Leaf	Wet, dry	Often
82	*Ulmaceae*	*Pteroceltis*	Qing tan	*Pteroceltis tatarinowii*	Leaf, flower	Wet, dry	Often
83	*Lauraceae*	*Cinnamomum*	Qing tan	*Cinnamomum burmannii*	Fruit	Wet	Few
84	*Lauraceae*	*Litsea*	Yin xiang	*Litsea glutinosa*	Leaf	Wet, dry	Fair
85	*Boraginaceae*	*Ehretia*	Shang si hou ke shu	*Ehretia tsangii*	Leaf, fruit	Wet, dry	Often
86	*Boraginaceae*	*Radermachera*	Cai dou shu	*Radermachera sinica*	Leaf	Wet, dry	Often
87	*Boraginaceae*	*Stereospermum*	Yu ye qiu	*Stereospermum colais*	Leaf	Wet	Often
88	*Boraginaceae*	*Oroxylum*	Mu hu die	*Oroxylum indicum*	Leaf	Wet	Often

(a) (b) (c) (d)

Fig. 7.2 Some of the plant species that the white-headed langurs feed on regularly (Photo by Liang Zuhong). (**a**) *Ficus microcarpa*, (**b**) *Croton euryphyllus*, (**c**) *Rhus chinensis*, (**d**) *Pteroceltis tatarinowii*

Fig. 7.3 Some of the plant species that white-headed langur consumed in large quantities (Photo by Liang Zuhong). (**a**) *Gouania javanica*, (**b**) *Cayratia japonica*, (**c**) *Pristimera arborea*, (**d**) *Thunbergia grandiflora*

belonging to pan-tropic distributing and tropical Asia distributing genera, as each category takes up 24.4% of the total diet. The species of old-world tropic distributing genera and tropical Asia–tropical Oceania distributing genera took up 17.9% and 11.5% of the total, respectively. The proportion of aggregate tropical distributing genera plants was 88.5% which did not differ markedly from the 91.5% we registered from observations in the early period of flora restoration (2002–2006) but was markedly higher than the 82.4% registered from observations in the intense human disturbances period (1996–2001). It was also higher than the proportion of tropical genera plants in the flora composition of the Nongguan Mountains. The fact that they would prefer tropical genera plants in an environment with adequate diet varieties and sufficient supplies of food could be further evidence indicating the tropical origin of white-headed langurs.

7.2.5 Diet Volume

As white-headed langurs would keep moving along their forages, it was not easy to determine precisely the diet volume of white-headed langurs in field observations, and it was rare that they fed on only one single species.

The cases cited in the following sections were among the few registers that had fully recorded the entire forage course of one species of plant that the white-headed langurs ate. We analyzed the diet volume of white-headed langurs from those field observation registers.

7.2.5.1 Accounts of White-Headed Langur Feeding on Yinhehuan (*Leucaena leucocephala*) Seeds and the Amount It Consumed

Yinhehuan (*Leucaena leucocephala*) of the *Mimosaceae* family can be a shrub or a small tree, 2–6 m in height, native to tropical America regions including Mexico. It was introduced initially into the Philippines before 1600 AD and later on introduced to other tropical regions like Indonesia, Hawaii, Mauritius, Northern Australia, etc. It was planted in the eighteenth century in Indonesia and Africa for nitrogen yielding and provided shade for economic crops like coffee, cocoa, cinchona, pepper, etc. In the 1960s, the species was introduced to China from Central America, and we cultivated it for its relatively high crude protein content and high yield rate. Presently it is being planted in relatively

Fig. 7.4 Some of the plant species that white-headed langurs consumed all year round (Photo by Liang Zuhong). (**a**) *Cansjera rheedii*, (**b**) *Ficus gibbosa*, (**c**) *Pueraria montana*, (**d**) *Sterculia nobilis*

large scales in Guangxi, Guangdong, Fujian, Hainan, Taiwan, Yunnan, Zhejiang, and Hubei in China.

We witnessed in the Nongguan Mountains the wide spreading of yinhehuan from a few trees and the rapid speed of their growth. In 1996, during the early years of our research, we found only one small yinhehuan growing by the pond in the Research Base. In 20 years, yinhehuan saplings are now everywhere in the Research Base except on the karst hills, and the tallest of them reach 12 m. Mature yinhehuan (Fig. 7.7) blossoms from March to April and August to September every year. Correspondingly, it has two seed ripening seasons from June to August and November to January in the following year. When the seeds ripen, the pods will split by itself, and the seeds will fall on the ground, germinating many saplings.

As the yinhehuan trees grew bigger and propagated, the white-headed langurs foraged on them more frequently. When November arrived and the yinhehuan beans turned gradually from green to reddish-brown, strips of them hung noticeably from the branches. It was a sign that the seeds inside were almost ripe, and they were going to be delicious food for the white-headed langurs. In the three following months, the transitory troop led by Yintangxiaotu and the "Wuque brothers troop" often came to forage for beans of the yinhehuans growing around the Research Base. It gave us opportunities to observe their forages and register how much they ate at a close distance.*Forage Register 1*

Date: December 23, 2014.

Hour: 16:10

The eldest of the brothers, 7-year-old Wuque was foraging on a yinhehuan. Swiftly it plucked a strip of beans and held it on its right hand, With its left hand holding the right arm, it ripped open a crack about 3–4 seeds in length along a side of the pod with its teeth and picked the exposed beans to eat. When finished, it nibbled open a similar length further down the side of the pod and ate the exposed beans, repeating until it finished the whole pod of beans. At 17:45, Wuque stopped feeding and left. It had consumed 170 pods of beans in 74 min (times spent on drinking, urinating, alerting, and other activities had been excluded). After Wuque had left, we collected 170 intact bean pods from the same tree it foraged, the empty pods, and the dropped beans it left behind and weighed them respectively. From them, we reckoned that Wuque had approximately consumed 421 g of yinhehuan beans in that feeding.

Forage Register 2

Date: December 25, 2014

Hour: 7:30

Fig. 7.5 Some of the plant species in which white-headed langur consumed flowers (Photo by Liang Zuhong). (a) *Delonix regia*, (b) *Pistacia weinmannifolia*, (c) *Argyreia capitiformis*, (d) *Gossampinus malabarica*

The 'Wuque brothers troop,' which consisted of 11 males and 1 immature female (the outlier female we named Wailaimei joined the troop lately), came to forage for yinhehuan beans by the pond. Amongst them, Wuque consumed a total of 216 pods of beans in 72 min. After they had gone, we weighed 216 intact pods of beans from the same tree and found it had 1253 g, and weighed the empty pods and dropped beans left behind by Wuque, and found it weighed total 850 g. From those figures, we reckoned Wuque consumed 403 g in this meal.

Forage Register 3

Date: December 28, 2014

Hour: 10:00

The 'Wuque brothers troop' again came to feed on the yinhehuan. The 5-year-old male Mengli had a feeding habit unlike that of its brother Wuque. It ripped open the entire pod and consumed only about half of the beans, dropping the other half all over the place. It had spent 55′40″ and foraged 235 pods of beans. The weight of the corresponding 235 pods of bean was 1458 g, the empty pods and dropped beans left behind by Mengli was 1050 g. Accordingly, we reckoned Mengli consumed 408 g.

From the above three registers, we worked out that the consumption speed of yinhehuan seeds by Wuque and Mengli were at 5.7 and 7.3 g/min, respectively. The difference probably is a result of the different ways of handling the pod of beans. Nonetheless, we noticed that regardless of their consuming speed, the total weight of yinhehuan seeds they consumed each time, 421, 403, and 408 g, were more or less the same.

Interestingly, we found that while the yinhehuan has seeds ripening twice in a year, white-headed langurs consume them in large quantities only during the dry months from November to January in the following year. There were even clashes between troops over the yinhehuan beans in that period. Nevertheless, in the rainy season from June to August, they do not consume yinhehuan beans. We suppose that the rapidly sprouting fresh and soft leaves of various other plant species in the rainy season had provided sufficient sustenance to the white-headed langurs. Hence, they left yinhehuan beans out of their chief diets. Whereas in the dry months when food was relatively deficient, there could hardly be any sprouting leaves, and hence yinhehuan beans which would be plentifully available at that time came to make up a significant portion in the white-headed langurs' chief diets. It was crucially important in the propagation of yinhehuan that white-headed langurs would not consume their beans in the rainy season. For the free fallen seeds nourished by rainwater would in just a few days germinate, sprout, and rapidly spread, becoming a small forest.

Fig. 7.6 Some of the plant species which white-headed langur consumed fruits of (Photo by Liang Zuhong). (**a**) *Dimocarpus longgana*, (**b**) *Cinnamomum burmannii*, (**c**) *Ehretia tsangii*, (**d**) *Ficus virens*

7.2.5.2 Accounts of White-Headed Langurs Feeding on Gou Trees (*Broussonetia papyrifera*) and the Amount Consumed

The gou tree (*Broussonetia papyrifera*) (http://frps.eflora.cn/frps/Broussonetiapapyrifera) is a deciduous tree in the *Moraceae* family, growing to a height of 10–20 m. It is extensively distributed over northern and southern China as well as in other tropical and temperate countries like Sikkim, Burma, Thailand, Vietnam, Malaysia, Japan, Korea, etc. It is also a widely distributed species in the Nongguan Mountains and a vital food resource for white-headed langurs. White-headed langurs feed on its leaves, flowers, and fruits all year round and even the old-grown leaves before they shed. In February and March, the budding blossoms and budding leaves of gou trees are delicacies for the white-headed langur (Fig. 7.8).

In the spring of 2015, we marked out seven gou trees (two males and five females) to observe the growth rate of leaves and flowers. We routinely picked ten random samples from each tree to measure their lengths and found that the growth rates of leaves sprouting were different between males and females. The growth of leaf buds on female gou trees were faster than those on males. Male inflorescences sprouted at the same time as its leaf buds but developed faster than male leaves, whereas the growth of inflorescences on female trees was evidently behind that of its leaves. We found in our field observations that in February and March, when gou trees just began to sprout, the white-headed langurs preferred to forage on male trees. They would not forage on female trees until the female inflorescences matured and turned red.

The two forage registers below may help us to understand the white-headed langurs' selections when feeding on the budding leaves and inflorescences of gou trees. The two registers were recorded in February and March 2015, the time when the "Yintangxiaotu transitory troop" was still roaming in Feijichang despite repulsions by the "Yuweng family troop." The individuals from the "Yintangxiaotu transitory troop" often encountered interferences from the Yuweng troop.*Forage Register 1*

Date: February 28, 2015

Hour: 17:40–18:20

(a) (b)

Fig. 7.7 (**a**) Yinhehuan in flowering season; (**b**) White-headed langurs foraging on ripen yinhehuan seeds (Photo by Liang Zuhong)

Yintangxiaotu was foraging on gou tree no. 5 at Feijichang. It was a male tree and had just begun to bud, and the leaf buds and the male inflorescence buds were at similar lengths. Time and time again Yintangxiaotu pulled the twigs to itself to nibble the leaf buds and inflorescence buds on it. After it had gathered a certain quantity in its mouth, it started chewing them altogether. We designate the process from pulling the stem to itself until complete chewing as a 'round.' We counted 82 rounds by Yintangxiaotu, consuming a total of 409 leaf buds and inflorescence buds in 40 min before it stopped and moved away upon hearing male langur shouts in the direction of Guoyuan Pass. Then we collected ten leaf buds and inflorescence buds from gou tree no. 5 to measure and weigh. The average length of each bud was 4.2 cm and weighed an average of 0.6 g. Accordingly, we reckoned Yintangxiaotu in this forage consumed a total of 245 g of leaf buds and inflorescence buds of gou trees, with an average rate of 6.1 g/min.

Forage Register 2

Date: March 10, 2015

Hour: 17:57–18:10

Four year old male Yiyi of the 'Yintangxiaotu transitory troop' was feeding on gou tree no. 5. The leaf sprouts were longer than they were 10 days ago, and the male inflorescences had swollen 2–3 times their former size. The manner and speed in which Yiyi consumed its food were different from its father Yintangxiaotu; it selectively ate only swollen male inflorescences and did not eat any sprouting leaves. In 10′50″, it consumed 79 male inflorescences, each having an average mass of 3.3 g at an average speed of 24.1 g/min, and the total mass consumed was 261 g. The others feeding on the nearby gou trees also consumed only male inflorescences, ignoring or even deliberately avoiding the sprouting leaves. The feeding was stopped by interference from the 'Yuweng family troop', which was on their way returning to the night shelter.

We can see from the above two registers that the feeding speed on the same diet varies significantly among individuals of different ages. From our observations at a close distance, we found out that adult male Yintangxiaotu, who came into the family in 2006 was now aging. Wounds from fighting over the years started to deform its forelimbs, and its teeth had become shaky, which could be dislodged at any time. Those factors could have affected its feeding speed. On the other hand, the youthful Yiyi was like a boy at his teens. It ate and swallowed ravenously. As mentioned earlier, the "Yintangxiaotu transitory troop" and "Yuweng family troop" were living on the same mountain tract, and they repelled each other. Feedings in both the registers were interrupted; hence, the diet volume of Yintangxiaotu and Yiyi might has been underestimated.

The meals of the white-headed langur did not harm the growth of the gou trees. Since, first, gou trees that grow at different environmental locations do not produce bud and sprout at the same time, so the langurs pursuing fresh leaves and budding inflorescences would roam all over the mountain tract, rather than feed fixedly on the same gou trees. Second, when the leaf buds and inflorescence buds on a gou tree could be identified to be male, the white-headed langurs would have a preference for feeding on the male inflorescences, and they ate almost no sprouting leaves. The sprouting leaves would proliferate, and in less than half a month, would have grown to their full sizes, beginning to gather cellulose and minerals and process photosynthesis. The white-headed langurs' preference for male inflorescences allowed the development of female inflorescences. Hence, the propagation of gou trees is not jeopardized to great extent.

From the registers above, we can see that regardless of the diet species, the consumption rate among individuals would be different, whereas the diet volume in each feeding was roughly the same.

In the register below, we can see even more clearly the diet volume of white-headed langurs.

Date: February 18, 2015

Hour: 10:30

We found a white-headed langur dead from an electric shock on a 10-m tall high voltage transformer pylon by the entrance of Eco-Park. From its body shape and hair color, it is estimated to be between 4 and 5 years of age and weighed 5920 g. The day was relatively hot, and the abdomen of the carcass was bloating, making cryopreservation undesirable. So laparotomy was processed without delay. Its stomach was filled with fine-textured

Fig. 7.8 White-headed langurs feeding on gou trees at various times (Photo by Liang Zuhong). (**a**) A white-headed langur consuming swollen male inflorescence. (**b**) A white-headed langur consuming ripened female inflorescence. (**c**) A white-headed langur consuming leaf buds between winter and spring. (**d**) A white-headed langur consuming leaves

> green chyme. It could be inferred that the accident happened 2 h after its morning meal. Contents in the stomach weighed 490 g.

From 421, 403, and 408 g of consumed yinhehuan, 245 and 261 g of consumed gou tree leaf buds and male inflorescences, and 490 g contents found in the stomach of the accidentally killed langur, we can get the white-headed langur's diet volume in an approximate range, despite that the data is different between individuals. We recollected the registers and statistical data obtained in our field observations over the years, and quantitatively analyzed the videos recorded. Then we came to know that the diet volume of a white-headed langur in Nongguan Mountains in each feeding was roughly between 300 and 500 g, with variations among different individuals, different ages, different genders, and in different seasons, as well as with different diet species.

Appendix: Nongguan Mountains Plant Directory

Family	Genus	Species (name in China)	Species (scientific name)	Remark[a] (previous name)
Euphorbiaceae	*Alchornea*	Shan ma gan	*Alchornea rugosa* (Lour.) M.A.	
		Hong bei shan ma gan	*Alchornea trewioides* (Benth.) Muell. Arg.	
	Antidesma	Wu yu cha	*Antidesma bunius* (L.) Spreng	Guang ye wu hua cha
		Bao Ye wu yu cha	*Antigesma* sp.	
	Breynia	Hei mian shen ○1	*Breynia fruticosa* (L.) HK.f.	
	Bridelia	Tu mi shu	*Bridelia monoica* (Lour.) Merr.	
	Cleidion	Bang bing hua	*Cleidion brevipetiolatum* Pax et Hoffm	
	Claoxylon	Bai tong shu ○2	*Claoxylon polot* (Burm.f.) Merr.	
	Cleidiocarpon	Hu die guo	*Cleidiocarpon cavaleriei* (Levl.) Airy Shaw	
	Croton	Shi shan ba dou○1○2	*Croton euryphyllus* W.W.Sm.	
	Drypetes	Wang mai he fuo mu○2	*Drypetes perreticulata* Gagn.	
	Excoecaria	Hong bei gui	*Excoecaria cochinchinensis* Lour	
	Flueggea	Bai fan shu ○1○2	*Flueggea virosa* (Willd.) Baill	
	Leptopus	Bao ye que she shu ○1	*Leptopus australis* (Zoll.et Merr.) Poiark	
	Bischofia	Wiu feng	*Bischofia javanica* Bl.	
	Mallotus	Cu kang chai ○1○2	*Mallotus philippinensis* (Lour.) H.-M.	
		Shi yan feng ○2	*Mallotus repandus* (Will.) M.-A.	
		Baii bei ye ○1	*Mallotus apelta* (Lour.) M.-A	
	Phyllanthus	Yu gan zi ○1	*Phyllanthus emblica* L.	
		Ye xia zhu	*P. uranaria* L.	
		Xiao guo ye xia zhu	*Phyllanthus reticulatus* Poir.	Guang ye long yan jing
	Sapium	Huilin wu jiu ○1	*Sapium sebiferum* (L.) Roxb.	
	Trigonostemon	Si geng san bao mu ○1	*Trigonostemon filipes* Y. T. Chang et X. L. Mo	
	Glochidion	Yuan guo suan pan zi	*Glochidion sphaerogynum* (Muell. Arg.) Kurz.	Li ye suan pan zi
Moraceae	*Artocarpus*	Mu bo luo	*Artocarpus heterophyllus* Lam	
	Broussonetia	Gou shu ○1○2	*Broussonetia papyrifera* (L.) Vent.	
		Chu ○1	*Broussonetia kazinoki* Sieb.	Teng gou
	Cudrania	Zhe shu ○2	*Cudrania cochinchinensis* (Lour.) Kudo et Masam.	
	Ficus	Xie ye rong ○1○2	*Ficus gibbosa* L.	
		Yin ye rong	*Ficus callosa* Willd.	
		Rong shu ○1○2	*Ficus microcarpa* L.	Xi ye rong
		Jiu ding rong ○1	*Ficus nervosa* Heyne	
		Shi shan rong ○1	*Ficus orthoneura* Levl. et Van.	
		Lu huang ge rong ○1	*Ficus virens* Aiton	
		Jian ye rong ○1	*Ficus henryi* Warb.	
		Gou shan rong	*Ficus altissima* Bl.	
		Chui ye rong	*Ficus benjamina* Linn.	
		Huang jin rong	*Ficus microcarpa* 'Aurea'	
		Zhi mai rong	*Ficus orthoneura* Lévl. et Vant.	
		Dou gou rong	*Ficus pisocarpa* Bl.	
		Yin du xiang pi rong	*Ficus elastica* Roxb. ex Hornem.	

(continued)

Family	Genus	Species (name in China)	Species (scientific name)	Remark[a] (previous name)
		Bai bei pa teng rong	*Ficus sarmentosa* Buch.-Ham. ex J. E. Sm. var. *nipponica* (Fr. et Sav.) Corner	
		Long zhou roung	*Ficus cardiophylla* Merr.	
	Malaisia	Niu gan teng ○1○2	*Malaisia scandens* (Lour.)	
	Morus	Sang ○1	*Morus alba* L.	
	Streblus	Mi yang ye	*Streblus tonkinensis* (Dub. et Eberh.) Corner	Mi nong ye
Rhamnaceae	*Berchemia*	Duo ye gou er cha	*Berchemia polyphylla* Wall.	
		Lao shu er ○2	*Berchemia racemosa* S. & Z.	
	Chaydaia	Bao ye mu ○2	*Chaydaia crenulata* H.-M.	
	Gouania	Ju qian ○1	*Gouania leptostachya* DC.	
		Mao ju qian ○2	*Gouania javanica* Miq.	Mao ye ju qian
	Paliurus	Tong qian shu	*Paliurus hemsleyanus* Rehd.	
	Rhamnus	Shu li	*Rhamnus davurica* Pall.	
		Xian xu shu li ○1	*Rhamnus nipalensis* Laws	
	Sageretia	Que mei teng ○1	*Sageretia theezans* (L.) Brongn.	
	Ventilago	Fu e yi he mu ○2	*Ventilago calyculata* Tul.	
		Hainan yi he guo ○1	*Ventilago inaequilateralis* Merr. et Chun	
	Ziziphus	Yin du zao ○2	*Ziziphus incurve* Roxb.	
Papilionaceae	*Christia*	Bian fu cao	*Christia vespertilionis* (L. f.) Bahn. F	
	Dalbergia	Teng huang tan ○1○2	*Dalbergia hancei* Benth.	
		Guangdong huang tan ○1	*Dalbergia benthamii* Prain	
		Qian huang tan ○2	*Dalbergia cavaleriei* Levl.	
		Huang tan ○1	*Dalbergia hupeana* Hance	
		Jiang xiang huang tan	*Dalbergia odorifera* T. Chen	
		Xie ye huang tan ○1	*Dalbergia pinnata* (Wall.) Benth.	
	Erythrina	Long ya hua	*Erythrina corallodendron* L.	Xiang ya hong
		Ji guan ci tong	*Erythrina crista-galli* L.	
	Canavalia	Hai dao dou	*Canavalia maritima* (Aubl.) Thou.	
	Derris	Mao yu teng ○2	*Derris elliptica* (Wall.) Benth.	
		Zhong nan yu teng	*Derris fordii* Oliv. var. *lucida* How	
		Yunnan yu teng	*Derris yunnanensis* Chun et How	
	Rhynchosia	Luhuo	*Rhynchosia volubilis* Lour.	
	Desmodium	E ma huang	*Desmodium multiflora* DC.	
		Dan ye ma huang	*D. zonatum* Miq	
	Dendrolobium	Jia mu dou	*Dendrolobium triangulare* (Retz.) Merr.	
	Indigofera	Tuo yuan ye mu lan	*Indigofera cassoides* Rottl. ex DC.	
	Lespedeza	Jie lie tie sao zhou	*Lespedeza cuneata* (Dum.-Cours.) G. Don	
	Zenia	Ren dou	*Zenia insignis* Chun	
	Millettia	Rong mao ji xue teng	*Millettia verlutina* Dunn.	
		Wang luo ji xue teng	*Millettia reticulata* Benth.Benth.	
		Hou guo ya dou teng ○2	*Millettia pachycarpa* Benth.Chun et T.Chen	
		Ya dou teng ○1	*Millettia* sp.	
	Pueraria	Ge teng ○1○2	*Pueraria lobata* (Willd.) Roem.	
	Phyllodium	Pa qian shu	*Phyllodium pulchellum* (Linn.) Desr.	
	Crotalaria	Zhu shi dou	*Crotalaria pallida* Ait.	
	Whitfordiodendron	Zhu yao dou	*Whitfordiodendron filipes* (Dunn) Dunn	

(continued)

Family	Genus	Species (name in China)	Species (scientific name)	Remark[a] (previous name)
Vitaceae	*Ampelopsis*	Guangdong she pu tao	*Ampelopsis cantoniensis* (Hook.etArn.) Pl.	
		She pu tao ○1	*Ampelopsis* sp.	
	Cayratia	Wu lian mei	*Cayratia japonica* (Thunb.) Gagnep.	
		Che suo teng ○1	*Cayratia* sp.	
	Tetrastigma	Guang ye ya pa teng ○1	*Tetrastigma nitida*	
		Ya pa teng	*Tetrastigma obtectum* (Wall.) Pl.	
		Bian dan teng ○1	*Tetrastigma planicaule*(HK.)Gagn.	
	Vitis	Ge lei pu tao ○1	*Vitis flexuosa* Thb	
		Ye pu tao ○1	*Vitis pentagona* Diesl.et Gilg.	
Caesalpiniaceae	*Bauhinia*	Xiao hua yang ti jia	*Bauhinia venustula* O.Htze.	
		Kuo li yang ti jia ○1	*Bauhinia corymbosa* Roxb.	
		Long xu teng ○1○2	*Bauhinia championi* Benth.	
	Caesalpinia	Ji zui lei ○1	*Caesalpinia tsoongii* Merr.	
		Su mu ○1	*Caesalpinia sappan* L.	
		Xiao ye yun shi ○1	*Caesalpinia milletii* HKet. Arn.	
	Gleditsia	Zao jia ○1	*Gleditsia sinensis* Lam.	
	Pterolobium	Lao hu ci	*Pterolobium punctatum* Hemsl.	
	Saraca	Zhong guo Wu you hua ○2	*Saraca chinensis Merr.*et Chun	
Asclepiadaceae	*Cryptolepis*	Gu gou teng ○1	*Cryptolepis buchananii* Roem.et Schult.	
	Cynanchum	E rong teng ○1	*Cynanchum* sp.	
	Gymnema	Shi geng teng ○1	*Gymnema alterniflorum* (Lour.) Merr.	
	Hoya	Qiu lan ○1	*Hoya carnosa* (L. f.) R. Br.	
	Tylophora	Dou hua wa er teng	*Tylophora floribunda* Miquel	Wa er teng
	Toxocarpus	Gong guo teng ○2	*Toxocarpus wightianus* HK.et Arn.	
	Secamone	Diao shan tao ○1	*Secamone sinica* Hand.-Mazz.	
Oleaceae	*Ligustrum*	Zi yao nu zhen ○1	*Ligustrum delavayanum* Hariot	
		Xiao ge	*Ligustrum sinense* Lour.	
	Jasminum	Qing teng zi ○1	*Jasminum nervosum* Lour.	Qing teng zhi
		Mo li hua	*Jasminum sambac* (L.) Aiton	Mao mo li
		Niu du teng ○1	*Jasminum elongatum* (Bergius) Willdenow	
		Ka fei su xin ○1	*Jasminum coffeinum* Buch.-Ham	
		Zhi hua su xin ○2	*Jasminum lang* Gagnepain	
		Bai e su xin	*Jasminum albicalyx* Kobuski	
	Osmanthus	Mu xi	*Osmanthus fragrans* (Thunb.) Loureiro	Gui hua
Ulmaceae	*Celtis*	Huanan pu ○1○2	*Celtis austro-sinensis* Chun	
		Zi dan shu ○1○2	*Celtis biondii* Pamp.	
		Pu shu ○1	*Celtis sinensis* Pers.	
		Shan hu pu ○2	*Celtis julianae* Schneid.	
		Jia yu gui	*Celtis timorensis* Span.	
	Pteroceltis	Qing tan ○1○2	*Pteroceltis tatarinowii* Maxim.	
	Trema	Shan huang ma	*Trema tomentosa* (Roxb.) Hara	
Anacardiaceae	*Mangifera*	Mango	*Mangifera indica* Linn	
		Tian tao mu	*Mangifera persiciforma* C.Y Wu et T.L.Ming	
	Pistacia	Huang lian mu ○1	*Pistacia chinensis* Bunge	
		Qing xiang mu ○1○2	*Pistacia weinmannifolia* J.Poiss	Xi ye jie mu
	Rhus	Yan fu mu ○1○2	*Rhus chinensis* Mill.	
Myrtaceae	*Eucalyptus*	Long yuan an	*Eucalyptus exserta* F. V. Mull.	
		Ning meng an	*Eucalyptus citriodora* Hook. f.	
		Che an ○1	*Eucalyptus camaldulensis* Dehnh.	
	Psidium	Guava ○1○2	*Psidium guajava* L.	

(continued)

Family	Genus	Species (name in China)	Species (scientific name)	Remark[a] (previous name)
	Syzygium	Mi mai pu tao	*Syzygium chunnianum* Merr.et Perry	
		Wu mo	*Syzygium cumini* (L.) Skeels	
Sterculiaceae	*Pterospermum*	Jie li chi zi shu	*Pterospermum truncatolobatum* Gagnep	Jie li ye fan bai ye shu
	Sterculia	Fen ping po	*Sterculia euosma* W. W. Smith	
		Ping po ○2	*Sterculia nobilis* Smith	
		Jia ping po ○2	*Sterculia lanceolata* Cav	
Bignoniaceae	*Oroxylum*	Mu Hu die ○1○2	*Oroxylum indicum* (L.) Vent.	
	Radermachera	Cai dou shu ○1○2	*Radermachera sinica* (Hance) Hemsl	
	Stereospermum	Yu ye qiu ○2	*Stereospermum chelonoides* (Oliv.) DC.	
Meliaceae	*Aglaia*	Mi zi lan	*Aglaia odorata* Lour.	
	Cipadessa	Hui mao jiang guo lian ○1	*Cipadessa cinerascens* (Pell.) H.-M.	
	Melia	Ku lian ○1	*Melia azedarach* L.	
	Walsura	Ge she shu ○1	*Walsura robusta* Roxb.	
Rutaceae	*Clausena*	Huang pi ○1	*Clausena lansium* (Lour.) Skeels	
	Evodia	San ya ku ○1	*Evodia lepta* (Spreng) Merr.	
	Micromelum	Xiao yun mu	*Micromelum integerrima* (B.-H.) Reem.	
	Murraya	Qian li xiang ○1	*Murraya paniculata* (L.) jack	Jiu li xiang
		Quangxi jiu li xiang ○2	*Murraya kwangsiensis* (Huang) Huang	
	Zanthoxylum	Le dang hua jiao	*Zanthoxylum avicennae* (Lam.) DC.	Hua mei tiao
		Dan mian zhen	*Zanthoxylum dissitum* Hemsl.	
		Ye hua jiao ○1	*Zanthoxylum simulans* Hance	
		Ya jiao	*Zanthoxylum schinifolium Sieb.* et Zacc.	
Mysinaceae	*Ardisia*	Lou san shu	*Ardisia quinquegona* Blume	
		Zi jin niu ○1	*Ardisia japonica* (Thunberg) Blume	
	Embelia	Wang mai suan teng zi ○1	*Embelia rudia* H.-M.	
		Lin pi kong suan teng zi	*Embelia scandens* (Lour.) Mez	
		Suan teng zi	*Embelia laeta* (L.) Mez	
	Maesa	Ding hua du jing shan	*Maesa balansae* Mez	Guang ye du jing shan
Cucurbitaceae	*Trichosanthes*	Jie ye gua lou	*Trichosanthes truncata* C. B. Clarke	
		Cao dian gua lou ○1	*Trichosanthes dunniana* Lévl.	
	Mukia	Mao er gua	*Mukia maderaspatana* (L.) M. J. Roem.	
	Thladiantha	Zhang ye che bo	*Thladiantha dubia* Bge	
	Solena	Mao gua	*Solena amplexicaulis* (Lam.) Gandhi	
	Gynostemma	Jiao gu lan	*Gynostemma pentaphyllum* (Thunb.) Makino	
Vernenaceae	*Clerodendrum*	Zheng tong ○2	*Clerodendrum japonicum* (Thunb.) Sweet	Zhuang yuan hong
	Duranta	Jia lian qiao	*Duranta repens* Linn.	
	Lantana	Ma ying dan ○1○2	*Lantana camara* L.	Wu se mei
	Premna	Huang mao dou fu chai ○2	*Premna fulva* Craib.	
		Shi shan dou fu chai	*Premna crassa* Hand.-Mazz.	
	Vitex	Guangxi mou jing ○1	*Vitex kwangsiensis* Pei	
		Mou jing ○1	*Vitex negundo* L.	
		Huang jing ○2	*Vitex negundo* L.var. *cannabifolia*	
Rubiaceae	*Canthium*	Tie shi mi ○2	*Canthium dicocum* (Gaertn.) Merr.	
	Mussaenda	Hainan yu ye jin hua ○1	*Mussaenda hainanensis* Merr.	
	Paederia	Ji xhi teng ○1	*Paederia scandens*(Lour.)Merr.	

(continued)

Family	Genus	Species (name in China)	Species (scientific name)	Remark[a] (previous name)
	Oxyceros	Ji jiao le	*Oxyceros sinensis* Lour.	
	Psychotria	Man jiu jie	*Psychotria serpens* L.	Chuan gen teng
	Catunaregam	Shan shi liu ○1	*Catunaregam spinosa* (Thunb.) Tirveng.	Zhu du le
	Tarenna	Wu kou shu	*Tarenna* sp.	
	Coptosapelta	Liu su zi ○2	*Coptosapelta diffusa* (Champ. ex Benth.) Van Steenis	Niu lao yao teng
	Uncaria	Xiao ye gou teng ○2	*Uncaria rhynchophylla* (Miq.) Jack	
Convolulaceae	*Argyreia*	Bai he teng ○1	*Argyreia acuta* Lour.	Yi pi chou
		Tou hua yin bei teng	*Argyreia capitiformis* (Poiret) van Ooststroom	
		Dong jing yin bei teng	*Argyreia pierreana* Bois	
	Cuscuta	Ri ben tu si zi ○1○2	*Cuscuta japonica* Choisy.	
	Ipomoea	Fan shu	*Ipomoea batatas.* (L.) Lamarck	
	Jacquemontia	Jia qian niu	*Jacquemontia* sp.	
Araliaceae	*Euaraliopsis*	Jia tong cao ○2	*Euaraliopsis ciliata* (Dunn) Hutch.	Jie mao zhang ye shu
	Schefflera	Qi ye lian ○1	*Schefflera arbo ricola* Hayata	
		Mi mai e zhang chai ○1	*Schefflera venulosa* (W.et A.) Harms.	
		e zhang cha	*Schefflera octophylla* (Lour.) Harms	
	Aralia	Huang mao cong mu	*Aralia decaisneana* Hance	
Lauraceae	*Cinnamomum*	Zhang shu	*Cinnamomum camphora* (L.) Presl.	
		Yin xiang	*Cinnamomum burmannii* (Nees) Bl.	
		Huang zhang ○1	*Cinnamomum parthenoxylon* (Jack) Meisner	Xiang zhang
	Litsea	Chan gao shu ○1○2	*Litsea glutinosa* (Lour) C.B.Rob.	
		Jia shi mu jiang zi	*Litsea monopetala* (Roxb.) Pers.	
	Persea	E li	*Persea americana* Mill	You li
	Actinodaphne	Mao huang rou nan	*Actinodaphne pilosa* (Lour.) Merr.	
Urticaceae	*Boehmeria*	Qing ye zhu ma	*Boehmeria nivea* (L.) Gaud.var. *tenacissima* Miq.	
		Zhu ma ○1	*Boehmeria nivea*(L.)Gaud.	
		Wei lu zhu ma ○2	*Boehmeria nivea*(L.)Gaud.var. *viridula* Yamamoto	
	Debregeasia	Shui ma	*Debregeasia orientalis* C. J. Chen	
	Oreocnide	Guangxi zi ma	*Oreocnide kwangsiensis* H.-M.	
		Zi ma	Oreocnide frutescens (Thunb.) Miq.	
	Pilea	Dian ru leng shui hua ○1	*Pilea glaberrima* (Bl.) Bl.	
		Hua ye leng shui hua	*Pilea cadierei* Gagnep. et Guill	Leng shui hua
Mimosaceae	*Adenanthera*	Hai hong dou	*Adenanthera pavonina* var. *microsperma* (Teijsm.et Binnend.) Nielsen	
	Albizia	Shan he huan ○1	*Albizia kalkora* (Roxb.) Benth.	
		Ying shu	*Albizia chinensis* (Osbeck) Merr.	
	Acacia	Yu ye jin he huan	*Acacia pennata* (L.) Willd.	She teng
		Taiwan xiang si	*Acacia confusa* Merr.	
		Teng jin he huan ○1	*Acacia sinuata* (Lour.) Merr.	
	Delonix	Feng huang mu	*Delonix regia* (Boj.) Raf.	
	Leucaena	Yin he huan	*Leucaena leucocephala* (Lam.) de Wit.	
	Mimosa	Han xiu cao	*Mimosa pudica* L.	
		Le zi shu	*Mimosa sepiaria* Benth.	
Sapindaceae	*Amesiodendron*	Long zhou xi zi long ○1	*Amesiodendron integrifoliolatum* H. S. Lo	
	Boniodendron	Huang li mu ○1	*Boniodendron minus* (Hemsl) T.Chen	
	Dimocarpus	Long an	*Dimocarpus longgana* Lour.	

(continued)

Family	Genus	Species (name in China)	Species (scientific name)	Remark[a] (previous name)
	Koelreuteria	Xiao ye luan	*Koelreuteria paniculata* Laxm	
	Litchi	Li zhi	*Litchi chinensis*	
Dioscoreaceae	*Dioscorea*	Ling shu yu	*Dioscorea bulbifera* L.	
		He bao shu yu ○1	*Dioscorea persimilis* Prain et Burk.	
Capparaceae	*Capparis*	Xiao lu ci	*Capparis urophylla* F.Chen	Wei ye chui guo teng
		Ma bin lang	*Capparis masaikai* Lev1	Shui bin lang
		Que tou ji	*Capparis versicolor* Griff	
		Shan gan teng ○1	*Cansjera rheedii* J. F. Gmel.	
Symplocaceae	*Symplocos*	Hua shan fan ○1	*Symplocos chinensis* (Lour.) Druce	
		Leng zhi shan fan ○1	*Symplocos howii* Merr.Et Chun ex H.L.LI	
		Bai teng	*Symplocos paniculata* (Thunb.) Miq.	
Magnoliaaceae	*Magnolia*	Zhang ye mu lan ○1	*Magnolia paenetalauma* Dandy	
	Michelia	Bai lan	*Michelia alba* Candolle	
		Han xiao hua	*Michelia figo* (Lour.) Spreng.	
		Huo li nan	*Michelia macclurei* Dandy	
Olacaceae	*Erythropalum*	Che cang teng ○1	*Erythropalum scandens* Bl.	
Ranunculaceae	*Clematis*	Tie xian lian	*Clematis florida* Thunb	
		Xiao mu tong ○1	*Clematis armandii* Franch.	
		Wei ling xian	*Clematis chinensis* Osbeck	
		Shan mu tong	*Clematis finetiana* Lévl. et Vant.	
		Mao zhu tian xian lian	*Clematis meyeniana* Walp.	
		Zhu guo tian xian lian	*Clematis uncinata* Champ.	
	Ranunculus	Mao lang	*Ranunculus japonicus* Thunb.	
Menispermaceae	*Diploclisia*	Cang bai cheng gou feng ○1	*Diploclisia glaucensens* (BL) Diels.	
		Cheng gou feng	*Diploclisia affinis* (Oliv.) Diels	
	Stephania	Fen ji du	*Stephania longa* Lour.	
	Tinospora	Qing niu dan	*Tinospora sagittata* (Oliv.) Gagnep.	
Theaceae	*Camellia*	Fusui jin hua cha ○2	*Camellia fusuiensis* S.Y.Liang etX.J.Dong	
		Cha	*Camellia sinensis* (L.) O. Ktze.	
Annonaceae	*Desmos*	Jia ying zhao ○1○2	*Desmos chinensis* Lour.	Ju bing ye
	Fissistigma	Jing hua gua fu mu	*Fissistigma caudiflorum* Hance	
		Guizhou gua fu mu	*Fissistigma wallichii* Merr.	
Bombacaceae	*Gossampinus*	Mu mian ○1	*Gossampinus malabarica* (DC.) Merr.	
Sapotaceae	*Sinosideroxylon*	Tie lan ○1○2	*Sinosideroxylon pedunculatum* (Hemsl.) H. Chuang	Mao ye tie lan
Illigeraceae	*Illigera*	Xiu mao qing teng ○1	*Illigera rhodantha* Hance var. *dunniana* (Lévl.) Kubitzki	Xiao ye qing teng, mao qing teng
Rosaceae	*Agrimonia*	Long ya cao	*Agrimonia viscidula* Bunge.	
	Eriobotrya	Pi pa	*Eriobotrya japonica* (Thunb.) Lindl.	
	Lauro-cerasus	Ci ye gui ying	*Lauro-cerasus spinulisa* (S et Z.) Schneid.	
	Pyrus	Tang li ○1	*Pyrus calleryana* Decne. var. *koehnei* (Schneid.) yii	
	Rubus	Mao mei	*Rubus parvifolius* L.	
		Yuen an xuan gou zi	*Rubus cochinchinensis* Tratt.	
	Potentilla	Chao tian wei ling cai	*Potentilla supina* L.	
Aquifoliaceae	*Ilex*	Dong qing ○2	*Ilex chinensis* Sims.	
		Ci dong qing ○1	*Ilex hylonoma* Hute Tang. var. *glabra* S.Y.Hu	
Celastraceae	*Euonymus*	Xi nan wei mao	*Euonymus hamiltonianus* Wall.	
		Bao ye wei mao ○1	*Euonymus* sp.	

(continued)

Family	Genus	Species (name in China)	Species (scientific name)	Remark[a] (previous name)
	Microtropes	Jia wei mao yi zhong ○2	*Microtropes* sp.	
Melastomataceae	*Memecylon*	Xi ye gu mu ○1	*Memecylon scutellatum* (Lour.) Naud.	
	Melastoma	Ye mu dan	*Melastoma candidum* D. Don	
Balsaminaceae	*Impatiens*	Feng xin hua ○1	*Impatiens balsamina*	
Araliaceae	*Alangium*	Ba jiao feng	*Alangium chinense* (Lour.) Harms	
Aristolochiaceae	*Aristolochia*	Quangxi ma dou ling ○1○2	*Aristolochia kwangsiensis* Chunet How.	
Santalaceae	*Osyris*	Shan zhen ○1	*Osyris wightiana* Wall.	
Burseraceae	*Garuga*	Yu ye bai tou shu ○1	*Garuga pinnata* Roxb.	
Compositae	*Artemisia*	Zhu mao hao	*Artemisia scoparia* Waldst.Et Kit.	
		Qing hao	*Artemisia caruifolia* Buchanan-Hamilton ex Roxburgh	
		Mou hao	*Artemisia japonica* Thunb.	
	Bidens	Gui zhen cao	*Bidens pilosa* Linn.	
	Eupatorium	Fei ji cao	*Eupatorium odoratum* L.	
	Senecio	Qian li guang ○1	*Senecio scandens* Buch.-Ham.	
	Wedelia	Pang qi ju	*Wedelia chinensis* (Osb.) Merr.	
	Vernonia	Ye xiang niu	*Vernonia cinerea* (L.) Less.	Jia xian xia cao
	Grangea	Tian je huang	*Grangea maderaspatana* (L.) Poir.	
	Crassocephalum	Ye tong hao	*Crassocephalum crepidioides* (Benth.) S. Moore	
	Tithonia	Zhong bing ju	*Tithonia diversifolia* A. Gray.	
Hipocrateaceae	*Pristimera*	Er zi dun ci guo	*Pristimera arborra* (Roxb.) A.C.Smith	
		Er zi bian shuo teng ○1	*Pristimera arborea* (Roxb.) A. C. Smith	
	Salacia	Leng zi teng	*Salacia sessiliflora* H.-M.	
		Wu ceng long ○2	*Salacia prinoides* (Willd.) DC.	
Icacinaceae	*Iodes*	Xiao guo wei hua teng ○1○2	*Iodes vitiginea* (hance) Gagn	
	Mappianthus	Tian guo teng ○2	*Mappianthus iodoides* H.-M.	
Solanaceae	*Solanum*	Ci tian jia ○1	*Solanum indicum* L.	
		Shui jia	*Solanum torvum* Swartz	
		Jia yan ye shu	*Solanum verbascifolium* L.	
Zingiberaceae	*Amomum*	Tu sha ren ○1	*Amomum* sp.	
Agavaceae	*Curculigo*	Da ye xin mao	*Curculigo capitulata* (Lour.) O. Kuntze.	
	Dracaena	Jian ye long xue shu ○1	*Dracaena cochinchinensis* Merr.	
		Hainan long xue shu	*Dracaena cambodiana* Pierre ex Gagnep.	Xiao long sue shu
Araceae	*Alocasia*	Hai yu ○1	*Alocasia odora* (Roxb.) C.Koch.	
	Pothos	Shi gan zi	*Pothos chinensis* (Raf.) Merr.	
Liliaceae	*Asparagus*	Tian mending	*Asparagus cochinchinensis* (Lour.) Merr.	
	Ophiopogon	Hou ye yan jie cao	*Ophiopogon corifolius* Wang et Dai	
		Kuo ye yan jie cao○1	*Ophiopogon platyphyllus* Merr.et Chun	
		Xia ye yan jie cao	*Ophiopogon stenophyllus* (Merr.) Rodrig.	
	Polygonatum	Da ye huang jing	*Polygonatum cyrtonema* Hua	
Loganiaceae	*Buddleja*	Zui yu cao	*Buddleja lindleyana* Lindley Summerlilic	
		Mi meng hua	*Buddleja officinalis* Max.	
Myristicaceae	*Horsfiedia*	Hainan Feng chui nan	*Horsfiedia hainanensis* Merr.	
Apocynaceae	*Alstonia*	Tang jiao shu	*Alstonia scholaris* (L.) R. Br.	
	Melodinus	Shan chen	*Melodinus suaveolens* (Hance) Champ. ex Benth.	
	Nerium	Jia zhu tao	*Nerium indicum* Mill	

(continued)

Family	Genus	Species (name in China)	Species (scientific name)	Remark[a] (previous name)
	Trachelospermum	Duan zhu luo shi	*Trachelospermum brevistilum* H.-M.	
		Luo shi	*Trachelospermum jasminoides* (Lindl.) Lem.	
	Urceola	Suan ye jiao teng	*Urceola rosea* (Hooker & Arnott) D. J. Middleton	
Ebenaceae	*Diospyros*	Xiao ye shan shi	*Diospyros dumetorum* W.W.Sm.	
		Shi	*Diospyros kaki* Thunb.	
		Luo fu ahi	Diospyros momrrisiana Hance	
Opiliaceae	*Cansjera*	Shan ju teng	*Cansjera rheedii* Grmel.	
	Lepionurus	Lin wei mu ○1○2	*Lepionurus latisquamus* Gagn.	
Tiliaceae	*Grewia*	Qing ma ye bian dan gan	*Grewia abutilifolia* Vent.Ex Tuss	
		Bian dan gan	*Grewia biloba* G. Don	
		Mao guo bian dan gan	*Grewia eriocarpa* Juss.	
		Xiao hua bian dan gan	*Grewia biloba* G. Don var. *parviflora* (Bunge) Hand. -Mazz.	
	Triumfetta	Mao ci shuo ma	*Triumfetta cana* Bl.	
		Ci shuo ma	*Triumfetta rhomboidea* Jacq.	
Malvaceae	*Abelmoschus*	Huang kui	*Abelmoschus moschatus* Medicus	
	Abutilon	Mo pan cao	*Abutilon indicum* (L.) Sweet [Sida indica L.]	
	Hibiscus	Diao dan fu sang	*Hibiscus schizopetalus* (Mast.) Hook. F	Diao deng hua
		Zhu jin	*Hibiscus rosa-sinensis* L.	
	Sida	Huang hua ren	*Sida acuta* Burm. F	
	Urena	De tao hua	*Urena lobata* L.	
Hypericaceae	*Cratoxylun*	Huang niu mu	*Cratoxylun ligustrinum* (Spach.) Bl	
Flacourtiaceae	*Bennettiodendron*	Shan gui hua	*Bennettiodendron brevipes* Merr.	
Boraginaceae	*Ehretia*	Shang si hou ke shu ○2	*Ehretia tsangii* I.M.Johnst.	
Combretaceae	*Combretum*	Ya xian feng che zi	*Combretum olivaeforme* var.*yaxianse* Y.R.Ling	
Escalloniaceae	*Itea*	Shu ci	*Itea chinensis* HK.et Arn.	
Acanthaceae	*Strobilanthes*	Ban lan	*Strobilanthes cusia*(Ness)O.Kuntze	Lan dian
	Thunbergia	Wan hua cao	*Thunbergia frangrans* Roxb.	Lao ya zui
		Shan qian niu ○2	*Thunbergia grandiflora* (Rottl. ex Willd.) Roxb.	Da hua lao ya zui
Passifloraceae	*Passiflora*	Long zhu guo	*Passiflora foetida* Linn.	
		Hu die teng	*Passiflora papilio* Li	
Hydrangeaceae	*Pilestegia*	Guan gai teng	*Pilestegia* sp.	
Sargentodoxaceae	*Sargentodoxa*	Da xue teng ○2	*Sargentodoxa cuneata* (Oliv) R.etW.	
Smalaceae	*Smilax*	Da ye ba qi	*Smilax china* L.	
		Fen ye ba qi	*Smalix corbularis* Kunth	
		Da guo ba qi	*Smilax macrocarpa* Bl.	
Connaraceae	*Rourea*	Hong ye teng	*Rourea microphylla* Pl.	
Polygonaceae	*Polygonum*	Huo tann mu	*Polygonum chinesis* L.	
Lythraceae	*Cuphea*	Xue jia hua	*Cuphea ignea* (syn.C.platycentra)	
	Lagerstroemia	Shi shan zi wei	*Lagerstroemia indica* L.	
		Nan zi wei	*Lagerstroemia subcostata* Koehne	
Nyctaginaceae	*Bougainvillea*	San jiao mei	*Bougainvillea spectabilis* Willd	
Gesneriaceae	*Rhynchotechum*	Ku ju tai	*Rhynchotechum formosanum* Hatus.	Jin yin hua
Caprifoliaceae	*Lonicera*	Ren dong	*Lonicera japonica* Thunb	
Guttiferae	*Garcinia*	Jin si li	*Garcinia paucinervis* Chun et How	
Commelinaceae	*Murdannia*	Shui zhu ye	*Murdannia triguetra* (Wal1.) Brijckn.	
	Pollia	Chuan du ruo	*Pollia omeiensis* Hong	
Amaranthaceae	*Achyranthes*	Tu niu xi	*Achyranthes bidentata* Bl.	
Pteridaceae	*Pteris*	Jian ye feng wei jue	*Pteris ensiformis* Burm.	
		Jing bian qian	*Pteris ensiformis* Burm.	

(continued)

Family	Genus	Species (name in China)	Species (scientific name)	Remark[a] (previous name)
		Wu gong cao	*Pteris vittata* L.	
		Ban bian qi	*Pteris semipinnata* L. Sp.	
Lygopodiaceae	*Lygopodium*	Xiao ye hai jin sha	*Lygopodium scandens* (L.) SW.	
Davaliaceae	*Nephrolepis*	Shen jue	*Nephrolepis cordifolia*(L.)Presl.	
Adiantaceae	*Adiantum*	You wei tie xian jue	*Adiantum caudatum* L.	
		Jia bian ye tie xian jue	*Adiantum malesianum* Ghatak	
Aspleniaceae	*Asplenium*	Zhang sheng tie jiao jue	*Asplenium prolongatum* L.	
Thelyteriaceae	*Parorthelypteris*	Jin xing jue	*Parorthelypteris glanduligera* (kunbze) ching	
	Cyclosorus	Hua nan mao jue	*Cyclosorus parasiticus* (L.) Farwell.	
		Jian jian mao jue	*Cyclosorus acuminatus* (Houtt.) Nakai	
Gnetaceae	*Gnetum*	Mai ma teng	*Gnetum montanum* Markgr.	
Chenopodiaceae	*Chenopodium*	Tu jing gai	*Chenopodium ambrosioides* L	
Cyperaceae	*Carex*	Tai cao yi zhung	*Carex* sp.	
Bambusoideae	*Monocladus*	Yun xiang zhu	*Monocladus amplexicaulis* Chia, Fung et. Y. L.Ya	
	Phyllostachys	Jin zhu	*Phyllostachys sulphurea* (Carr.) A.et C.Riv.	
	Bambusa	Fen dan zhu	*Bambusa chungii* McClure	
	Pleioblastus	Ku zhu	*Pleioblastus amarus* (Keng) Keng f.	
	Dendrocalamus	Ma zhu	*Dendrocalamus latiflorus* Munro	
Agrostidodeae	*Apluda*	Xhui zhe cao	*Apluda mutica* L.	
	Arthraxon	Jin cao	*Arthraxon hispidus* (Thunb.) Makio	
	Arundinella	Shi mang cao	*Arundinella nepalensis* Trin.	
	Brachiaria	Bi xing cao	*Brachiaria villosa* (Lamk.) A.Camus	
	Cyrtococcum	Gong guo shu	*Cyrtococcum patens* (L.) A.Gamus	
	Isachne	Liu ye ruo	*Isachne globosa* (Thunb.) Kuntze	
	Imperata	Bai mao	*Imperata cylindrica* var.*major* (Ness)	
	Microstegium	Er xing you zhu	*Microstegium biforme* Keng	
		Man san you zhu	*Microstegium vagans* (Nees) A.Camus	
	Miscanthus	Wu jie mang	*Miscanthus floridulus* (Labvill.) Warb.	
	Neyraudia	Lei lu	*Neyraudia reynandiana* (Kunth) Keng	
	Rottboellia	Tong zhou mao	*Rottboellia cochinchinensis* (Loureiro) Clayton	Lou shi cao
	Saccharum	Ban mao	*Saccharum arundinaceum* R.et Z.	
	Themeda	Huang bei cao	*Themeda triandra* var. *japonica* (Willd.) Makino.	
	Thysanolaena	Zong ye lu	*Thysanolaena maxima* (Roxb.) Ktze.	
	Zoysia	Xi ye jie lu cao	*Zoysia pacifica* (Goudswaard) M. Hotta & S. Kuroki	Taiwan cao
Palmaceae	*Arenga*	Guang lang	*Arenga pinnata* (Kze.) Merr.	
	Daemonorops	Huang teng	*Daemonorops mangaritae* (Hance) Becc	
	Guihaia	Shi shan zong	*Guihaia argrata* S.K.Leo & F.N.Wei.J.Dransfida	
	Rhapis	Cu zong zhu	*Rhapis robusta* L.	
Cycadaceae	*Cycas*	Kuan ye su tie	*Cycas balansae* Warburg	Yunnan su tei
		Su tie	*Cycas revolute* Thunb.	
Polypodiaceae	*Drynaria*	Hu jue	*Drynaria fortune* (Kze) J.SM.	
	Lemmaphyllum	Fu shi jue	*Lemmaphyllum microphyllum* C. Presl	
		Xi ye gu pai jue	*Lemmaphyllum diversum* (Rosenst.) De Vol et C. M. Kuo	Pai gu jue
	Lepidogrammitis	Bao shi lian	*Lepidogrammitis drymoglossoides* (Baker) Ching	
	Pseudodrynaria	Ya jiang jue	*Pseudodrynaria coronans* (Wall. ex Mett.) Ching	
Umbelliferae	*Centella*	Ji xue cao	*Centella asiatica* (L.) Urban	Ting qian cao
Loranthaceae	*Macrosolen*	Qiao hua	*Macrosolen cochinchinensis* (Lour.) Van Tiegn.	
	Taxillus	Guang ji san	*Taxillus chinensis* (DC.) Danser	
Actinidiaceae	*Actinidia*	Ge ye mi hou tao o1	*Actinidia rubricaulis* Dunn var. *coriacea* (Fin. & Gagn.) C.F. Liang	

(continued)

Family	Genus	Species (name in China)	Species (scientific name)	Remark[a] (previous name)
Caricaceae	*Carica*	Fan mu gua ○1	*Carica papaya* L.	
Juglandaceae	*Engelhardia*	Huang qi	*Engelhardia roxburghiana* Wall.	
Lamiaceae	*Salvia*	Li zhi cao	*Salvia plebeia* R. Br.	
Hamamelidaceae	*Liquidambar*	Feng xiang shu	*Liquidambar formosana* Hance	
Blechnaceae	*Blechnum*	Wu mao jue	*Blechnum orientale* L.	

Remarks: (○1) Species included in white-headed langur diet composition in intense human activity disturbances period (1990–2001) (Yan Lijie 2009). (○2) Species included in white-headed langur diet composition in the early period of flora restoration (2002–2006) (Xiangdong et al. 2008).
[a]Species names in "remark" column were names formerly known by new names adapted from *Flora Republicae Popularis Sinicae, web version*

References

Chengming H, Ruyong S, Yuegui X, Suling W, Youbang L. Research on white-headed langur diet and its forage hour allocations (白头叶猴食谱与觅食时间分配的研究). Anthrol J. 2000;19(1):65–72.

Hongying L. Plant resources survey and research in white-headed langur researh base in Chungzuo City (崇左市白头叶猴研究基地植物资源的调查研究). Degree thesis. Guangxi: Life Science and Technology College, Guangxi University; 2010.

Kuanzhao H. Dictionary of seeding plant genera and families in China (中国种子植物科属词典). Beijing: Science Press; 1982.

Lantian S, Hanhua L. White-headed langur of Guangxi (广西的白头叶猴). Guangxi Normal Univ J (Nat Sci Ed). 1982;1982:27–33.

Li Z, Wei Y, Rogers E. Food choice of white-headed langurs in Fusui, China. Int J Primatol. 2003;24(6):1189–205.

Lijie Y. Male white-headed langur (*Trachypithecus leucocephalus*) reproduction tactics – a research on infanticide and paternal fostering (雄性白头叶猴的繁殖策略 - 对杀婴行为和雄性照料的研究). Doctoral thesis. Beijing: Peking University; 2009.

Qiang X, Liren Lu, Chengming H. Research on white-headed langur ecology – the eating habits of wild white-headed langurs (白头叶猴环境生态的研究——白头叶猴的野生食性). Acad J Guangxi Normal Univ (Nat Sci). 1994;12:1.

Xiangdong L, Shiyang L. Preliminary analysis on the vegetative diet of white-headed langurs (白头叶猴采食植物初析. 基因组学与应用生物学). Genomics Appl Biol. 2005;24(02):153–60.

Xiangdong L, Lijie Y, Li jun Liu, et al. Further researches on the vegetative diet of white-headed langurs (白头叶猴采食植物的再研究. 基因组学与应用生物学). Genomics Appl Biol. 2008;27:1.

Yuchang S. Animal ethology (动物行为学). Beijing: Peking University Press; 2005.

Yuemei L. The living habits of white-headed langur, rearing and reproducing (白头叶猴的生活习性和饲养繁殖). Wild Anim. 1987;38(4):23–4.

Zhengyi W. Phytogeographic attributes of plants in China (中国种子植物属的分布区类型). Acta Bot Yunnan (Suppl Ed). 1997;4:1–190.

Zhou Q, Tang Z, Li Y, et al. Food diversity and choice of white-headed langur in fragmented limestone hill habitat in Guangxi, China. Acta Ecol Sin. 2013;33(2):109–13.

8 Population Restoration and Social Structure of White-Headed Langurs in the Nongguan Mountains

Abstract

The restoration of the white-headed langur population in the Nongguan Mountains has undergone several phases. In February 1997, only six family troops and three all-male troops survived the anthropogenic calamities (Chap. 4), with an aggregate of 105 white-headed langurs. In 2002, the population increased to 12 family troops and seven all-male troops, with a population density of 42.5 langurs/km^2. From 2006 to 2015, the population increased to above 800, and the all-male troops started to migrate and disperse all over the Nongguan Mountains. We divide the lifespan of a white-headed langur into six age stages: infant ($\leq$0.5 years old), yellow-haired pup (0.5–1.3 years old), juvenile (1.3–2.5 years old), adolescent (2.5–4 years old), sub-adult (4–5 years old), adult (>5 years old), and old (♂ >13 years; ♀ >25 years). The social structure of the white-headed langurs could be summarized into four basic systems: ante-family troop, family troop, transitory troop, and all-male troop. The white-headed langur does not show obvious sexual dimorphism, but the canine teeth of adult males are noticeably larger than in females. During the family troop phase, the white-headed langurs are strictly polygynous, but in their lifetime, they are polygamous.

Keywords

Population density · Social structure · Polygamy · Polygyny

In the late 1990s, poaching was escalating in the Nongguan Mountains as well as in other white-headed langurs habitats. Traps targeting at white-headed langurs could be found everywhere in the distribution regions. For a time, this preciously rare species was made into trendy tonics and meat on the feasting tables.

On April 10, 1998, as I was walking to Feijichang, I could see from afar that there was something like a white headed langur on the branch above the cave. As I got nearer, I found it was a white-headed langur with its head hanging down in midair, motionless. Through the binoculars, I saw one of its hind legs clamped by a "Steel Cat" that was tightly roped on a tree. The tormented look of its dying struggle remained frozen on its contorted face. That was the first time I came face to face with the tragic death of a white-headed langur. I was indignant that such a vicious act had happened in the core district where we conducted our research every day.

In those days, the white-headed langur in the Nongguan Mountains seemed doomed for a shaky future. There were only a small number of white-headed langurs that remained alive, and they were threatened to disappear anytime, which is what happened to the populations in the other four geographical distributions. I pleaded with the Chungzuo County Party Committee to take immediate action to enhance the patrolling of the region to protect the white-headed langurs in the Nongguan Mountains. After the government took a series of measures to stop poaching, illicit tree cutting, and blasting of hills, the Nongguan Mountains white-headed langur population finally had the luck to survive and restore.

The restoration of the Nongguan Mountains white-headed langur population consists of three phases; in the beginning, it was the troop-founding phase, then the multiplication phase, and finally the saturation phase.

8.1 Troop-Founding Phase

The troop-founding phase of the Nongguan Mountains white-headed langur population had taken 10 years (1996–2006). To detail the restoration process, we shall divide the phase into two periods.

W. Pan, *The Population Ecology of White-Headed Langur*, https://doi.org/10.1007/978-981-33-4118-0_8

8.1.1 Founding Troops

From the day I arrived at the Nongguan Mountains in November 1996, I spent more than 2 months searching for the few white-headed langurs that remained in the region. I found in the end nine troops survived after the calamities: they were six family troops and three all-male troops with an aggregate of 105 white-headed langurs distributed in disconnected areas.

Since then, the population started to increase, and the langurs started to spread over extended areas. The population multiplied by way of "grafting" or "splitting." Three young females from the founding population (FJC Big Cave troop) dispersed to Fangshaoshan (FSS) in 2000 and by way of "grafting" initiated the "Fangshaoshan family troop." In 2002, another three adult females with three immature females in the "FJC Big Cave troop" split from the family to inhabit in Xishan, thus by way of "splitting" initiated the "Xishan family troop" (FJC-XS).

8.1.2 Gradual Increase of Population

In 2002, the Nongguan Mountains white-headed langur population increased to 19 troops, including family troops with fixed home ranges, transitory troops, and all-male troops without fixed home ranges. The distributions of the troops are shown in Fig. 8.1. While the population had over time increased, the home range for each troop had been reduced. For the 12 family troops with fixed home ranges, their average home range had been reduced from 0.42 m^2/troop in the early days to 0.31 m^2/troop in this period.

From 1996 to 2002, the utilized habitat by white-headed langurs in the Nongguan Mountains had changed a bit. In 1996, it was about 5 km^2, and by 2002, it was expanded to cover 5.84 km^2. At the same time, the population density had increased from 29.4 langurs/km^2 in 1998 to 42.5 langurs/km^2 in August 2002.

8.2 Multiplication Phase

Field observation registers had indicated that the number of white-headed langurs in the Nongguan Mountains started to increase significantly in 2006, and it is continuing this trend up to the present.

We shall again take the example of the FJC Big Cave troop. The matrilineal inherited home range of the female-centered family troop had not been enlarged, while the number of family individuals keeps increasing (Table 8.1). In September 2006, Yintangxiaotu successfully fought its way into the family and became the resident alpha male, initiating the "Yintangxiaotu troop I" ("Yintangxiaotu family troop"). The family was split in July 2012, and the "Yintangxiaotu troop II" (the "Yintangxiaotu transitory troop") was initiated. In March 2015, 15 male langurs left the transitory troop to become the "Yintangxiaotu troop III" (the "Yintangxiaotu all-male troop"). In spring of 2015, the "Yintangxiaotu troop III" migrated southward, departing its former home range.

The multiplication of the "Yintangxiaotu troop" could be taken as a paradigm of the growth of the white-headed langur population density and distribution expansion in the Nongguan Mountains. It started 9 years ago and is still continuing. We shall divide the progression into three stages.

8.2.1 Stage 1: 6 Years with Rapid Growth in Number

In September 2006, our cameras caught the images of a robust male langur with a small lump of protruding flesh between its eyebrows (we named it Yintangxiaotu, Fig. 8.2) attacking the "α-Gonghou family troop," which had its fixed home range in FJC Cave. It won and took over the seven adult females and six premature females and began a new family which we named the "Yintangxiaotu family troop" (an outlier immature female came and joined the troop in 2010.08). α-Gonghou left FJC Big Cave with its four immature sons and became an all-male troop inhabiting temporarily at Bokyueshan and Fanghoushan, where our research base is.

By July 2010, the "Yintangxiaotu family troop" had multiplied to include 26 females and 16 males. Together with Yintangxiaotu, it had become a big family with 43 langurs (one daughter died of sickness in January 2010, an outlier adolescent female came and joined). However, the home range of the "Yintangxiaotu family troop" had not become wider; their night shelter remained at the Big Cave and Small Cave in FJC, and their roaming territory was still bounded within the area from the eastern half of FJC to the Taohuagu and Fanghoushan areas around our Research Base.

8.2.2 Stage 2: 3 Years of Rapid Increase of Population Density

In July 2010, an outlier male langur attacked the "Yintangxiaotu family troop" ("Yintangxiaotu troop I"), and Yintangxiaotu had to relinquish the Big Cave night shelter. It withdrew with only three adult females, four immature daughters, and 13 sons to FJC Small Cave and began a new troop in this "split." We named the "Yintangxiaotu troop" at this stage the "Yintangxiaotu transitory troop" ("Yintangxiaotu troop II"). In the spring of 2014, three pups (2♂ 1♀) were born to this troop.

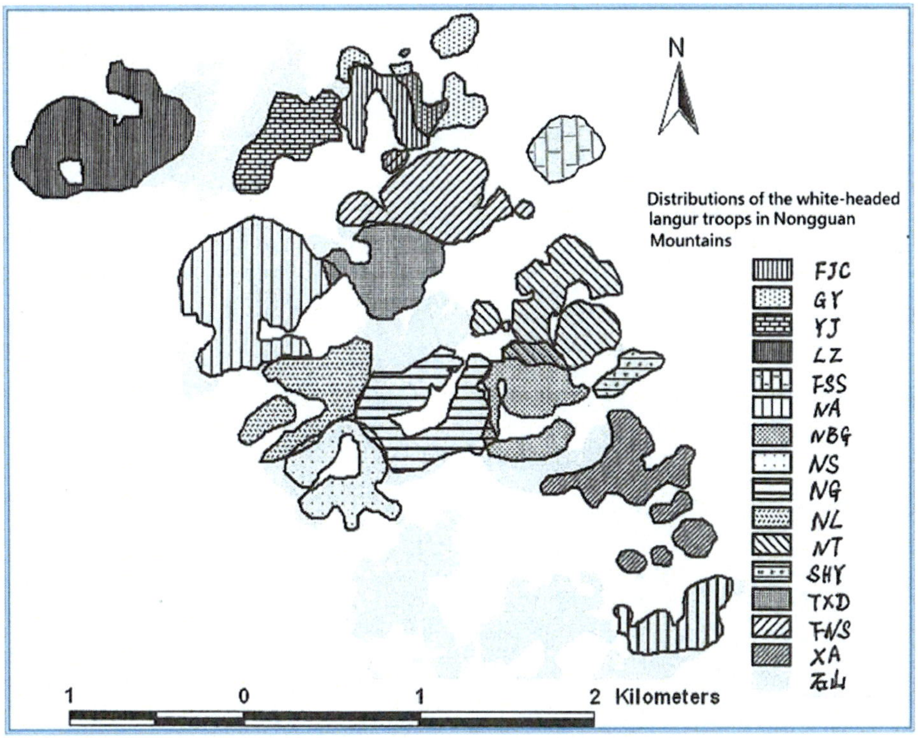

Fig. 8.1 White-headed langur troops distributions in Nongguan Mountains in the period 1996–2002

From July 2012 to July 2013, the FJC Big Cave troop had been taken over successively by six adult male langurs in three combinations (Duanwei + Romeo, Shengge + Shengdi, Yuweng + Xiaoyuweng), and each had occupied the night shelter for a short period of time. Eventually, the adult male langur Yuweng dominated FJC Big Cave troop.

Yuweng took over the 11 adult female langurs of the FJC Big Cave troop and the four immature daughters they had born to Yintangxiaotu. Together with one outlier adolescent female that came and joined, Yuweng began a bigger family consisting of 17 reproductive langurs (including Yuweng), and we named the troop "Yuweng family troop." By December 2015, there were 15 pups born in the "Yuweng family troop," and the family became a big one with 33 langurs in less than 2 years. Yet, their home range became smaller, and their daily roaming area was limited to the hills at the front and back of FJC Big Cave and areas around the mountain pass at Guoyuan. At the same time, their occupied territory was also small. It was bounded in the south by the boundary with the "Yintangxiaotu troop II" at the Tianxian mountain pass, in the west by the boundary with the "Xishan family troop" at a short distance from the mountain pass at Guoyuan, and the northern boundary was just a little further north of the Guoyuan pass. It can be foreseen that the number of langurs in the "Yuweng family troop" would continue to grow, and the density would become even higher.

8.2.3 Stage 3: Formation of an All-Male Troop and Migration to Distant Territories

In March 2015, the outlier male langur Laoxie fought into FJC Small Cave and took over the three wives and four daughters of Yintangxiaotu, and together with two outlier females which came and joined, started the "Laoxie family troop." Yintangxiaotu then led its 14 sons, leaving their FJC Small Cave night shelter, and became "Yintangxiaotu troop III" ("Yintangxiaotu all-male troop"). They inhabited somewhere around the Research Base for a short period of time, and then dispersed to the south.

Before long, the eldest son of Yintangxiaotu, Wuque (so named for there were no scars on both its ears, Fig. 8.3a) started a family troop of its own with 13 females inhabiting on a mountain tract not far from FJC. The second son Zuoque (so named for there was a nick on its left ear, Fig. 8.3b) and third son Youque (so named for there was a nick on its right ear, Fig. 8.3c) also left their old father and younger brothers to strive to establish their respective realms. Presently, Yintangxiaotu is still leading the remaining 11 sons and roaming through the hinterland of the Nongguan Mountains to further places. It has to help its sons to find target troops to take over.

Assuming the multiplication trend continues, we can predict that as the natural habitat in the Nongguan Mountains becomes better and better under enhanced conservational and

Table 8.1 White-headed langur family troops and all-male troops engendered by "graft" or "split" from FJC Big Cave troop from 1996.11 to 2015.12

Sequence	Duration	Troop	Night shelter	Langurs in the troop	Remark
1	1994.07–1998.04	Queque family troop	FJC Big Cave, FJC Small Cave, Xishan, Guoyuan, Yangjuan	18 (1♂ 8♀ 10 pups)	1♀ was killed by poaching. Believe 2 pups had been killed in infanticide
2	1998.04–2002.07	Archeng family troop	FJC Big Cave, FJC Small Cave, Xishan, Guoyuan	20 (1♂ 7♀ 12 pups)	1 pup was killed in infanticide 3 pups died in rain storm
3	2000.06–2000.07	Queque transitory troop	Guoyuan, Bokyueshan, Fangshaoshan	10 (6♂ 4♀)	"Queque" was killed in battle in 2000.07 1 adult ♀ died from miscarriage 3 daughters initiated new family troop with an outlier ♂ in Fangshaoshan
4	2000.07–2001.11	Queque sons all-male troop	Bokyueshan	3♂	The 3 sons of Queque
5	2002.07–2006.06	α-Gonghou family troop	FJC Big Cave, FJC Small Cave	21 (1♂ 7♀ 13 pups)	1 pup was killed in infanticide
6	2002.07–2002.10	Archeng transitory troop	Guoyuan	1♂ 3♀ 8 pups	–
7	2002.10–2006.06	*β-Gonghou* family troop	Xishan	1♂ 6♀ ? pups	–
8	2006.06–2012.07	Yintangxiaotu family troop	FJC Big Cave, FJC Small Cave	44 (1♂ 13♀ 29 pups, +1 adolescent ♀joined)	Pups (3♂ 1♀) disappeared or were killed in infanticide 1 pup died from sickness
9	2006.06–2007	α-Gonghou all-male troop	Fanghoushan, Bokyueshan	5♂	–
10	2013.07–present	Yuweng family troop	FJC Big Cave	32 (1♂ 16♀ 15 pups)	–
11	2012.07–2015.03	Yintangxiaotu transitory troop	FJC Minor Cave, Research Base	23 (12♂ 8♀ 3 pups; 1 adolescent ♀joined)	2 pups were killed in infanticide
12	2015.03–present	Laoxie family troop	FJC Small Cave	12 (1♂ 9♀ 2 pups; 1 adult ♀; and 1 adolescent ♀ joined)	–
13	2015.03–present	Yintangxiaotu all-male troop	Without permanent night shelter	15♂	The 3 eldest sons had left the troop one after another

managerial efforts, the number of white-headed langurs in the region would continue to increase. When white-headed langurs have distributed all over the tropical karst monsoon rainforest habitat, their population growth would have reached its saturation and will maintain itself at a stabilized state.

8.3 Identification of White-Headed Langurs

To register the white-headed langurs' interactive behaviors, their social relationships underlying their gregarious lives, it is necessary to identify the individual langurs. We used high-resolution binoculars for detailed identification, and capture their images with high-definition cameras and camcorders for repeated comparisons. We used the following method to identify the individual white-headed langurs.

8.3.1 Identification of Age by External Appearances

The hair color of a white-headed langur and the distribution of black and white hair changes with age (Fig. 8.4). Newborn pups have golden yellow hair all over their body. At the age of 1 year, streaks of black hair start to appear amidst the yellow hair on its body and limbs, giving it an overall light brown or brownish black color, while the hair on the head, shoulders, and tail begin to turn white. Around the age of 2 years, hair on the body will be mostly black, but traces of brown could still be noted, whereas on the head, shoulders, and tail, white hair would notably replace hair of other colors while the hair on the back of hands and feet will have all turned white. At the adolescent age of 3 years, the overall hair colors will rapidly become distinctly contrasting like those appearing on adults, as the head, shoulders, and tail will be covered with white hairs, while the body will be covered by

Fig. 8.2 (a) Full photo portrait of Yintangxiaotu (2009); (b) Facial photo portrait of Yintangxiaotu (2013)

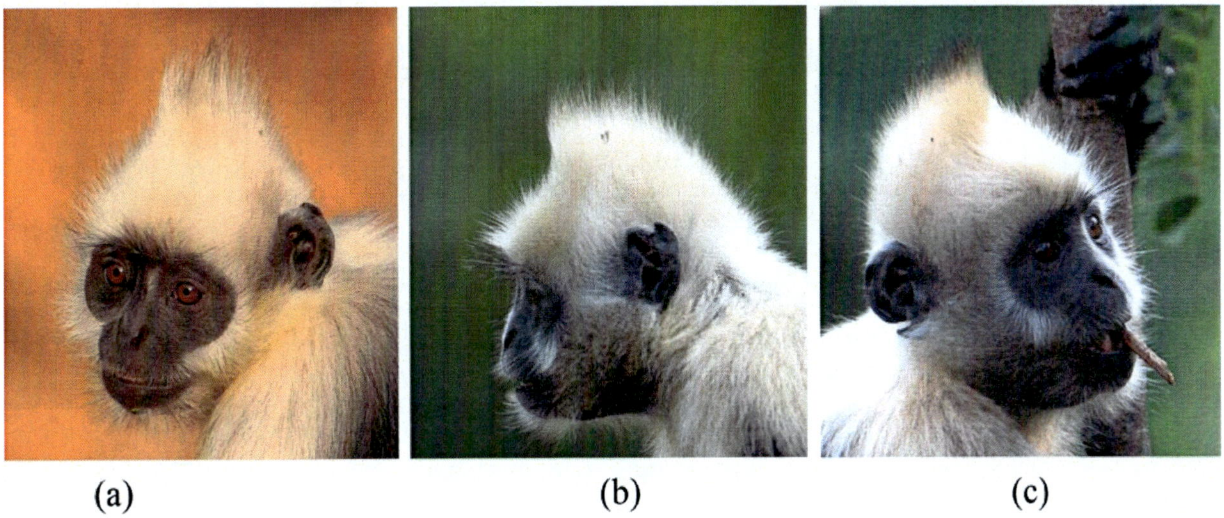

Fig. 8.3 The three eldest sons of Yintangxiaotu and their individualistic features: (a) Wuque; (b) Zuoque; (c) Youque

black hairs. The body size would still be smaller than that of an adult.

We divide the lifespan of a white-headed langur into six age stages: newborn (≤0.5 years), yellow-haired pup (0.5–1.3 years old), juvenile (1.3–2.5 years old), adolescent (2.5–4 years old), sub-adult (4–5 years old), adult (>5 years old), and old (♂ >13 years; ♀ >25 years).

8.3.2 Identification of Individual Females

We found there is a white triangular skin patch on the groins of female white-headed langurs resembling the sex skin on all other female primates. It functions to transit reproductive signals of the female. That small patch of skin will remain white through its lifetime with black blotches distributed on it, like birthmarks on humans whose pattern stay unchanged through lifetime (Fig. 8.5). Nonetheless, the sizes, shapes, and distributions of the black blotches are varied, and we used this feature to identify the different individuals. Table 8.2 shows the photos of the individualistic features on the triangular skin patch on the groins of some of the "FJC Big Cave troop" female langurs taken with HD cameras and camcorders.

Fig. 8.4 Hair color changes over various age stages of white-headed langur. (a) Newborn (yellow-haired infant) and mother (adult female). (b) Yellow-haired pup (younger brother) and juvenile (elder sister). (c) Adolescent ♂ and sub-adult ♂. (d) Adolescent ♀ (Wailaimei). (e) Adult ♂ (Yintangxiaotu). (f) Old ♀ (Dahei)

8.3.3 Identification of Individual Males

There is no white skin patch on the groins of male white-headed langurs, but nearly all adult males have permanent scars on their bodies from fighting for reproduction rights, as shown in photos in Table 8.3. Like hot iron brands, they become the outward markings, which we could use to identify the male individuals.

8.4 The Social System of the Nongguan Mountains White-Headed Langur Population

White-headed langurs are social animals that live in troops centered on females. Most female langurs live their entire lives in their matrilineal inherited home range, whereas young male langurs would follow their defeated father and leave their birthplace, their mothers, aunts, and sisters to look for mates in the broad expanse of the white-headed langur society. Therein lies the white-headed langur reproductive tactics: males compete against each other for breeding mates while females decide on the most suitable male for reproduction in rounds and rounds of male competitions.

There are few academic papers detailing the social system and reproductive behavior of white-headed langurs. The statistics we are using below was gathered from our observation registers on the white-headed langurs in the Nongguan Mountains in the last 20 years. They are primarily registers from our in-depth research on the "Feijichang (FJC) Big Cave troop" from 2008 to 2015. Some relevant data are adapted from the respective academic degree thesis by eight Peking University postgraduates in the dates from 1998.03 to 2008.06 (Lei 1998; Yingyi 2002; Lijie 2008; Wang 2004; Wenzhong 2003; Qing 2005; Lizhong 2002; Tong 2008).

The different forms of white-headed langur social systems are invariably based on core families. During the 20 years of our research, the white-headed langur society in the Nongguan Mountains was an "enclosed" population. They could only evolve on the lineal track exclusive to its own population. We have been regarding the Nongguan Mountains as one laboratory of nature, with one single research object in an insular environment.

From November 1996 to December 2015, we had 63 counts of tracking on socially structured white-headed langur troops to observe and register their social systems. We considered a troop socially structured only if it had more than two langurs in the troop (we encountered more than ten

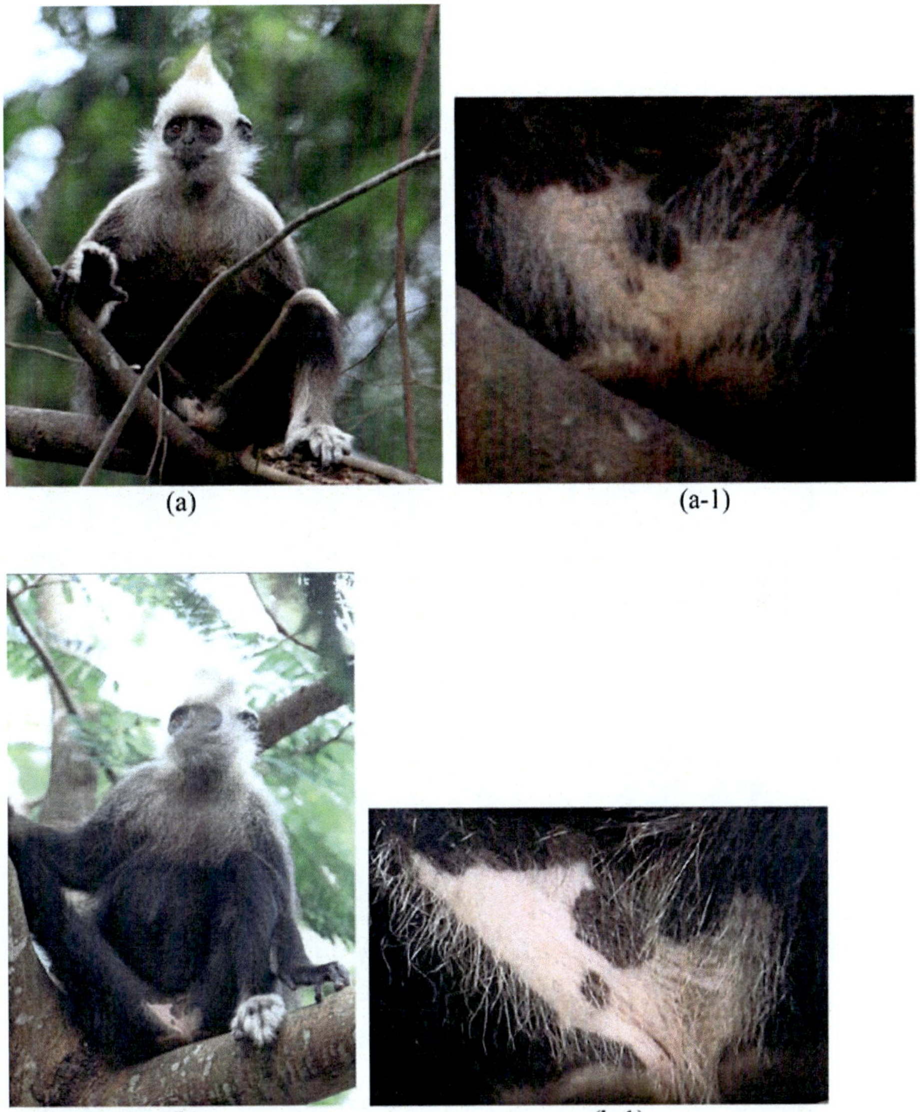

Fig. 8.5 The outward appearance of the female langur Gaga at various age stages and the individualistic blotches on the triangular skin patch on her groin (Photo by Liang Zuhong). (**a**) The full body appearance at 1 year and 11 months old. (**a-1**) The individualistic blotches on the triangular skin patch on her groin. (**b**) The full body appearance at 7 years and 6 months old. (**b-1**) The distinct blotch distribution on the triangular skin patch on her groin

Table 8.2 The individualistic blotches on the triangular skin patch on the groins of some of the "FJC Big Cave troop" female langurs

Name	*Archen*	*Yingying*	*Lanqi*	*Pingping*	*Lanlan*	*Tiantian*
Individualistic blotches on the triangular skin patch on the groin		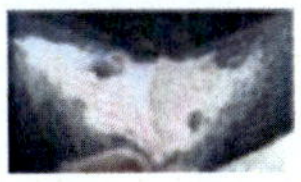		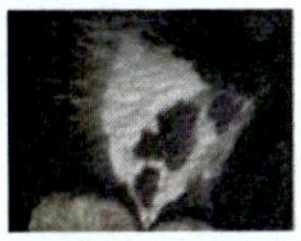		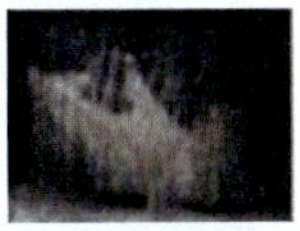
Name	*Yaxi*	*Yumei*	*Wailaimei*	*Qingming*	*Gaga*	*Xiaoying*
Individualistic blotches on the triangular skin patch on the groin	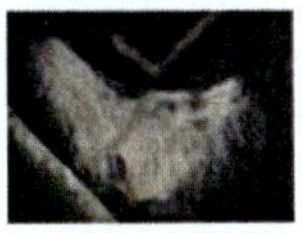	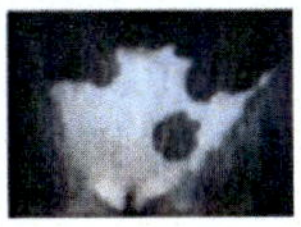	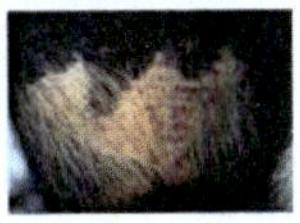	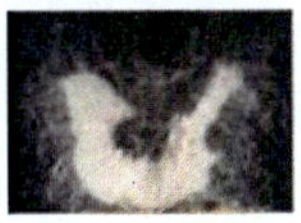		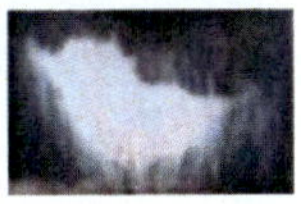

occasions of solitary males or solitary females, but those were not considered registers of the social system). We had registered a total of ten social systems from observations on socially structured troops (Table 8.4), but in essence, they could be summarized into four basic social systems: ante-family troop, family troop, transitory troop, and all-male troop.

Table 8.3 The individualistic outward identifications on individual male white-headed langurs

Name	*Yintangxiaotu*	*Shengge*	*Yuweng*	*Duanwei*
Individualistic facial feature				
Specifics	Lump between eyebrows	Chipped right ear	Scar on left face	Wound scar on the left face
Individualistic ear feature				
Specifics	Chipped right ear	Chipped left ear	Chipped left ear	Perfect ears
Individualistic tail feature				
Specifics	Crooked tail	Crooked tail	[illegible]	[illegible]

Table 8.4 The social systems of the white-headed langurs in the Nongguan Mountains (1996.11–2015.12)

No.	System	Social structure	Troop designation	Observation period	Site	Individuals in the troop	Observation and registration by
1	Ante-family troop	1 adult ♂ + *n*-numbers adult ♀	FNS	1996.11–1997.02	Fangniushan	A♂1: Tutu (TT) A♀8	Pan Wenshi
2			SHY	1999.12–2000.03	Shihuiyao	A♂1: SHY A♀3: Daimei-F1, Ermei-F2, Sanmei-F3	Zhao Qing, Pan Wenshi
3		1 adult ♂ + *n*-numbers adult ♀ + *n*-numbers adopted immature ♀	FJC II	1998.04–1998.11	Feijichang	A♂1: Archeng (ACH) A♀6: Dahei, Archen, Jiajia, Lingling, Bibi, Beike SA♀1: Xiaoxiao	Yin Lijie, Qin Dagong, Pan Wenshi
4			NT-X	2000.06–2000.10	Nongtong	A♂1, A♀3, SA♀3	Ran Wenzhong, Qin Dagong, Pan Wenshi
5			FJC III A	2002.07–2003.01	Feijichang	A♂1: α-Gonghou A♀4: Dahei, Archen, Jiajia, *Xiaoxiao* SA♀3: Yingying, Pingping, Lanlan	Jin Tong, Wand Dezhi, Pan Wenshi
6			FNS II	2002.02–2002.05	Fangniushan	A♂1: Waiwai (WW) A♀3 SA♀3	Yin Lijie, Qin Dagong, Pan Wenshi
7			FJC III B	2002.07–2003.06	Feijichang, Xishan	A♂1: *β*-Gonghou A♀3: Lingling, Bibi, Beike SA♀1: Ziao-W	Jin Tong, Wand Dezhi, Pan Wenshi
8			FNS III A	2006.03–2008.03	Fangniushan	A♂1: Aiai (AA) A♀3, SA♀7	Yin Lijie, Qin Dagong, Pan Wenshi
9			FNS III B	2007.06–2008.03	Fangniushan	A♂1: Bangbang (BB) A♀2, SA♀4	Yin Lijie, Qin Dagong, Pan Wenshi
10			FJC-Y	2006.09–2007.10	Feijichang	A♂1: Yintangxiaotu A♀2: Archen, Dahei, Jiajia, Xiaoxiao, Yingying, Pingping, Lanlan J♀1: Lanqi E♀5: Tiantian, Aryu, A-Xi, Arlu, Beibei	Pan Wenshi, Liang Zuhong, Liu Lijun
11			FJC-W	2013.07.14–2013.12.05	Feijichang Big Cave	A♂1: Yuweng A♀11: Archen, Dahei, Jiajia, Xiaoxiao, Pingping, Tiantian, Aryu, A-Xi, Arlu, Beibei, Yumei SA♀5: Zhenzhu, Qiqi, Mimi, Da-S, Wenwen J♀1: Xiaojiu	Gu Tieliu, Feng Chunguang, Li Xinyang, Pan Yue, Zhao Yi, Meng Hao, Yang Yanping, Huang Meiyang, Luo Zuoye, Wei Meijiao, Qin Jiajia, Zheng Xinma, Liang Zuhong, Pan Wenshi
12			FJC-X	2015.03.20–2015.11.17	Feijichang Small Cave	A♂1: Laoxie A♀3: Yingying, Lanlan, Lanqi, Panma SA♀3: Qingming, Gaga, Wailaimei E♀2: Xiaoying, Xiaolan	Gu Tieliu, Feng Chunguang, Li Xinyang, Pan Yue, Zhao Yi, Yang Yanping, Huang Meiyang, Luo Zuoye, Wei Meijiao, Zheng

(continued)

Table 8.4 (continued)

No.	System	Social structure	Troop designation	Observation period	Site	Individuals in the troop	Observation and registration by
							Xinma, Liang Zuhong, Pan Wenshi
13		2 adult ♂ + n-numbers adult ♀ + n-numbers adopted premature ♀	FJC-DL	2012.07.22–2012.12.25	Feijichang Big Cave	A♂2: Romeo, Duanwei A♀10: Archen, Dahei, Jiajia, Xiaoxiao, Pingping, Tiantian, Aryu, A-Xi, Arlu, Beibei SA♀5: Zhenzhu, Qiqi, Mimi, Da-S, Yumei J♀1: Wenwen E♀1: Xiaojiu	Gu Tieliu, Feng Chunguang, Li Xinyang, Pan Yue, Zhao Yi, Long Yu, Meng Hao, Yang Yanping, Huang Meiyang, Luo Zuoye, Wei Meijiao, Qin Jiajia, Zheng Xinma, Liang Zuhong, Pan Wenshi
14			FJC-SS	2012.12.15–2013.03.10	Feijichang Big Cave	A♂2: Shengge, Shengdi A♀10: Archen, Dahei, Jiajia, Xiaoxiao, Pingping, Tiantian, Aryu, Arxi, Arlu, Beibei SA♀5: Zhenzhu, Qiqi, Mimi, Da-S, Yumei J♀1: Wenwen Xiaojiu	Gu Tieliu, Feng Chunguang, Li Xinyang, Pan Yue, Zhao Yi, Meng Hao, Yang Yanping, Huang Meiyang, Luo Zuoye, Wei Meijiao, Qin Jiajia, Zheng Xinma, Liang Zuhong, Pan Wenshi
15			FJC-YX	2013.07.14–2013.09.07	Feijichang Big Cave	A♂2: Yuweng, Xiaoyuweng A♀11: Archen, Dahei, Jiajia, Xiaoxiao, Pingping, Tiantian, Aryu, Arxi, Arlu, Beibei, Yumei SA♀5: Zhenzhu, Qiqi, Mimi, Da-S, Wenwen J♀1: Xiaojiu	Gu Tieliu, Feng Chunguang, Li Xinyang, Pan Yue, Zhao Yi, Yang Yanping, Huang Meiyang, Luo Zuoye, Wei Meijiao, Qin Jiajia, Zheng Xinma, Liang Zuhong, Pan Wenshi
16		1 adult ♂ + n-numbers adopted premature ♀	FSS	2000.07–2000.08.28	Fangshaoshan	A♂1: AZ SA♀3: Dagege, Ergege, Sangege	Zhang Yingyi, Zhu Lizhong, Pan Wenshi
			FSS	2000.08.29–2001.05	Fangshaoshan	A♂1: W SA♀3: Dagege Ergege Sangege	Zhang Yingyi, Zhu Lizhong, Pan Wenshi
17	Family troop	1 adult ♂ + n-numbers adult ♀ + n-numbers premature offspring	FJC I	1996.11–1997.02	Feijichang	A♂1: Queque (QQ) A♀7: Heimama, Dahei, Archen, Jiajia, Lingling, Bibi, Beike SA♀1: Weiming E♀1: Dagege, Xiaoxiao E♂1: Daarge I♂2: Erarge, Sanarge I♀3: Ergege, Sangege	Pan Wenshi
18			YJ	1996.11–1997.02	Yangjuan	A♂1, A♀6, SA♀1, I 2	Pan Wenshi
19			LZ	1996.11–1997.02	Leizhai	A♂1, A♀6, SA♀3, I 4	Pan Wenshi
20			NBG 1	1996.11–1997.02	Nongbaga 1	A♂1, A♀9, I 9	Pan Wenshi
21			NT	1996.11–1997.02	Nongtong	A♂1, A♀7, SA♀5, I 6	Pan Wenshi
22			FJC I	1997.03–1998.04	Feijichang	A♂1: Queque (QQ) A♀8: Heimama, Dahei, Archen, Jiajia,	Pan Wenshi, Zhang Lei

(continued)

Table 8.4 (continued)

No.	System	Social structure	Troop designation	Observation period	Site	Individuals in the troop	Observation and registration by
						Lingling, Bibi, Beike, Weiming J♀2: Dagege, Xiaoxiao J♂1: Daarge E♂2: Erarge, Sanarge E♀2: Ergege, Sangege I♂2: Siarge, Xiaoarge	
23			NT	1997.03–2000.05	Nongtong	A♂1, A♀5 SA♂2, SA♀4 J♂4, J♀2 I♀2	Zhang Lei, Zhang Yingyi
24			NA	1998.02.05–2000.06	Nongan	A♂1, A♀6 J♂3, J♀1 I♂2	Ran Wenzhong, Pan Wenshi
25			NBG 1	1997.03–2000.12	Nongbaga 1	A♂1, A♀5 SA♂3, SA♀2 J♂2, J♀2, I♂3	Zhang Lei, Ran Wenzhong, Pan Wenshi
26			TXD	1998.01–2000.10	Taoxindong	A♂1, A♀4 J♂1, J♀1 I♂2	Ran Wenzhong, Pan Wenshi
27			FNS	1997.03–2002.02	Fangniushan	A♂1: Tutu (TT) A♀8, immature offspring 16	Yin Lijie, Qin Dagong, Pan Wenshi
28			FJC II	1998.12–2002.07	Feijichang	A♂1: Archeng (ACH) A♀6: Dahei, Archen, Jiajia, Lingling, Bibi, Beike SA♀1: Xiaoxiao, immature offspring: ♂8, ♀6,? 1: Yingying, Arying, Nianying, Pingping, Arping, Nianping, Qingqing, Arqing, Feifei, Arfei, Lanlan, Arlan, Xinxin, Yiyi, Aryi	Jin tong, Wand Dezhi, Yin Lijie, Qin Dagong, Pan Wenshi
29			SHY	2001.01–2002 summer	Shihuiyao	A♂1: SHY A♀3: Damei F1 Ermei F2 Sanmei F3 I♂2, I♀1	Zhao Qing, Pan Wenshi
30			NGK	2000.06–2002.06	Nongguankou	A♂1, A♀3 SA♂2, SA♀2 J♂3, J♀2 I♀2	Zhang Yingyi, Pan Wenshi
31			NA-X	2000.07–2002.06	Nongan	A♂1, A♀2 J♂3, J♀1 I♂2	Ran Wenzhong, Pan Wenshi
32			NL	2001.02–2001.03	Nonglao	A♂1, A♀6 J♂5, I♂3 I♀2	Ran Wenzhong, Pan Wenshi
33			NBJ	2001.02–2001.03	Nongbajiao	A♂1, A♀5 J♂1 I♂1, I♀1	Ran Wenzhong, Pan Wenshi
34			FSS	2001.06–(still existed by 2015.12)	Fangshaoshan	A♂1: W A♀3: Dagege, Ergege, Sangege I♂2	Pan Wenshi

(continued)

Table 8.4 (continued)

No.	System	Social structure	Troop designation	Observation period	Site	Individuals in the troop	Observation and registration by
35			FNS II	2002.05–2006.03	Fangniushan	A♂1: Waiwai A♀12 Immature offspring 21	Yin Lijie, Qin Dagong, Pan Wenshi
36			FJC III A	2003.02–2006.09	Feijichang	A♂1: α-Gonghou A♀7: Dahei, Archen, Jiajia, Xiaoxiao, Yingying, Pingping, Lanlan Immature offspring ♂7 ♀6: Yixi, Feilu, Lanqi, Xinyu, Tiantian, Baobao, Beibei, Arxi, Arlu, Aryu, Arqi, Artian, Arbao	Jin Tong, Wang Dezhi, Pan Wenshi
37			XS	2003.07–2006.06	Feijichang, Xishan	A♂1: β-Gonghou A♀4: Lingling, Bibi, Beike, Xiao-W Immature offspring: Yingfu, Arfu, Pingan, Weiming, Qingsi, Arsi, Shenshen	Jin Tong, Wang Dezhi, Pan Wenshi
38			FJC-Y I	2007.11–2012.07	Feijichang	A♂1: Yintangxiaotu A♀13: Dahei, Archen, Jiajia, Xiaoxiao, Yingying, Pingping, Lanlan, Lanqi, Tiantian, Arxi, Arlu, Aryu, Beibei J♂3: Wuque, Zuoque, Youque J♀6: Qiqi, Mimi, Yumei, Zhenzhu, Da-S, Bandazai E♂9: Mengli, Yiling, Gualai, Hanlin, Xiaoliu, Chongsheng, Xiaohei, Shishi, Yiyi E♀5: Wenwen, Xiaojiu, Qingming, Gaga, Chuliu I♂4: Xiaolu, Xiaobei, Xiaozhen, 007 I♀3: Yutu, Xiaolan, Xiaoying	Gu Tieliu, Feng Chunguang, Lin Chen, Zhang Lan, Li Xinyang, Pan Yue, Zhao Yi, Meng Hao, Yang Yanping, Huang Meiyang, Luo Zuoye, Wei Meijiao, Zheng Xinma, Liang Zuhong, Pan Wenshi
39			FJC-SS	2013.03.11–2013.07.14	Feijichang Big Cave	A♂2: Shengge, Shengdi A♀11: Archen, Dahei, Jiajia, Xiaoxiao, Pingping, Tiantian, Aryu, Arxi, Arlu, Beibei, Yumei SA♀4: Zhenzhu, Qiqi, Mimi, Da-S J♀2: Wenwen, Xiaojiu I♂1: Xiao-M (was killed on 2013.03.31) I♀1: Xiao-P (was killed on 2013.06.13)	Gu Tieliu, Feng Chunguang, Li Xinyang, Pan Yue, Zhao Yi, Meng Hao, Yang Yanping, Huang Meiyang, Luo Zuoye, Wei Meijiao, Qin Jiajia, Zheng Xinma, Liang Zuhong, Pan Wenshi
40			FJC-W	2013.12.06–(still existed by 2015.12)	Feijichang	A♂1: Yuweng A♀16: Archen, Dahei, Jiajia, Xiaoxiao, Pingping, Tiantian,	Gu Tieliu, Feng Chunguang, Li Xinyang, Pan Yue, Zhao Yi, Yang

(continued)

Table 8.4 (continued)

No.	System	Social structure	Troop designation	Observation period	Site	Individuals in the troop	Observation and registration by
						Aryu, Arxi, Arlu, Beibei, Yumei, Mimi, Zhenzhu, Wenwen, Da-S, Xiaojiu I♂8: Xiaoxuan, Huanghunhou, Mizai, Beckham, Arweng, Xiaoyuer, Xiao-S, Nannan I♀7: Nuhanzi, Xiaosanba, Xiaowen, Tianniu, Zhenni, Zhuli, Ningning	Yanping, Huang Meiyang, Luo Zuoye, Wei Meijiao, Zheng Xinma, Qin Jiajia, Liang Zuhong, Pan Wenshi
41			FJC-X	2015.11.18–(still existed by 2015.12)	Feijichang Small Cave	A♂1: Laoxie A♀3: Yingying, Lanlan, Lanqi, Panma SA♀3: Qingming, Gaga, Wailaimei E♀2: Xiaoying, Xiaolan I 2: Chaosheng, Yousheng	Gu Tieliu, Feng Chunguang, Li Xinyang, Pan Yue, Zhao Yi, Yang Yanping, Huang Meiyang, Luo Zuoye, Wei Meijiao, Zheng Xinma, Liang Zuhong, Pan Wenshi
42	Transitory troop	1 adult ♂ + *n*-numbers adult ♀ + *n*-numbers premature daughters + *n*-numbers sons	FSS-QQ	2000.06–2000.07	Fangshaoshan	A♂1: Queque (QQ) SA♂1: Daarge SA♀1: Dagege J♂2: Erarge, Sanarge J♀2: Ergege, Sangege	Pan Wenshi, Zhang Yingyi
43			FNS	2002.02–2002.05	Fangniushan	A♂1: Tutu (TT) A♀4: male1, male2, male3, male4 Immature offspring: ♂12 ♀1	Yin Lijie, Qin Dagong, Pan Wenshi
44			FJC-Ach	2002.07–2002.09	Feijichang Xishan	A♂1: Archeng (Ach) A♀3: Lingling, Beibei, Beike E♂1: Feifei I♂4: Xinxin, Arqing, Arlan, Nianping I♀2: Aryi, Arfei	Jin Tong, Wang Dezhi, Pan Wenshi
45			FNS II	2006.04–2006.09	Fangniushan	A♂1: Waiwai A♀9, SA♂4, SA♀1, I♂5, I♀4	Yin Lijie, Qin Dagong, Pan Wenshi
46			FJC-Y II	2012.07–2015.03	Feijichang Small Cave	A♂1: Yintangxiaotu A♀3: Yingying, Lanlan, Lanqi J♂3: Wuque, Zuoque, Youque E♂9: Mengli, Yiling, Gualai, Hanlin, Xiaoliu, Chongsheng, Xiaohei, Shishi, Yiyi E♀3: Qingming, Gaga, Wailaimei I♂3: 007, Wangzai, Liushaotou I♀3: Xiaolan, Xiaoying, Qinu	Gu Tieliu, Feng Chunguang, Li Xinyang, Pan Yue, Zhao Yi, Yang Yanping, Huang Meiyang, Luo Zuoye, Wei Meijiao, Zheng Xinma, Qin Jiajia, Liang Zuhong, Pan Wenshi

(continued)

Table 8.4 (continued)

No.	System	Social structure	Troop designation	Observation period	Site	Individuals in the troop	Observation and registration by
47		2 adult ♂ + *n*-numbers adult ♀ + *n*-numbers premature daughters + *n*-numbers sons	GY	1998.04–2000.05	Guoyuan	A♂2:Queque, Liaoya A♀1: Weiming (died from miscarriage) SA♂1: Daarge J♂2: Erarge, San-a-ge J♀3: Dagege, Ergege, Sangege	Zhang Yingyi, Zhu Lizhong, Pan Wenshi
48			BYS-QQ	2000.05–2000.06	Bokyueshan	A♂2:Queque, Liaoya SA♂1: Daarge J♂2: Erarge, Sanarge J♀3: Dagege, Ergege, Sangege	Zhang Yingyi, Zhu Lizhong, Pan Wenshi
49		1 adult ♂ + *n*-numbers premature offspring (Male + premature langurs troop)	TXD-X	2000.11–2002.06	Taoxindong	A♂1, SA♂2, J♂2, J♀1	Ran Wenzhong, Pan Wenshi
50			FNS-X	2013.09–2015.05	Fangniushan	A♂1, SA♂3, J♂3, J♀1	Gu Tieliu, Feng Chunguang, Li Xinyang, Yang Yanping, Huang Meiyang, Liang Zuhong, Pan Wenshi
51			BYS-II	2014.08–(still existed by 2015.12)	Bokyueshan	A♂1, A♀1, SA♂5, J♂5, J♀1, I♂1, I♀1	Feng Chunguang, Pan Yue, Zhao Yi, Yang Yanping, Luo Zuoye, Wei Meijiao, Liang Zuhong, Pan Wenshi
52	All-male troop	1 adult ♂ + *n*-numbers premature♂	TXD	1996.11–1997.02	Taoxindong	A♂1, SA♂5	Pan Wenshi
53			NA	1996.11–1997.02	Nongan	A♂1, SA♂4	Pan Wenshi
54			NBG 2	1996.11–1997.02	Nongbaga 2	A♂1, SA♂7	Pan Wenshi
55			FNS-TT	2002.06–2005.05	Fangniushan	A♂1: Tutu (TT) SA♂ + J♂ + I♂ = 12	Yin Lijie, Qin Dagong, Pan Wenshi
56			FNS-WW	2007.06–(still existed by 2015.12)	Fangniushan	A♂1: Waiwai (WW) SA♂ + J♂ + I♂ = 8	Yin Lijie, Qin Dagong, Pan Wenshi
57			FHS	2006.07–2008.10	Fanghoushan	A♂1: α-Gonghou J♂2: Yixi, Baobao I♂2: Arbao, Arqi	Pan Wenshi, Liang Zuhong, Liu Lijun
58			BYS-I	2013.10–2014.05	Bokyueshan	A♂1 SA♂6 ~ 7	Gu Tieliu, Feng Chunguang, Li Xinyang
59			FJC-NL	2015.04–(still existed by 2015.12)	Feijichang—Research Base—Nonglao	A♂1: Yintangxiaotu SA♂7: Mengli, Yiling, Gualai, Hanlin, Xiaoliu, Chongsheng Xiaohei J♂3: Shishi, Yiyi, 007 I♂1: Wangzai	Gu Tieliu, Yang Yanping, Lu Haibin, Pan Wenshi
60		*n*-numbers premature♂	NX	2000.02–2000.03	Nongxiang	SA♂2	Ran Wenzhong, Pan Wenshi
61			BYS	2000.07–2001.11	Bokyueshan, Fanghoushan	SA♂1: Daarge J♂2: Erarge, Sanarge	Zhu Lizhong, Pan Wenshi
62			SHY-II	2001.02–2001.03	Shihuiyao	SA♂3	Ran Wenzhong, Pan Wenshi
63			GY-MK	2012.04.17–(still existed by 2015.12)	Gongyuan	SA♂2 (one of which was knocked down by vehicle on the highway outside the park and died)	Pan Wenshi, Liang Zuhong, Lu Jintong

Notes: ♂ male, ♀ female, *A* adult langur, *SA* sub-adult langur, *J* adolescent langur, *E* juvenile langur, *I* pup. Numeral denotes the number of langurs in that category

Fig. 8.6 On the evening of May 24, 2012, "Yintangxiaotu" and its 41 wives and pups returned to the night shelter at FJC Cave. The family had a quiet rest, only the dutiful alpha-male was still sitting in the "nose" bush above the Big Cave to keep watch (Photo by Liang Zuhong)

Fig. 8.7 In July 2012, Yintangxiaotu relinquished FJC Big Cave, took its three wives, four daughters and 13 sons, and withdrew to a small cave on the precipice at FJC east, starting the "Yintangxiaotu transitory troop" (Photo by Liang Zuhong)

- Ante-family troop: Made up by a newly incoming male together with the adult females and immature females from the former family. No pups.
- Family troop: A stable troop consisting of one resident alpha male, adult females, pups born of the alpha male, and sub-adult females (Fig. 8.6).
- Transitory troop: Troop split from the former family troop, consisting of the defeated male, some adult females, some immature daughters and all the sons, all from the former family troop (Fig. 8.7).
- All-male troop: Consists of the defeated old male (or old male relinquished the family troop) and its sons.

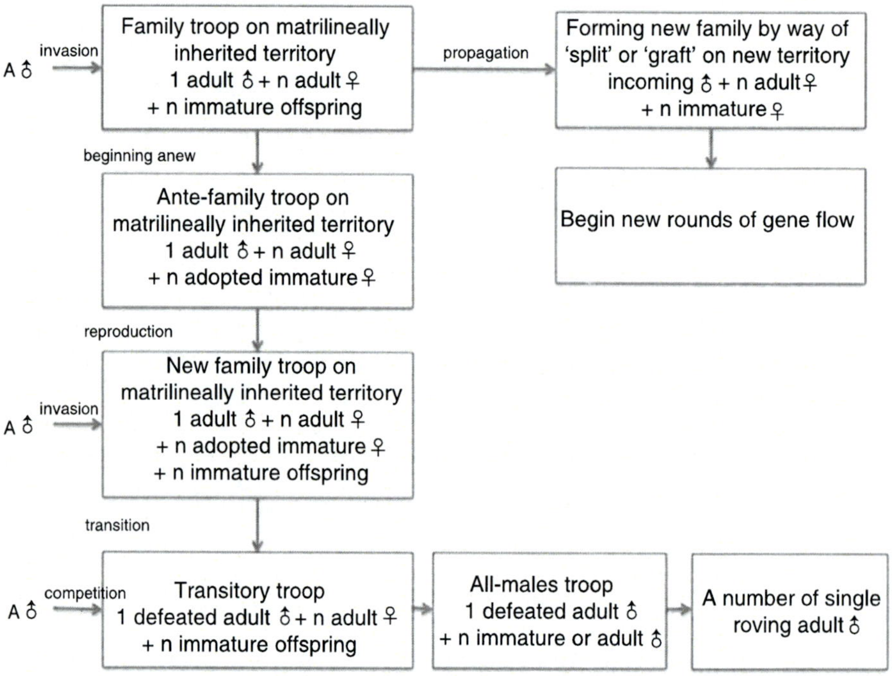

Fig. 8.8 Gene flows in white-headed langur societies

Fig. 8.9 On March 25, 2015, "Yintangxiaotu" and its sons sat on a kapok tree with red flowers in full bloom, and left their original birthplace in FJC. The troop carrying the genes of its own ethnic group will flow to every corner of the Nongguan Mountain area (Photo by Yang Yanping)

Alternatively, the troop may consist of only brothers. Troops of this structure have no permanent night shelters.

When an outlier male langur successfully fights into and takes over a family, it introduces heterogeneous genes into that matrilineal family. Such gene flow gets replaced every 4 years and with each replacement the white-headed langur family would be refreshed (Fig. 8.8, Fig. 8.9).

8.5 Reproductive Behaviors of White-Headed Langurs

Different sexes of white-headed langurs have different reproductive tactics. An adult male white-headed langur must compete fiercely with other males to possess females while females would decide on whether to stay behind to reproduce with the new incoming male. In the following paragraph, we shall give an account of the conjugal relationships in the eight family and transitory troops that successively inhabited FJC Big Cave, as well as study the causes of different tactics in engaging reproductive mates by the different sexes.

8.5.1 The White-Headed Langur Is Essentially Polygamous Through Its Lifetime

We have since 1996 observed a total of eight white-headed langur families in FJC Big Cave and have kept tracking the life histories of the individual reproductive males and females (Table 8.5). We found that a male white-headed langur has only one period of reproduction opportunity (4 years) in its entire life. During that time, it will mate with every adult female in the family, making the relationship polygynous. On the other hand, every female in a family troop will through its lifetime mate with and bear the babies of successive incoming males. In that sense, the relationship is polyandry. So while adult white-headed langurs appear to be polygynous (in every reproduction cycle), in their lifetimes, they are polygamous.

We suppose there are three factors underlying the conjugal relationships of white-headed langurs:

First, it is related to the habitat environment. Places that allow for shelter at night and are suitable for raising young ones are scantly available on karst hilly territory. And, in the early days (through Pleistocene to Holocene, until 50 years ago), there were numerous predacious animals foraging in karst hilly territories. To avoid being preyed upon, female

Table 8.5 Registers on the conjugal relationship of individual ♂ and ♀ in the seven white-headed langur family troops and three transitory troops in FJC

Sequence	Troop structure	Name of ♂	Duration	Number and name of conjugal mates	Remark
1	Queque family troop	Queque	1994.10–1998.3	A♀8: Heimama, Dahei, Jiajia, Archen, Lingling, Bibi, Beike, Weiming.	Heimama died from poaching. There were totally 9 pups born to the family
2	Queque transitory troop	Queque, Liaoya	1998.3–2000.7	A♀1: Weiming	8 immature langurs were stay-behinds from the last family 2000.6: Weiming died from miscarriage 2000.7: Queque died in fighting in Fangshaoshan Liaoya disappeared
3	Archeng family troop	Archeng	1998.4–2002.7	A♀7: Dahei, Jiajia, Archen, Lingling, Bibi, Beike, Xiaoxiao	12 pups were born to the family
4	Archeng transitory troop	Archeng	2002.7–2002.10	A♀3: Lingling, Bibi, Beike	11 immature langurs were stay-behinds from the last family
5	α-Gonghou family troop	α-Gonghou	2002.8–2006.6	A♀7: Dahei, Jiajia, Archen, Xiaoxiao, Yingying, Lanlan, Pingping	13 pups were born to this family
6	*β*-Gonghou family troop	*β*-Gonghou	2002.7–2006.6	A♀4: Lingling, Bibi, Beike, Xiao-W	7 pups were born to this family
7	Yintangxiaotu family troop	Yintangxiaotu	2006.6–2012.7	A♀13: Dahei, Jiajia, Archen, Xiaoxiao, Yingying, Lanlan, Pingping, Lanqi, Tiantian, Beibei, Arxi, Arlu, Aryu	1 outlier female langur came and joined the family 29 pups were born to this family
8	Yintangxiaotu transitory troop	Yintangxiaotu	2012.8–2015.3.20	A♀3: Yingying, Lanlan, Lanqi SA♀1: Wailaimei	There were adult and immature (sons) altogether 14 ♂ in the family. 3 pups were born to this transitory troop
9	Yuweng family troop	Yuweng	2013.7–(still existed by 2015.12)	A♀16: Dahei, Jiajia, Archen, Xiaoxiao, Pingping, Tiantian, Beibei, Arxi, Arlu, Aryu, Yumei, Zhenzhu, Mimi, Wenwen, Da-S, Xiaojiu	Until 2015.12, there were 15 pups born to this family
10	Laoxie family troop	Laoxie	2015.3.20–(still existed by 2015.12)	A♀4: Yingying, Lanlan, Lanqi, Panma SA♀2: Wailaimei, Qingming EA♀2: Gaga, Xiaoying, Xiaolan	Until 2015.12, there were 2 pups born to this family

white-headed langurs had to stay together in the limitedly available safe places at night. This trait has since instilled "gene memories" in white-headed langurs through evolution, even when carnivorous predators have almost disappeared from the territory. Every night when the sky darkens, they will climb up the steeply inclined precipice on karst hills and rest there for the night.

Second, there is the food-related factor. There are northern tropical forests scattered all around the disjointed karst hills, and they are thriving in all seasons. The profuse vegetation there, the leaves, shoot sprouts, flowers, and fruits provide white-headed langurs with endless food resources. Hence, the individual female langurs are able to live together in relatively secluded small areas for safety, which allows an incoming male white-headed langur to claim multiple females for its own.

The third factor is the most important of all. That is the genetic pressure, which is innate in white-headed langurs. Every adult male langur in its 4 years being the resident alpha male of a family would beget 3–4 litters of offspring. When its daughters and sons of the first litter reach sub-adult age (about 4–5 years of age), they would become restive and enter into reproductive age very soon. The sexual stress between the resident male and its daughter and that between the sons and its mother would then become very discomfiting. To avoid inbreeding, the family would split up at this time. The old father langur and its sons have to leave the family, and another outlier male would come to take the place of the resident alpha male, bringing in new genes and starting a new reproduction cycle. Thus, every white-headed langur, regardless of gender, is in essence polygamous in its lifetime.

8.5.2 Is There Sexual Dimorphism in White-Headed Langurs?

Sexual dimorphism in animals is basically a result of natural selection, an inevitable outcome of multiple males competing for mating. Sexual dimorphic traits nevertheless are considered secondary characteristics of natural selection. "Sexual selection" drives males and females to develop distinctly different outward appearances, and the sexual dimorphic feature usually increases the competitiveness and sexual charm of the male. While some dimorphic features, such as the beautiful tail feathers of a peacock and the antler of a stag, sometimes would reduce the survival chance of the male, they nonetheless increase the male's mating chances.

Sexual dimorphism occurs in many animals. Males of carnivorous species usually have bigger-built bodies. The adult male polar bear weighs 36% heavier than the female, male black bears weigh 29–41% heavier than females, and male giant pandas weigh 18% heavier than females. Many species in primate order have bigger-built males than females as well. Some examples are gorillas, baboons, and macaques. They all display sexual dimorphism characteristics (Table 8.6).

We found no essential differences in the outward appearance between males and females from the pup stage to the adolescent stage, as the physical data of 3-year-old male and female langurs in Table 8.7 illustrate. In adulthood, males weigh about 15% more than females, as illustrated in Table 8.8. Nevertheless, no significant sexual dimorphic features could be observed in white-headed langurs in field observations, as illustrated in Fig. 8.10.

However, we noticed that the size of the canine teeth on male and female white-headed langurs would gradually come to be different with the growing age. The shape of canine teeth on a 1-year-old male white-headed langur appears to be no different from that of a same age female. At the juvenile age of around 2 years, there is still very little difference between males and females on the shape and size of their canine teeth. But when male langurs reach adolescence at 4 years old, the canine teeth would grow rapidly, reaching 2 cm in length and would continue to grow to 2.5 cm or more in adulthood, whereas canine teeth in females of corresponding ages show no apparent growth in size (Fig. 8.11).

Additionally, the triangular white skin patch on the groin of adult female langurs would turn moist and red during ovulation, male langur would usually sit solitarily high up on a tree with its legs spread out displaying its genitals, and different sexes of white-headed langurs generally rest at different positions in the night shelter. Hence, despite that the sexual dimorphism is not distinct between male and female white-headed langurs in respect to body weight, body length, and hair color, it is still apparent on the sizes of their canine teeth, the skin color on their groins, and in particular the behaviors.

8.5.3 Kinship of Females in a White-Headed Langur Family

Through 20 years of observations, we found that the females in a white-headed langur family are essentially a closely blood-related troop. Most female white-headed langurs stay in their birthplace, while male langurs must fight to gain control of these reproduction resources. The females would in every 4 years make a new reproductive family with an incoming outlier male langur who fought and succeeded in becoming the resident alpha, while the other adult male langurs roving around would always attempt as much as they could to invade the family and possess the females. For that reason, in white-headed langur societies, the male langurs often fight each other to contest for possessing a

Table 8.6 Sexual dimorphism characteristics in 17 primate species (MacDonald 2006)

Species	Body length (cm)	Tail length (cm)	Average body weight (kg) ♂	Average body weight (kg) ♀	♂ Heavier than ♀ by (%)	Distinctness of sexual dimorphism
Golden snub-nosed Monkey	47–78	48–88	16.5	9.9	40	Very distinct
Black snub-nosed monkey	51–83	49–72	15	9	40	Very distinct
Gray snub-nosed monkey	64–73	85–97	14.6	7.8	46.6	Very distinct
Black langur	47–69	77–90	7.6	5.7	25	Moderately distinct
Silver langur	43–59	63–84	5.5	4.9	10.9	Moderately distinct
White-headed langur	40–45	85–95	10.1	8.5	15.8	Moderately distinct
Gray langur	42–60	55–84	6.5	–	–	–
Nepal gray langur	41–78	61–108	14.5	–	–	–
Macaque	53–68	19–38	11.3	6.9	38.9	Very distinct
Japanese macaque	47–64	19–30	7.7	5.4	29.9	Distinct
Baboon	56–79	42–60	21	11.6	44.8	Very distinct
Black crested gibbon	45–64	–	5.7	–	–	–
White-handed gibbon	45–64	–	5.5	5.7	−3.6	Non-evident
Baboon	♂ 73–83 ♀ 70–76	–	39	30	20.5	Moderately distinct
Common chimpanzee	♂ 77–92 ♀ 70–85	–	40	30	25	Moderately distinct
Lowland gorilla	♂ 170–180 ♀ 150	–	160	90	43.8	Very distinct
Orangutan	♂ head and body 97: height 137 ♀ head and body 78: height 115		75	45	40	Very distinct

Table 8.7 Survey findings of weight and measurement parameters respective of 3-year-old male and female white-headed langurs

Parameters	Sex Male	Sex Female	Remark
Head length (cm)	5	5	The female had suffered body wounds for 5–7 days and weight could have been reduced by body consumption
Body length (cm)	37	36	
Tail length (cm)	64	59	
Weight (kg)	3	2.5	

Table 8.8 Survey findings of weight and other parameters respective of male and female adult white-headed langurs

Parameters	Sex Male	Sex Female	Remark
Head length (cm)	–	–	The female was died from poaching Male figures are averages of findings on two male langurs
Body length (cm)	44.5	43	
Tail length (cm)	90	87	
Weight (kg)	10.1	8.5	

family of females. While sometimes fights would be formalistic, sometimes they could also be ferocious. Whenever a male langur successfully invades and takes over a family, it will undertake infanticide (killing the suckling pups of the previous male langur) and take possession of the immature females (the immature daughters of the previous male langur).

Fig. 8.10 Comparative appearances and body builds of female (Archen, *above*) and male (Yintangxiaotu, *below*) white-headed langur (Photo by Liang Zuhong). (In field observations, the sex of white-headed langur could be identified by whether there is a triangular white patch on the groin, but its sexual dimorphic features are difficult to be discerned from a distance)

The FJC Big Cave troop, like every other family troop, is female-centered. By making reference to the ages of the pups in the troop, we estimated that the troop was started in 1994. Since November 1996, we had recorded reproductions by Jiajia, one of the female founders of the FJC Big Cave troop, and reproductions by her female progenies, which stayed behind in the "Big Cave troop" and mated with successive incoming resident alpha male langurs. When every sex-matured female langur in the FJC Big Cave troop has mated with the resident alpha male langur and bore its babies, the family becomes one big closely blood-related family centered on females.

My observations in November 1996 found that the troop led by resident alpha male Queque was a polygynous family troop with offspring. In 1998, resident alpha male Queque was replaced by the incoming outlier male langur Archeng and the latter mated with the female langurs in the family and reproduced offspring until 2002. In next 4 years, another outlier male α-Gonghou fought into FJC Big Cave and replaced Archeng to be the resident alpha male. At this time, the family split. Three adult females followed Archeng and began a new troop at FJC Xishan, while three other adult females stayed in FJC Big Cave and became seniors in the FJC Big Cave family troop, bearing babies for each successive incoming male langur until today. In 2006, outlier male langur Yintangxiaotu replaced α-Gonghou, becoming the resident alpha male of the FJC Big Cave family troop and reproduced 29 offspring. In 2013, outlier male langur Yuweng won in the fight for reproduction rights and successfully took over the family troop. By this time, the family was a big troop composing of five generations of females. We separate the offspring respectively born by the three senior female langurs into the "Dahei lineage," "Archen lineage," and "Jiajia lineage." In the following, we will take the blood ties in the "Jiajia lineage" (Figs. 8.12 and 8.13) to exemplify the close kinship in the family.

From the fact that we have observed in the wild, most of the female white-headed langurs live in their home area inherited from their ancestors. However, males must participate in the competition for the mating rights and migrate. When an old and seemingly "defeated" male langur takes all his sons and leaves the family to enter a broader space to start a new life, his wives and daughters would stay in the same old place to welcome the new invading males and start a new round of reproductive cycle. It is this behavior mode that ensures the restoration and continuity of the population of white-headed langurs in Nonguan Mountains. This is their life, and this is their ethics.

(a) (b) (c) (d)

Fig. 8.11 Comparing the canine teeth of male and female white-headed langurs (Photo by Liang Zuhong). (**a**) 007 (♂) at 1 year; (**b**) Xiaolan (♀) at 1 year; (**c**) Adult "Youque" (♂); (**d**) Adult "Fanma" (♀)

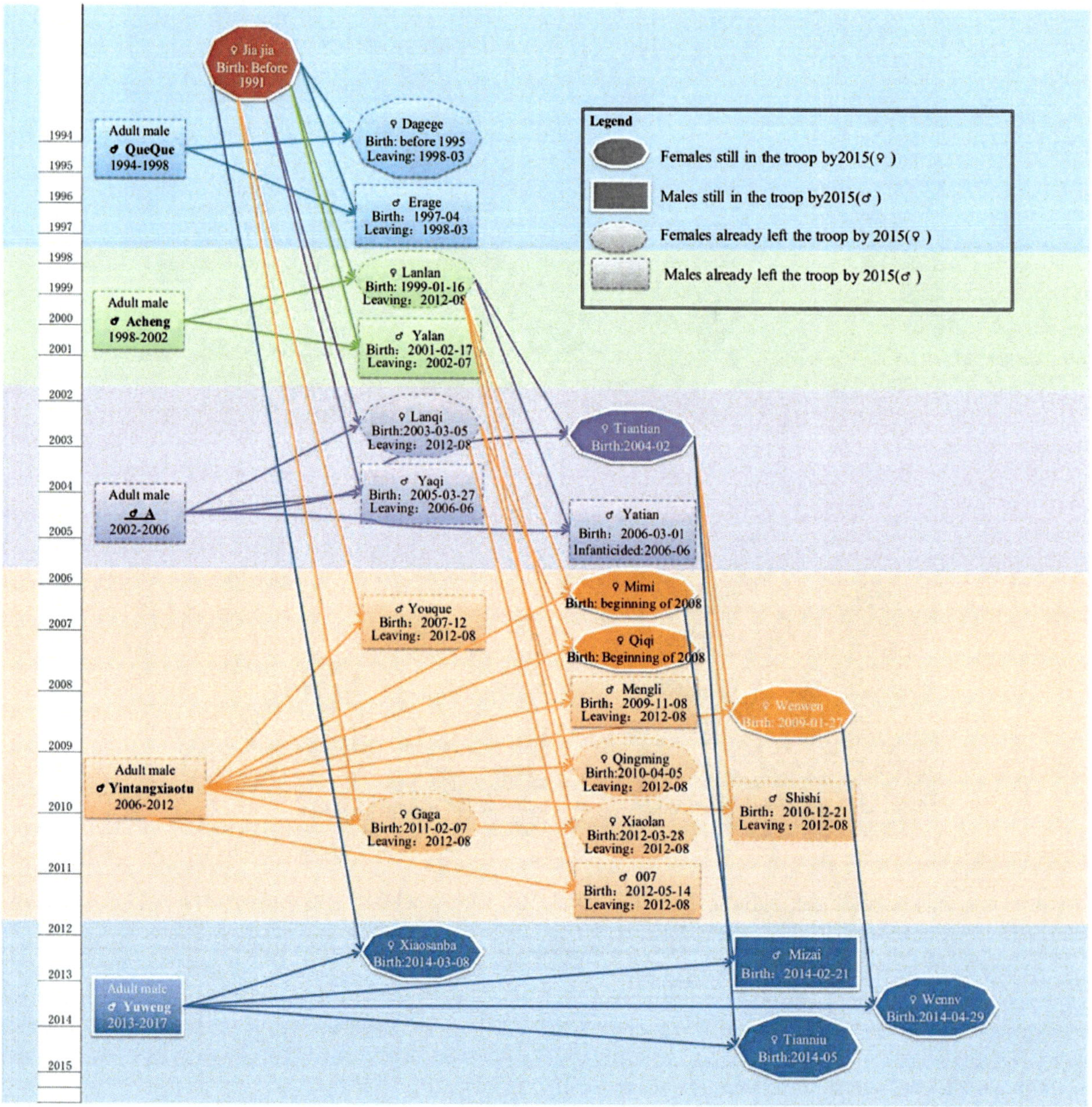

Fig. 8.12 Jiajia lineage chart (1994–2015) (Drawn by Pan Yue). Showing the reproductions of Jiajia, one of the founders of the "FJC Big Cave troop" with successive resident alpha males starting from November 1996 as registered from observation, as well as reproductions by those Jiajia female progenies, which stayed behind through successive resident alpha males periods

Fig. 8.13 Some of those in Jiajia lineage (Photo by Liang Zuhong). (**a**) Jiajia (about 25 years old by December 2015). (**b**) Jiajia and Gaga (daughter of Jiajia and Yintangxiaotu). (**c**) Lanqi (daughter of Jiajia and α-Gonghou) and 007 (son of Lanqi and Yintangxiaotu, grandson of Jiajia). (**d**) Youque (son of Jiajia and Yintangxiaotu). (**e**) Lanlan (daughter of Jiajia and Archeng) and Xiaolan (daughter of Lanlan and Yintangxiaotu, granddaughter of Jiajia). (**f**) Wenwen (daughter of Tiantian and Yintangxiaotu, great-granddaughter of Jiajia) and Shishi (son of Tiantian and Yintangxiaotu, great-grandson of Jiajia)

References

Lei Z. A preliminary research on the white-headed langur population at Banli township in Chungzuo County of Guangxi Province, and its Social System (广西省崇左县板利乡白头叶猴种群及社会结构的初步研究). Master's thesis. Beijing: Peking University; 1998.

Lijie Y. Male white-headed langur (*Trachypithecus leucocephalus*) reproduction tactics – a research on infanticide and paternal fostering (雄性白头叶猴的繁殖策略 – 对杀婴行为和雄性照料的研究). Doctoral thesis. Beijing: Peking University; 2008.

Lizhong Z. The home range utilization of an all-males wild white-headed langur troop and the relationships between individuals in the troop and relationship between troops (野生白头叶猴全雄群的巢域利用以及群内群间关系). Master's thesis. Beijing: Peking University; 2002.

MacDonald DW. The encyclopedia of mammals (new edition). New York, NY: Facts on File; 2006.

Qing Z. White-headed langur (*Trachypithecus leucocephalus*) mother-pup relationship and behavioral development in pups (白头叶猴的母幼关系和幼仔的行为发育). Doctoral thesis. Beijing: Peking University; 2005.

Tong J. Wild female white-headed langur (*Trachypithecus leucocephalus*) reproduction tactics – its social relationships, mating and reproduction, and surrogate behavior (野生白头叶猴的雌性繁殖策略 – 社会关系、交配繁殖与拟母亲行为). Doctoral thesis. Beijing: Peking University; 2008.

Wang D. A composing unit in wild white-headed langur (*Trachypithecus leucocephalus*) population – the research and tracking of a white-headed langur family troop in Nongguan Mountains, Guangxi (野生白头叶猴社群的结构单元 – 对广西崇左弄官山区一个白头叶猴家庭的跟踪研究). Doctoral thesis. Beijing: Peking University; 2004.

Wenzhong R. The present state of wild white-headed langur (*Trachypithecus francoisi leucocephalus*): an analysis on its distributions, populations numbers, habitat use and population viability analysis (野生白头叶猴的现状:分布、数量、栖息地利用和种群生存力分析). Doctoral thesis. Beijing: Peking University; 2003.

Yingyi Z. Researches on the social system of wild white-headed langur: its composition, the social relationship within a troop and between troops, and changes in its population (野生白头叶猴的社会组织:有关其社会组成、群内群间社会关系及社群变化的研究). Doctoral thesis. Beijing: Peking University; 2002.

9 The Behavior Patterns of a White-Headed Langur Family

Abstract

In 2012, we worked in the bush for 357 days and recorded the all-day activities of a white-headed langur family with high-definition video cameras, in both "Sampling All Occurrences of Some Behaviors" method and "Focal-Animal" method. Activities of white-headed langurs follow a constant 24 h diurnal cyclical pattern. In this chapter, we present a detailed summary of their diurnal activities. White-headed langurs periodically adjust the timing of their activities and rests according to seasonal cycles. Their inherent circadian clock keeps biological activities synchronized with the seasonal photoperiod changes in the Northern Hemisphere. We also researched on the "roles" of individuals in a family troop with respect to different sexes and ages and found that the duty of defending the family rested primarily on the resident alpha male, while sub-adult male and adolescent male served as assistance to their father. All females acted in roles of reproducing or nursing pups. Yellow-haired pups and infants spent a lot of time playing with family members.

Keywords

All-day activities · Circadian clock · Social roles

In September 2006, when Yintangxiaotu was at a robust age (about 7 years old), it invaded the "FJC Big Cave troop" and became the resident alpha male of the family. That gave us a chance to look into the white-headed langur's way of living. Our advanced equipment enables us to conduct research more meticulously. From 2006 to 2016, we had the luck to record everything we observed without interruption.

9.1 A Brief History of the "Yintangxiaotu Family Troop"

As it had happened, the living sphere of Yintangxiaotu and its family through the 10 years had covered large parts of the Nongguan Mountains. The boundary starts at Feijichang, goes south through Nongbaga to Nonglao, then turns westwards to Nongsha and hinterland. Before we give an account of their way of life, we have to look back briefly into the founding of the family troop, as well as its developments over the years.

9.1.1 Yintangxiaotu's Invasion into the "FJC Big Cave Troop"

In September 2006, adult male langur Yintangxiaotu invaded the "FJC Big Cave troop," replaced α-Gonghou as the resident alpha male langur, and started a new family cycle with the seven adult female langurs and their six immature daughters that stayed behind, initiating the "Yintangxiaotu ante-family troop" (Table 9.1).

9.1.2 Founding of the "Yintangxiaotu Family Troop"

2007.11–2008.3, three sons and five daughters were born to the Yintangxiaotu family, turning the troop into the "Yintangxiaotu family troop" with 22 family members.

2009.1.27, another daughter was born to the family, increasing the number of family members to 23.

2009.11–2010.5, a further seven sons and three daughters were born to the family, and the number of family members

W. Pan, *The Population Ecology of White-Headed Langur*, https://doi.org/10.1007/978-981-33-4118-0_9

Table 9.1 Family members in "Yintangxiaotu ante-family troop" initiated by Yintangxiaotu invading over the "FJC Big Cave troop"

Period	Night shelter	Adult ♂	Adult ♀	Offspring
2006.09–2007.10	Big cave, small cave	Yintangxiaotu	Adult ♀: Arzhen, Dahei, Jiajia, Xiaoxiao, Pingping, Yingying, Lanlan Immature ♀: Lanqi, Tiantian, Arxi, Arlu, Aryu, Beibei	–

Table 9.2 Family members in "Yintangxiaotu family troop"

Period	Night shelter	Adult ♂	Adult ♀	Offspring	Adolescent ♀	Remark
2007.11–2012.7	Big cave, small cave	Yintangxiaotu	Adult ♀: Dahei, Jiajia, Xiaoxiao, Pingping, Yingying, Lanlan, Lanqi, Tiantian, Arxi, Arlu, Aryu, Beibei	♂16: Wuque, Zuoque, Youque, Mengli, Yiling, Gualai, Hanlin, Xiaoliu, Chongsheng, Xiaohei, Shishi, Yiyi, Xiaolu, Xiaobei, Xiaozhen, 007 ♀14: Qiqi, Mimi, Yumei, Zhenzhu, Bandazai, Wenwen, Xiaojiu, Qingming, Gaga, Chuliu, Yutu, Xiaolan, Xiaoying	Da-S	Daughter Bandazai died of sickness Da-S was an outlier came joined the family

Table 9.3 Family members in the short-lived "Duanwei and Romeo pair ante-family troop" in FJC

Period	Night shelter	Adult ♂	Adult ♀	Offspring	Remark
2012.07–2012.12.23	Big cave	Duanwei, Romeo	Adult ♀: Arzhen, Dahei, Jiajia, Xiaoxiao, Pingping, Yingying, Lanlan, Lanqi, Tiantian, Arxi, Arlu, Aryu, Beibei Premature ♀: Qiqi, Mimi, Yumei, Zhenzhu, Da-S, Wenwen, Xiaojiu	–	Yutu, Xiaolu, Xiaobei and Xiaozhen were killed in infanticide

drastically increased to 32 (one daughter born in 2008 died from sickness).

2010.12–2011.2, two more sons and two more daughters were born to the family, increasing the number of family members to 36.

2012.1–2012.5, four sons and three daughters were born to the family, enlarging the family to 43 members, making it the biggest family troop ever recorded in the Nongguan Mountains (Table 9.2).

9.1.3 Invasion by Outlier Adult Male White-Headed Langurs

Between 2012.7 and 2012.12, a pair of outlier male langurs Duanwei and Romeo invaded the FJC Big Cave and formed an ante-family troop with the ten adult females and their seven immature daughters from the Yintangxiaotu family. It had not reproduced any offspring but killed four infant pups, three sons, and one daughter of Yintangxiaotu (Table 9.3).

9.1.4 Founding of the "Yintangxiaotu Transitory Troop"

During 2012.07–2015.03, Yintangxiaotu, taking along three adult females, four daughters, and 13 sons, inhabited in FJC Small Cave and became the "Yintangxiaotu transitory troop." In this period, two sons and one daughter were born to the troop (Table 9.4).

Table 9.4 Family members in the "Yintangxiaotu transitory troop"

Period	Night shelter	Adult ♂	Adult ♀	Offspring	Remark
2012.7–2015.3	Small cave	Yintangxiaotu	Adult ♀: Yingying, Lanlan, Lanqi	♂15: Wuque, Zuoque, Youque, Mengli, Yiling, Gualai, Hanlin, Xiaoliu, Chongsheng, Xiaohei, Shishi, Yiyi, 007, Wangzai, Liushaotou ♀5: Qingming, Gaga, Xiaolan, Xiaoying, Qinu	Daughter Xiaoliu disappeared upon family split

Table 9.5 Members in "Yintangxiaotu all-male troop"

Period	Night shelter	Adult ♂	Adult ♀	Offspring	Remark
2015.3	Research base	Yintangxiaotu	–	♂14: Wuque, Zuoque, Youque, Mengli, Yiling, Gualai, Hanlin, Xiaoliu, Chongsheng, Xiaohei, Shishi, Yiyi, 007, Wangzai	Upon family split, Qinu was killed in infanticide, Liushaotou was seriously wounded and died

9.1.5 Forming of the "Yintangxiaotu All-Male Troop"

By the end of March 2015, outlier male langur Laoxie invaded FJC Small Cave, and Yintangxiaotu had to leave. It gradually withdrew from FJC territory, leading its 14 sons, started roving in the Nongguan Mountains and became "Yintangxiaotu all-male troop" (Table 9.5). As time went by, the eldest son Wuque, the second son Zuoque, and the third son Youque left the troop one after another to develop territories of their own. The aged Yintangxiaotu is still leading the rest of its 11 sons roving in the southernmost tip of the Nongguan Mountains, which is 6 km away from FJC.

We shall excerpt from some of the video-recorded activities of the "Yintangxiaotu family troop," the "Yintangxiaotu transitory troop," and the "Duanwei + Romeo troop" during their short invasion period in FJC Big Cave as example cases. The activities in the recordings are paradigmatic and could well illustrate the living patterns of white-headed langurs in the Nongguan Mountains.

9.2 Research Method

While employing wireless tracking on animals with individualistic territories is a workable method, we think it is not suitable for the white-headed langurs, which live gregariously in troops. On the one hand, white-headed langurs are often in crowd, they would be so close to each other that the researcher would not be able to match the auditory signals with the corresponding transmitters worn by the animals. On the other hand, white-headed langurs live on precipitous precipices where researchers are unable to reach. Thereby, they could not observe them at a close distance. In addition, the observer could have no idea of the relationships of an individual's actions to its peers, nor could they sense the animal's moods only from the signals received.

In December 2008, we built a 35m-tall observation platform, which we called the treehouse (Fig. 9.1a–c). It is 50 m from FJC Big Cave, and we installed a high-definition video recording monitor system on it to observe and record the lives of the white-headed langurs. In 2012, we had recorded over 300 h of video data.

Since December 2008, we set up eight work platforms on precipices and in forests along the regular routes of the FJC Big Cave troop (Fig. 9.1d–f). In 2012 alone, one of our teams stationed on those work platforms had worked there for 357 days and recorded 14,745 mins of white-headed langurs' activity with high-definition video cameras.

Our analysis in this chapter are largely based on the video recordings of the three white-headed langur troops inhabiting FJC Big Cave and Small Cave taken in 2012. In addition, there are field observation registers on the "Yintangxiaotu family troop" in November and December in 2009 and in 2010 and 2011.

9.3 The Activity Cycles of White-Headed Langurs

We used both the "Sampling All Occurrences of Some Behaviors" method and "Focal-Animal" method in our field video recordings. "Sampling All Occurrences of Some Behaviors" means to scan and record the maximum possible activities of all animals. While "Focal-Animal" means to focus on a stochastically appeared animal, whose activity is worth noticing, and document the beginning and ending time of its behavior. Our analysis on the white-headed langur's activity cycles is based on the following field video recordings and observation registers:

1. Data of "Yintangxiaotu family troop" registered in 2012.
2. Video recordings on the three white-headed langur troops that successively inhabited FJC Small Cave taken in 2012. They were the "Duanwei + Romeo troop," "Shengge + Shengdi troop," and "Yintangxiaotu transitory troop."
3. Observation registers made between 2014.4 and 2014.5, and between 2015.12 and 2016.2 on the daytime activities of the FJC Big Cave's "Yuweng family troop," FJC Small Cave's "Laoxie family troop," and FJC "Xishan troop."
4. Registers from tracking the full day activities of the "Bokyueshan transitional troop" from 2015.12 to 2016.2.

The video recordings from the monitoring system through the entire year of 2012 and our field registers through 2015–2016 had revealed that white-headed langurs would come out from their night shelter every day before dawn, rest for a while in the shrub thickets close to the night shelter, and then move to forage at daybreak. Usually, they would only return to their night shelter at dusk, and then quickly move inside to rest and sleep. Below is the analysis:

Fig. 9.1 Fieldwork and facilities (photo by Feng Chunguang, Liang Zuhong). (**a**) Field observation platform—treehouse (daytime); (**b**) field observation platform—treehouse (nighttime); (**c**) Professor Pan Wenshi observes Big Cave group at Treehouse at night; (**d**) Professor Pan Wenshi observes at No. 1 work platform; (**e**) Researchers observe at No. 6 work platform; (**f**) Track and observe in the field; (**g**) Photographing at No. 7 work station

05:30–07:30: Wake up in the night shelter.

06:00–07:30 or 08:30: Come out from night shelter, rest in the thicket around the night shelter or on the craggy walls of the night shelter.

06:00–09:30: Start daytime activities. The resident alpha male langur sits up on a high place to scan around. Some adult female langurs would suckle their yellow-haired infants, and the rest would either remain resting close to the night shelter or have slight activities.

08:30–10:30: The whole troop starts entering into their first daytime activity peak. They start leaving the night shelter area and forage as they move while the resident alpha male langur keeps being vigilant as it forages and moves along.

10:30–12:00: Move to lush forest near hilltops or at mid-hills to take a rest. Some individuals would continue foraging.

12:00–14:30 or 15:00: Time for a mid-day break. Yellow-haired pups and infants would play a little or be suckled by their mothers.

14:30–17:00: The second daytime activity peak. They start moving and forage along the way.

17:00–18:00: The troop moves in the direction of the night shelter.

18:00–19:30: The langurs appear in the thickets on the craggy walls near the night shelter. One after another they climb up the precipice to move inside the night shelter whilst the resident alpha male langur sits up high while scanning all around for danger. It usually is the last one to move inside the night shelter.

19:30–20:00: Sitting inside the night shelter, the langurs embrace or groom each other, expose their teeth, scramble to cuddle the yellow-haired pups, and display other behaviors such as presenting their anogenital areas, copulating, suckling yellow-haired pups, etc.

20:00–05:00: Sleep. (Between 23:00–23:30 and 04:30–05:00 there may be intermittent excretion behavior of minute volume).

05:00–05:30: Stay sleeping or resting in a state of calmness.

The precise schedule of white-headed langurs that follows sunrises and sunsets proves they are typically diurnal animals.

Vegetation in the Feijichang habitat in the Nongguan Mountains had since 2006 been gradually restored and now provides an abundant and steady food supply to the white-headed langurs. We noticed that unlike some other orangutan species in the *primate* order, which would collect branches of trees to make comfortable dens every night, white-headed langurs do not make dens. Their home is nothing but crannies or depressions on craggy walls of the precipices. Only in the coldest nights in winter (December–February) do they move inside the small caves on the precipice to shelter from the cold. There were times when they slept sprawling on the windowsill or high up on the rooftop of the Research Base. The openness of their behaviors has made it easy for us to observe them at nighttime.

Table 9.6 Time statistics of the day and night activities of "Yintangxiaotu family troop" 2012.01–2012.12

Month	Number of days observed	Time coming out from night shelter in morning	Time day activities started	Time back at night shelter area at dusk	Daytime activity hours (including rest)	Hours of rests in night shelter area at night (including slight movements)
January	5	5:52 (5:00–7:04)	8:11 (7:23–9:30)	18:07 (17:27–18:48)	9 h 56 min	14 h 4 min
February	5	7:54 (7:11–8:43)	8:14 (7:30–9:02)	18:50 (17:49–19:20)	10 h 36 min	13 h 24 min
March	5	6:34 (5:49–6:58)	7:41 (7:02–8:27)	18:57 (18:21–19:17)	11 h 16 min	12 h 44 min
April	8	5:55 (5:34–6:15)	6:58 (6:30–7:20)	19:04 (18:42–19:21)	12 h 6 min	11 h 54 min
May	6	6:06 (5:38–6:40)	7:00 (6:20–7:38)	19:15 (19:00–19:26)	12 h 15 min	11 h 45 min
June	7	5:42 (5:29–6:10)	6:35 (6:14–7:10)	19:16 (18:03–19:48)	12 h 41 min	11 h 19 min
July	8	5:49 (5:26–6:15)	6:29 (6:17–6:55)	19:11 (18:49–19:32)	12 h 42 min	11 h 18 min
August	8		6:16 (6:00–6:30)	18:53 (18:24–19:20)	12 h 37 min	11 h 23 min
September	6	6:09 (6:00–6:14)	6:39 (6:30–7:00)	19:06 (18:57–19:14)	12 h 27 min	11 h 33 min
October	4	6:22 (6:11–6:36)	7:00 (6:38–7:20)	19:08 (18:25–19:37)	12 h 8 min	11 h 52 min
November.	6	6:23 (6:13–6:32)	7:31 (6:54–8:55)	18:28 (18:21–18:34)	10 h 57 min	13 h 3 min
December	7		8:15 (7:30–9:10)	17:59 (17:03–18:23)	9 h 44 min	14 h 16 min
Earliest hour		5:00	6:00	17:03		
Latest hour		8:43	9:30	19:48		
Shortest duration					9 h 44 min	11 h 18 min
Longest duration					12 h 42 min	14 h 16 min

White-headed langurs sleep soundly through the night, with only intermittent slight activities like scratching itches or raising their tails to excrete waste down the precipice. They would immediately fall asleep again after these activities. While such slight activities would also exhaust energy, they are slight, brief, and within an arm's length. Hence, we still regard the time interval of these activities as resting in our statistics.

We found the white-headed langurs have two activity peaks in the daytime: one in the third hour after daybreak, and another in the second hour before dusk. Between the two activity peaks, they would take breaks in the afternoon that can be either brief or last as long as 3–4 h. We also noticed that they rest in obscure places that are difficult to find. Their states of rest are different, though they all stay under the shade of dense trees. During rest time, there will still be restless ones, with slight activities, such as moving to rest on another branch, or eating leaves, etc.

9.4 The Activity Rhythm of White-Headed Langurs in Relation to the Seasons

White-headed langurs will periodically regulate the timing of their activities and rest according to season cycles, but what they do through the 24 h always remain the same. We are citing two instances in the following for illustration.

Instance 1: The day and night activity rhythm of the "Yintangxiaotu family troop"

Through 2012.1–2012.12, we tracked the "Yintangxiaotu family troop" to observe their day and night activities and documented them, as listed in Table 9.6.

The statistics in Table 9.6 shows that the timing of daytime activities by the "Yintangxiaotu family troop" varies slightly according to the seasons. It had the longest hour of daytime activities in July, which took up 12 h and 42 min, while the shortest hour of daytime activities was in December, which took up 9 h and 44 min. The difference

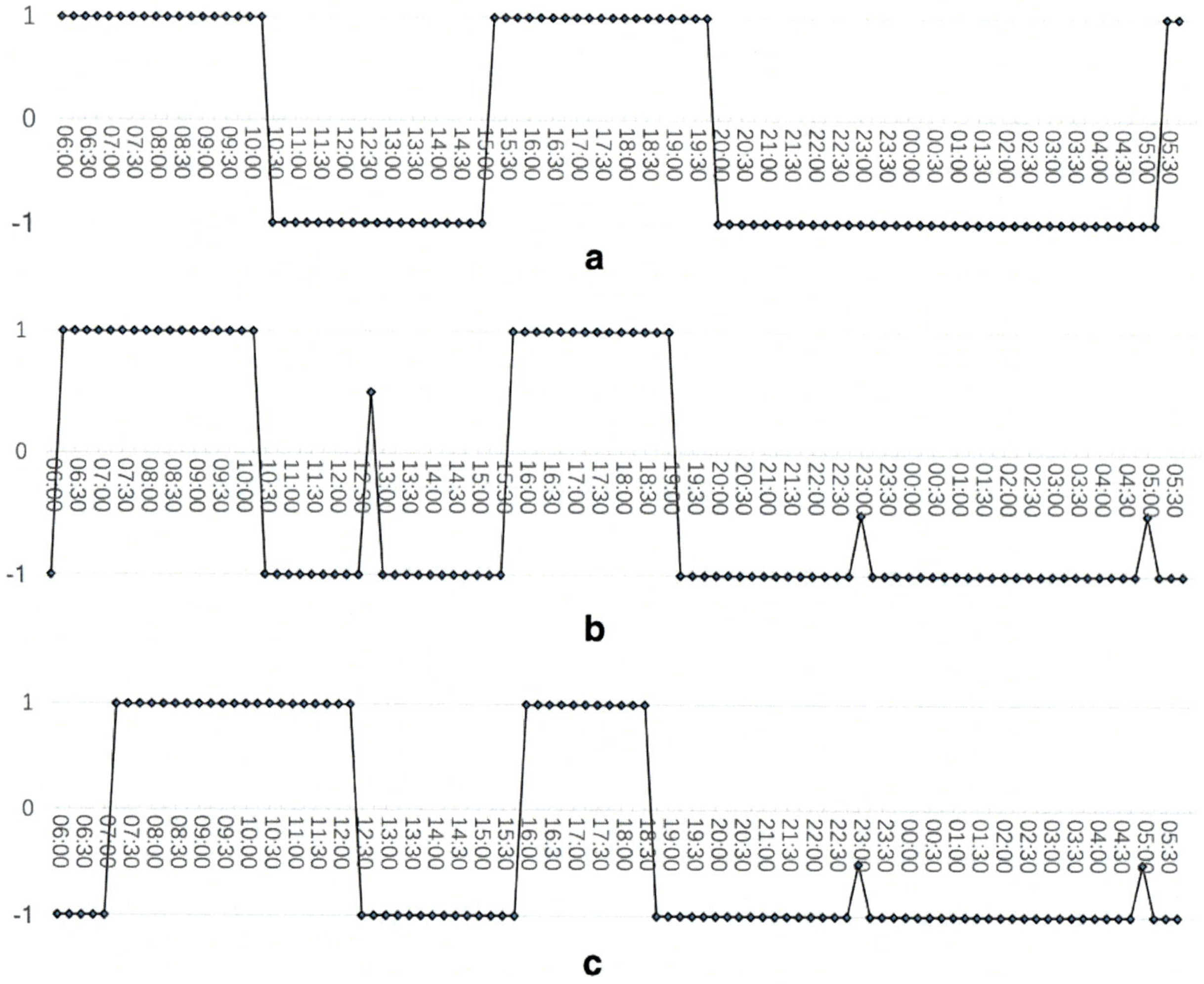

Fig. 9.2 The day and night activity rhythm chart of "Wuque brothers troop". (**a**) The day and night activity rhythm of "Wuque brothers troop" on 2014.06.06. (**b**) The day and night activity rhythm of "Wuque brothers troop" on 2014.09.25. (**c**) The day and night activity rhythm of "Wuque brothers troop" on 2014.12.31

between the longest and shortest daytime activity hours was 2 h and 58 min.

Instance 2: The day and night activity rhythm of the "Wuque brothers troop"

After Yintangxiaotu led some of its wives and offspring to retreat to FJC Small Cave and began the "Yintangxiaotu transitory troop" in July 2012, its eldest son Wuque frequently led its 11 brothers and half-brothers and got apart from their parents and sisters (we called the contingent team "Wuque brothers troop"). Their territory was within an area roughly at a straight-line distance of 500 m from the southeast of FJC Small Cave. They often came to Fanghoushan and up on the rooftop of our building. We wondered whether the activity rhythm of the all-male troop would be different from that of a family troop inhabiting in the same area.

We tracked the "Wuque brothers troop" for days and nights in June, September, and December of 2014. We registered the state of the troop in every 15 min. That gave us 96 activity registers in one day and night cycle. We then processed the data by computer analysis to get their activity rhythm (Fig. 9.2).

Figure 9.2a shows that in June 2014, the "Wuque brothers troop" had daytime activities of 12 h and 40 min, which differed by only 2 min from the July register of "Yintangxiaotu family troop" in Table 9.6. Figure 9.2b shows that the daytime activity hours of the "Wuque brothers troop" in September 2014 was 12 h and 30 min, differing from the September register of "Yintangxiaotu family troop" in Table 9.6 by just 3 min. In December 2014, daytime activity hours of the "Wuque brothers troop" lasted 9 h and 50 min (Fig. 9.2c), differing from the December register of the "Yintangxiaotu family troop" in Table 9.6 by only 6 min. It appears that the activity rhythm of the "Wuque brothers troop" and the "Yintangxiaotu family troop" is almost identical. Although the time length of daytime activities varied slightly in different seasons, they all had a period of rest between two activity peaks. The period of rest averaged between 3 and 4 h, at times with slight activities in between.

Comparing the registers in Instance 1 and Instance 2, it reveals that there were common traits in the daytime and nighttime activities of the "Yintangxiaotu family troop" and the "Wuque brothers troop":

1. White-headed langurs are active in daytime, but at night, they have almost no activities other than sleeping.
2. There are two activity peaks in a day with a period of rest in between, but there were always individuals being slightly active during the period of rest.

How do white-headed langurs strike a balance between activity and rest? Since 2012, white-headed langurs have been living in a habitat that has steady food resources. Their regular daily forages and rests are in line with their physiological needs. We shall excerpt in below several of our registers from observations in the wilderness to illustrate the white-headed langur activities of forages, rests, etc.

Register 1: Time taken for food to pass through the digestive tracts in three adolescent male white-headed langurs

Date: November 13, 2014

Time: 18:20

> We found the 3 younger brothers of Wuque eating the beans of a shanhehuan (Albizia kalkora) by the Research Base. Since the beans were pretty shriveled, we estimated that the pods might be difficult to fully digest. Fortuitously, they were using the rooftop of Research Base building as their night shelter, so we decided to keep tracking and observing them through the night. We collected their feces on that night at 22:00 and in next morning at 02:30 and 06:00. We found in the first two feces specimens residues of leaf cellulose, but pods of shanhehuan in the last feces specimen.

Register 2: Time taken for food to pass through the digestive tracts in a juvenile male white-headed langur

Date: February 26–27, 2015

Time: 18:00 (on the 26th) to 06:00 (on the 27th)

> We found the little male langur 007 was eating the figs of a xieyerong (Ficus tinctoria) at the neighborhood of the Research Base. By chance, it was also sleeping on the rooftop of the Research Base building that night, so we kept tracking it and observed it through the night. We noticed it defecated at 22:30 that night and at 02:00 and 05:50 on the next day. We collected the feces to examine and found only residues of leaf cellulose in the first two feces specimens, but there were fig coats of xieyerong in the third specimen.

Register 3: Length of digestive tract in an adolescent male white-headed langur

Date: February 18, 2015

Time: 12:30

> We dissected an adolescent male white-headed langur, which was found dead from an electric shock while it moved among the trees, to measure the length and capacity of its digestive tracts. Its esophagus was found to have a length of 20 cm, stomach about 20 cm, duodenum 12 cm, small intestine 331 cm, appendix 14 cm, colon 88 cm, and rectum 5.5 cm. The total length of its digestive tract reached 490.5 cm. The stomach of the male adolescent langur at the time of death was fully filled with dark green chyme, weighing a total of 580 g while the chyme alone weighed about 490 g. The intestines were empty.

Register 4: Estimating the weight of food needed to fully fill the stomach of an average male white-headed langur in each forage.

Date: December 22, 2014

Time: 10:00

> We observed at near distance the 'Wuque brothers troop' on a yinhehuan tree (Leucaena leucocephala) picking the seeds (beans) out from the pods to eat. After about 1 h, they stopped feeding, which indicated that their stomachs were full. We estimated each one of them had consumed about 400 g of seeds.

The above registers help to explain why white-headed langurs have two forage peaks in the daytime. The first peak is when the white-headed langurs move out from their night shelter. For 2 h they feed as they move until their stomachs are full. Then they rest under shade of dense trees for 3–4 h while the chyme in the stomach moves to the small intestine for further digestion, then to the large intestine. The entire digestion process involves passing through the digestive tract of a length of about ten times that of the body (45–50 cm), and it takes 11–12 h to complete. Hence, digestive residues of leaves consumed at 10:00 in the morning would only be excreted at 22:00 at the night of the same day, and food consumed at 18:00 in the afternoon would, after digestion, be excreted around 06:00 on the next day. The timing agrees with the hours when we found yinhehuan seed coats and xieyerong seed coats in white-headed langur feces (Fig. 9.3c, d).

All creatures have to look for food and rest every day. Carnivores generally take a long time foraging but spend little time on feeding. Ruminants spend about 40% of their time each day on feeding while non-ruminant herbivores have to spend 50% or more of their daily time on feeding. The giant panda lives on bamboos, and it has to spend more than 50% of its total day and night time on foraging and feeding. Table 9.7 lists the time distribution between activities and rests in one day and night of eight animal species.

As shown in Table 9.7, feeding takes up 16.62% of the time of white-headed langurs, which is relatively close to the 12.5% with mountain gorillas in Dian Fossey's study. The two species are high-order old-continent primates and are anatomically very similar. We know that mutation had taken place in the foreguts of white-headed langurs, as there

Fig. 9.3 Forage record of white-headed langur (Photo by Liang Zuhong, Feng Chunguang). (**a**) The dried pods of *Albizia kalkora*; (**b**) the hard fruits of *Ficus obliquus*; (**c**) the dried feces samples of the white-headed langur; (**d**) carefully check the food residue through the digestive tract of the white-headed langur, and collect the seed coat of *Leucaena leucocephala* (in the red circle), so as to calculate the time of food staying in the gastrointestinal tract of the white-headed langur

Table 9.7 The time distribution between activities and rests in the day and night of eight animal species (in percentages)

Species/ activity	White-headed langur (subject of research)	Gorilla (Fossey and Harcourt 1977)	Impala (Jarman and Jarman 1973)	Burchellii zebra (Gogan 1973)	African elephant (Wyatt and Eltringham 1974)	Giant panda (Schaller et al. 1985)	Polar bear (Stirling 1974)	Lion (Schaller 1972)
Feeding	16.62	12.5	40.0	56.0+/–	74.2	55.0	2.3	3.1
Resting	55.31	76.5	41.7	29.0+/–	13.2	40.9	42.0	84.0
Walking (or migrating)	22.87	7.2	7.5	15.0+/–	11.3	2.1	29.0	8.3
Foraging	–	–	–	–	–	–	24.6	
Others	5.2	3.8	10.5	No data	1.3	2.0	2.0	4.6+/–

are different sections that developed in its stomach. Symbiosis of protists and bacteria is found in some sections, which helps to digest hermicellulose and cellulose, enabling it to live on leaves. On the other hand, with zebras and elephants, the mutation happened in the hindgut, and even though symbiotic microorganisms are found, the digestion process is not as effective as in ruminant species such as cows and sheep, whose foreguts had mutated to a profound extent through evolution. As a result, zebra and elephants have to spend more time on feeding. African elephants spend 74.2% of their time on feeding, and they consume astounding volumes. The total weight of leaves, grass, barks, and roots consumed by an elephant in a day reaches 300 kg, though more than 50% of it is excreted without being completely digested. The zebra spends 56% of its time on feeding, which is about the same with that of a giant panda (55%). While the time spend on feeding by white-headed langurs is somewhere between impalas and gorillas. The impala is a ruminant species with an extraordinary stomach that has a specialized chamber with symbiotic microorganisms in it that digest cellulose. Anatomies of it indicate that its digestive system is more adapted to grass eating than primates. The gorilla spends only 12.5% of its time each day on feeding while spending most of its time (76.5%) on resting. The time it spends on other activities, including walking, takes up merely 11%. On the whole, the time distribution with white-headed langurs is not far from those of gorillas except that white-headed langurs spend 4.12% more time than gorillas on feeding, 21.19% less time on resting, and 17.07% more time on walking and other activities.

Do the figures indicate that primates do better in achieving a balance between being active and resting than non-primate mammals? Because in the phylogeny of primates, there was no digestive system that was more adaptive to vegetative diets than ruminants, we conjecture that the high efficiency in feeding and foraging by primates came as a result of brain developments which eventually turned them into animals with higher intelligence.

9.5 The Biorhythm of the White-Headed Langur

John Welsh, in studying the biorhythm of crustaceans and insects, proposed the "circadian clock" concept in 1934, but it received no particular attention (Shang Yuchang 2005). In 1948, Brown believed that living things, by drawing upon cues from the environment, have senses of time and will know when to do what (Brown and Webb 1948). It was only since then that the circadian clock concept was widely acknowledged. The sense of time inherent in living beings keeps the organic functions from being completely influenced by environmental changes. While the inherent circadian clock can also draw cues from the changes in the environment to adjust itself.

As early as 140 BC during the Qin Dynasty and Han Dynasty, the Chinese had mapped the sun's celestial path and determined the 24 solar terms, thereby creating a calendar that would signal the time of year for agricultural activities. That is the *Taichu Calendar*. The 24 solar terms denote the different positions of the sun on its ecliptic orbit, which bring about the different seasons. The season changes obviously influence the weather in the Nongguan Mountains, which situates in the transitional zone between tropical and subtropical climate areas.

As we depict the year of 2012 in solar term sequences along with the activity and rest hours of the "Yintangxiaotu family troop" in each month (there are two solar terms in a month) in Table 9.8 and Fig. 9.4, we note that the periodic monthly activity rhythm changes of white-headed langurs correspond to the seasonal photoperiod changes in the Northern Hemisphere. In every summer, the Northern Hemisphere has longer periods facing the sun, giving rise to longer photoperiods in the months of June, July, August, September, and October. During that season, white-headed langurs have their daytime activity hours lengthened, waking up earlier and sleeping later than they do at any other point in the year. However, in the months of November, December, January, and February when the Northern Hemisphere enters winter, the nights become longer and white-headed langurs wake up later but sleep earlier and accordingly as their daytime activity hours become shortened.

We can see from Table 9.8 and Fig. 9.4 that:

1. On the spring equinox on March 20, 2012 and on the autumn equinox on September 22, 2012, when the sun was directly overhead, making the equator area have an equal length of day and night in the Nongguan Mountains, the daytime activities and nightly rests of the white-headed langurs also came to be of equal time lengths.
2. On the summer solstice on June 21, 2012, when the sun was directly overhead the Tropic of Cancer, it started in the Northern Hemisphere as described in the context of solar terms, "as the sun arrives the North, days become long and sun shadows become short." As the Nongguan Mountains is situated right next to the Tropic of Cancer, that day had the longest daytime and shortest night of the year. And the white-headed langurs correspondingly had the longest daytime activity hours (12 h 42 min) and shortest night rest hours (11 h 18 min) on that day.
3. On the winter solstice on December 21, 2012, when the sun was almost exactly overhead the Tropic of Capricorn, it started in the Northern Hemisphere as described in the context of solar terms, "as the sun arrives the South, days become short and sun shadows become long." In the Northern Hemisphere (including Nongguan Mountains), the winter solstice was the day with the shortest daytime and longest nighttime. On that solar term date, the white-headed langurs also happened to have the shortest daytime activity hours (9 h 44 min) and longest nightly rest hours (14 h 16 min).

From Table 9.8, we can work out that the average distribution is about 11 h for daytime activities and about 13 h for nightly rests.

9.6 Social Activities Among White-Headed Langurs

Although there had been many researchers tracking the social activities of primates in tropical forests since the 1930s, today our understanding of the social behaviors of primates, as well as that of other mammals, is still in a primary stage. Our research on the social structure and behaviors of white-headed langurs is as well at the beginning stage. Before we discuss white-headed langurs' social behaviors, we shall first delineate some relevant concepts and specific terms.

9.6.1 "Sex–Age Group"

The structure of an animal society will vary as the individual animals in it grow up. In order to make accounts of it correctly, it is necessary that accounts be made when the animal society is at a stable stage. In analyzing the white-

Table 9.8 Changes in the activity and rest hours of "Yintangxiaotu family troop" as regulated by their circadian clock with accord to the changes of photoperiods in 2012

Solar term	Date	The sun's incidence, its effect on Nongguan Mountains	Hours of activity and rest by the white-headed langurs: Activity starting time Average daily activity hours	Time start to sleep Average daily resting hours
Moderate cold	Jan 6		08:11 ($N = 5$) 9 h 56 min	18:07 ($N = 5$) 14 h 4 min
Severe cold	Jan 21			
Spring commences	Feb 4		08:14 ($N = 5$) 10 h 36 min	18:50 ($N = 5$) 13 h 24 min
Spring showers	Feb 10			
Insects waken	Mar 5		07:41 ($N = 5$) 11 h 16 min	18:57 ($N = 5$) 12 h 44 min
Spring divide (vernal equinox)	Mar 20	Sun overhead the equator		
Bring and clear	Apr 4	Equal length of day and night	06:58 ($N = 8$) 12 h 6 min	19:04 ($N = 8$) 11 h 54 min
Rain comes for corn	Apr 20			
Summer commences	May 5		07:00 ($N = 6$) 12 h 15 min	19:15 ($N = 6$) 11 h 45 min
Corn flourishes	May 20			
Corn on ear	Jun 5		06:35 ($N = 7$) 12 h 41 min	19:16 ($N = 7$) 11 h 19 min
Summer arrives (summer solstice)	Jun 21	Sun overhead the Tropic of Cancer		
Moderate heat	Jul 2	The longest day of the year	06:29 ($N = 8$) 12 h 42 min	19:11 ($N = 8$) 11 h 18 min
Great heat	Jul 22			
Autumn commences	Aug 7		06:16 ($N = 8$) 12 h 37 min	18:53 ($N = 8$) 11 h 23 min
Heat recess	Aug 23			
White dew	Sep 7		06:39 ($N = 6$) 12 h 27 min	19:06 ($N = 6$) 11 h 33 min
Autumn divide (autumnal equinox)	Sep 22	Sun overhead the equator		
Cold dew	Oct 8	Equal length of day and night	07:00 ($N = 4$) 12 h 8 min	19:08 ($N = 4$) 11 h 52 min
Frost	Oct 23			
Winter commences	Nov 7		07:31 ($N = 6$) 10 h 57 min	18:28 ($N = 6$) 13 h 3 min
Light snow	Nov 22			
Heavy snow	Dec 7		08:15 ($N = 7$) 9 h 44 min	17:59 ($N = 7$) 14 h 16 min
Winter arrives (winter solstice)	Dec 21	Sun overhead the Tropic of Capricorn		
		The shortest day of the year		

headed langur social structure, we adopted the "sex–age group" concept put forward by renowned primatologist Clarence Ray Carpenter (1940) in his study of Lar gibbons (*Hylobates lar*) in the 1930s and 1940s in Chiangmai, Thailand.

9.6.2 "Role"

The "role" concept in biology is adapted from sociology. In our following report, we use "role" to imply a recurrent social activity by a white-headed langur relative to its "sex–age group." In the broadest sense, the mounting act by a male in copulation is a "role" and the act of taking care of pups by a female (whether it is a mother or not) is also a "role."

9.6.3 The "Sex–Age Groups" and "Roles" in the "Yintangxiaotu Family Troop"

The "Yintangxiaotu family troop" had in May 2012 grown to consist of 43 individuals, but in July the family split and some langurs had to leave and became the "Yintangxiaotu transitory troop." Despite that the troop is transitional in nature, it continued to breed. In our idea, the "Yintangxiaotu transitory troop" was similar to the "Yintangxiaotu family troop" in that both were polygynous and with offspring, which was a

Fig. 9.4 Circadian clock regulated activity cycles of the "Yintangxiaotu family troop" in various solar terms and months in year 2012

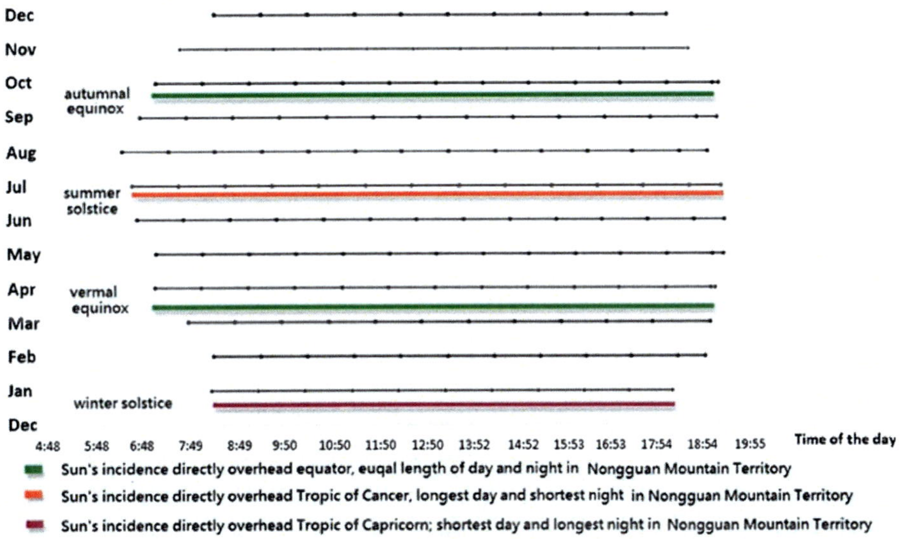

typical core structure in white-headed langur society, and that they were both at a stable stage. We are listing in Table 9.9 the "sex–age groups" of the individuals in the "Yintangxiaotu family troop" and their respective "roles" in July 2012.

From the field registers listed in Table 9.9, we can categorize the individuals in the "Yintangxiaotu family troop" with reference to their sexes, ages, and "roles" into the following groups:

Male (♂) groups consist of resident alpha ♂, sub-adult ♂, adolescent ♂, juvenile ♂, and old ♂. The duty of defending the family rested primarily on the resident alpha ♂. Sub-adult ♂ and adolescent ♂ assisted their father while old ♂, yellow-haired ♂ pups, and yellow-haired ♂ infants played no part in defending the troop.

Female (♀) groups consist of old ♀, adult ♀, sub-adult ♀, adolescent ♀, and juvenile ♀, and nearly all of them had the roles of reproducing or nursing pups (regardless if it was one's offspring, brother, or sister) and played little to no part in defending the family. A major trait in the behaviors by ♀ is that adolescent ♀ and juvenile ♀ had the helper "role," that is to help adult ♀ in nursing yellow-haired pups.

Yellow-haired pups and infants spent a lot of time acting "roles" of play. The term "act" is used here on the pups' playing in order to imply that "playing" is the "role" for yellow-haired pups and infants.

9.6.4 "Seasons" in White-Headed Langur Reproduction Ecology

We had registered 98 pup births in FJC Big Cave and Small Cave during the period from 1996 to 2015 (see Chap. 10), among which 70 pups were born in the months of November, December, January, and February, accounting for 71.4% of a given year's births. We regard those 4 months as the reproduction season of the Nongguan Mountains white-headed langurs. Among the 70 pups born in that season, 39 pups were born in December and January, which equaled 39.8% of the year's births and accounted for 55.7% of births in the reproduction season. It clearly indicated that the peak period of white-headed langur reproduction is in December and January.

We also registered in the period from 2010 to 2015 that there were in the Nongguan Mountains a total of 468 counts of female white-headed langurs presenting their anogenital area and male white-headed langurs performing an act of mounting (see Chap. 10). Of these cases, 219 counts happened in the summer to autumn season of July, August, September, and October, accounting for 46.8% of the total. However, behaviors of presenting anogenital area and mounting happened mostly in July and August, a total of 178 times, which was 38% of the registered total and equaled 81.3% of the registers in the summer to autumn period. From that we would surmise that July and August are the peak months when white-headed langurs' "actual copulations" take place.

The registers from 1996 to 2015 have shown that the birth rate in March was still lingering at 13.3%, whereas it descended drastically in the following 3 months (Table 10.3). The most characteristic white-headed langur family troop activity in the entire spring season of March, April, May, and June is that every individual in the family troop is obliged to receive and foster infant pups born in the "reproductive cycle" of the year (from November to June/July in the next year). During the spring months, about 94%

Table 9.9 The "roles" of individuals in the respective "sex–age groups" in "Yintangxiaotu family troop"

Names of the individuals	Sex–age group		Activity in the family (role)				Remarks
	Sex	Age	Defense	Foster[a]	Family helper[b]	Play	
–	Aged ♂	>13					No register
Yintangxiaotu (resident alpha male)	Adult ♂	12	+++	–	–	–	Adult, sub-adults and adolescents took up defense duties
Wuque, Zuoque, Youque	Sub-adult ♂	4.0–5.0	+++	–	–	–	
Mengli	Adolescent ♂	2.5–4.0	+	–	–	–	
Hanlin, Yiling, Gualai, Xiaoliu, Xiaohei, Chongsheng	Juvenile ♂	1.0–2.5	+	+	+	+++	
Shishi, Yiyi	Yellow-haired infant pup ♂	0.5–1.0	–	+	+	+++	
Xiaolu, Xiaobei, Xiaozhen, 007	Yellow-haired infant pup ♂	≤0.5	–	–	–	+	
Dahei	Aged ♀	>25	–	+	++	–	Juvenile ♀ and adolescent ♀ were the main helpers in taking care pups
Arzhen, Jiajia, Xiaoxiao, Yingying, Pingping, Lanlan, Lanqi, Tiantian, Beibei, Arxi, Arlu, Aryu	Adult ♀	≥5	+	+++	++	–	
Qiqi, Mimi, Yumei, Zhenzhu, Da-S	Sub-adult ♀	4.0–5.0	–	+	+	–	
Wenwen	Adolescent ♀	2.5–4.0	–	+	++	–	
Xiaojiu, Qingming	Juvenile♀	1.0–2.5	–	++	+++	+	
Gaga, Chuliu	Yellow-haired pup ♀	0.5–1.0	–	+	+	++	
Tutu, Xiaolan, Xiaoying	Yellow-haired infant ♀	≤0.5	–	–	–	+	

Notes: "+" denotes the intensity acted in the role; "+" = slight, "++" = relatively intense, "+++" = very intense; "–" denotes not acting that role
[a]Embracing yellow-haired pup acts other than suckling
[b]Carrying around or nursing pups in the family

of the pups would be yellow-haired pups of 6 months or less, while the remaining 6% would be newborns.

On the basis of the yearly cycles of reproduction described above, we divide a year of the lives of white-headed langurs in the Nongguan Mountains into the following ecological seasons:

- "Ecological season of mating" (July, August, September, October)
- "Ecological season of parturition" (November, December, January, February)
- "Ecological season rearing pups" (March, April, May, June)

9.6.5 Definitions of Behaviors

Theoretically, the characteristic behaviors of an animal are related to the ecologic environment it lives in and its development of growth. Making references to the age and sex of each individual in "Yintangxiaotu family troop" and their contributive "roles" in the family in the first half of 2012, we divide their behaviors into five categories. Below we list the behaviors under each category with specifications.

First category: Defense behaviors (Fig. 9.5)

1. Protection—Behavior of grown individuals staying close and watching over young individuals.
2. Vigilance—Behavior of male langurs (individual female langur sometimes do this too) keeping watch on outlier langurs distance away or gazing far away.
3. In alert—Reflective behavior by male langurs upon possible dangers to the family. Behaviors include looking in every direction and intermittently barking monosyllabic alarms. Whenever a male langur barks an alarm, adult females in the family would rapidly collect the pups to their sides and nervously look around.
4. Intimidation—Behavior of an adult male langur that dashes and leaps about in the forest or on craggy walls, forcefully shaking the branches and at the same time

Fig. 9.5 Defense behaviors (Photo by Liang Zuhong). (**a**) Intimidation; (**b**) In alert

making fierce grunts, in order to threaten away any outlier male langur intending to invade the family.

5. Driving away—Behavior of a resident alpha male langur dashing about or engaging in a fight to drive away invading male langurs. Female langurs in the troop sometimes do the same to drive away outlier female langurs.

Second category: forage behaviors (Fig. 9.6)

6. Recess—Brief breaks during forages.
7. Forage—The entire process from approaching, picking, handling, chewing to swallowing the food.
8. Drink—Approaching the water, stooping down, and drawing in water.
9. On the move—the upward and downward climbing of trees and craggy walls.

Third category: foster behavior (Fig. 9.7)

10. Carry—By initiative, take a yellow-haired pup in embrace or allow a yellow-haired pup to cling to the chest, clasping the hairs in order to move as one. This behavior mostly happens when the yellow-haired pup is in the first month of birth.
11. Look after—Approaching yellow-haired pups with the purpose of taking care of and protecting them.
12. Snatch—To snatch a yellow-haired pup from the embrace of another.
13. Suckle—Yellow-haired pups, infants, or juvenile langurs suckling milk from the mother langur. This often comes alongside the embracing behavior of the mother langur (it may also refer to the mother langur milking the pup or juvenile langur).
14. Lead—The behavior of an individual leading the troop's movement, or a mother langur or other individual leading the way for yellow-haired pups.
15. Embrace—Picking up single-handedly or with both hands a pup to clasp it in its bosom, sit it quietly on the knees, or dandle it. Generally refers to an adult female langur holding a pup on her chest.
16. Nestle—Yellow-haired pup sleeping in the embrace of an individual. The term generally refers to adult female langur nestling a yellow-haired pup.

Fourth category: social behavior (Fig. 9.8)

17. Affection—Behavior of the resident alpha male langur sitting peacefully with its wives and offspring.
18. Groom—Using hands or mouth to comb through hairs in order to remove dirt or dead skin, etc. It generally happens when the troop is at rest and at most time happens simultaneously with embracing. It can sometimes pacify the tense relationship between individuals.
19. Cuddle—Two individuals facing each other and spreading their arms while holding each other. Often takes place before grooming.

Fig. 9.6 Forage behaviors (Photo by Liang Zuhong). (**a**) Drinking; (**b**) Foraging

Fig. 9.7 Foster behaviors (Photo by Liang Zuhong). (**a**) Look after; (**b**) Snatch

20. Showing teeth in appeasement—The behavior of one individual opening its mouth, curling its lips, and showing its teeth and gums when getting near another dominant individual.
21. Showing teeth in resentment—Targeting at an individual or a direction, repeatedly flapping the lips up and down quickly, and showing its teeth and gums. The behavior could last a long or short period, and the frequency could be different each time. A resident alpha male langur always behaves this way toward an invading male langur far away, sometimes accompanied by low and deep "Ga..." growls emitted from the peritoneal cavity.
22. Offering place—Giving up a more secure and snug position to another individual by initiative.
23. Holding place—Not giving up the more secure and snug position to another.
24. Mounting—One mounting on another with its abdomen on the back of the other. The one on top has its hind legs on the ground and forelimbs on the back of another whilst the other one is hunched with all four limbs on the ground. The behavior lasts only for a short while. The behavior may happen between individuals of the same sex in the family.
25. Intimate behaviors—Besides between mother and pup, intimate behaviors like caressing, holding each other, etc. may happen between other individuals in the troop.
26. Kissing—Putting the mouth on another's body parts such as head, tail, and mouth.
27. Present anogenital area—Behavior of female langur hunching with all four limbs on the ground, raising the tail or flicking it sideward, displaying its anogenital area toward the male langur. Sometimes, it will turn its head to see the male langur or the other individuals.
28. Copulation—the male langur mounting on the female langur, their genitals come into contact and the male langur makes pushes and presses with its buttock. The behavior refers to the entire process from the invitation by female, mounting by male, inserting the penis into the vagina, rapid in and out, and until ejaculation. It can be ascertained by semen traces left on the genitals of the participating parties whether ejaculation had taken place, but it is difficult to empirically discern. We can only determine whether it occurred by the length of time it had taken. Male langurs generally in one occasion behave multiple rounds of mounting and copulation, and each time last from a few seconds to a little more than 20 s.

(a) (b)

Fig. 9.8 (**a**) Grooming and (**b**) mounting

Fig. 9.9 January 2009, Youque who just reached juvenile age was learning defense skill from its father (Photo by Liang Zuhong)

29. Play—Playing behaviors include chasing one another, jumping, dragging, tumbling, wrestling, feign biting, etc., mostly among immature individuals.

Fifth category: other behaviors

30. Climbing—Moving up or down a precipice.
31. Leaping—The act of leaping from the ground and falling on another object, often performed between branches, as well as between precipices and branches.
32. Resting—Dozing or taking respite on tree or precipice.
33. Sleeping—With closed eyes, entering into peaceful sleep in a night shelter.

9.7 Behavioral Traits of the "Yintangxiaotu Family Troop"

In our study of the white-headed langur's social behaviors, we emphasize on the relevance of the "roles" taken by each "sex-age group" in the family to the white-headed langur's reproduction ecological season. We also made comparisons between the sex–age groups the probability of occurrence of each behavior to find the group's level of social involvements and its interrelationship with the other sex–age groups in the family.

We sorted the behaviors observed in the "Yintangxiaotu family troop" under the 33 social behavior definitions and found the time length each behavior lasted. Then, on the basis of "sampling all occurrences of some behavior," we calculated the occurrence frequency of each behavior in time unit of 15 min. After that, we classify the behaviors according to its purpose (intended on itself or on others) into the five activity categories.

In Table 9.10, 9.14, 9.15, 9.16, 9.17, 9.18, and 9.19, we assembled the probabilities of the different "roles" acted by the six relatively stable "sex-age groups" in the "Yintangxiaotu family troop" in the three reproduction ecological seasons of 2012.

9.7.1 Trait 1: Only the Resident Alpha Male, Sub-adult, and Adolescent Males Take Up Roles of Defense

Resident alpha male langur Yintangxiaotu and its sons that were over 3 years old took up home defending roles (Fig. 9.9). They often guard the family territory jointly at the boundaries with exaggerative ways of barking, leaping, and stomping on tree branches to ward off intruders, whether it is an outlier male langur or other animal or human, from entering into family territory. If the intruder keeps approaching and gets to a distance that poses a threat to the females and pups in the family, Yintangxiaotu would lead its sons to strike. Sometimes the fight would be so fierce that wounds occur. Yintangxiaotu would always be the last one to withdraw in order to keep the intruder away from its wives and offspring.

On January 4, 2014, Gu Tieliu, leading his filming team, recorded an incident in which Yintangxiaotu and its sons jointly defended the family territory.

09:00: The nine sons of Yintangxiaotu led by their eldest brother Wuque came to forage beans on the yinhehuan tree (Leucaena leucocephala) at the southeast corner of the pond in front of the Research Base. This was their winter forage ground.

10:00: A troop of seven roving male langurs which had split from the 'Fangniushan troop' appeared in the direction of the Convention Center at the southern Research Base. Judging from their body sizes, they were a bit older than the 'Wuque brothers troop.' The nine younger male langurs of the 'Wuque brothers troop' began to withdraw to the direction of Fanghoushan where their 'old home base' was, but the eldest brother Wuque stayed on a tree in the southern Research Base, which was the outpost of the 'Wuque brothers troop'.

10:10: At the foothill between Feilaishi and Fanghoushan, the old dad of the brothers Yintangxiaotu was sitting high up on a tree and keeps watching, safeguarding its wives and offspring which were taking recess in the forest at its back.

10:40: The third brother, Youque, which we considered the most faithful, most intelligent, and most brave, had disappeared for several days and suddenly appeared. It showed itself on the flatland between the Research Base and Wudong, then rapidly moved up the trees at the hilltop of Wudongshan, going back and forth the Convention Center and Bokyueshan to issue warnings. The alert lasted 1 h 20 min during which Yintangxiaotu, Wuque and Youque made a triangular joint defence formation.

12:00: The eldest brother Wuque withdrew from the direction of Convention Center, limping, and reached the rooftop of Wudong, joining its younger brothers. Afterwards, they remained in the tree clusters at the foothill of Wudongshan, and kept watching towards the direction of Convention Center.

16:50: Youque appeared on the rooftop of the Research Base main building, still in a state of alert. It kept looking towards the direction of Feilaishi and Bokyueshan.

18:00: The 'Yintangxiaotu troop' disappeared into the midst of dusk.

When we put on a bar chart the time spent on defensive activities by the ♂ groups and ♀ groups in the three reproduction ecological seasons in the year as listed in Table 9.10 (Fig. 9.10), we can easily see:

The contribution to family defense by resident alpha male langur Yintangxiaotu: its defensive activities made up 39.82% (the average of 32.52%, 43.90%, and 43.04% in the three reproduction ecological seasons) of its total activities in the year. Its young sons also helped in defending the family territory, which accounted to 12.86% (the average of 13.33%, 13.79%, and 11.46% in the three reproduction ecological seasons) of their total activities in a year.

Contributions by ♀ groups toward defense was insignificantly low (adult ♀ 1.19%, sub-adult ♀ and adolescent ♀ 2%, juvenile ♀ 1.3%), which had clearly indicated that defending the family was not the "role" for female langurs.

9.7.2 Trait 2: Adult, Sub-adult, and Adolescent Females Take Up Roles of Nursing Pups

The adult females are the center of a family. Together with sub-adult and adolescent and juvenile females, they take up the role of nursing the pups. Mothers share a solid relationship with their infants. Immature females in the family helping foster the pups could improve the genetic fitness of the family. Like any mother of old-continent primate species, a mother white-headed langur would carefully take care of its newborn infant. In the first few days of the infant's birth, the mother langur would not pass it to the hands of inexperienced females or allow others to touch it. Then, after about 1 week, the adolescent and juvenile female white-headed langurs would help to take care of the newborn yellow-haired pup. In fact, all the young female helpers are half-sisters or cousins of the yellow-haired pup, if not agnate sisters.

From the registers in Table 9.10, we estimate that the proportion of foster activities in a full year by ♀ groups out of their total activities are: adult ♀ 46.63%, sub-adult and adolescent ♀ 23.4%, juvenile ♀ 22.3%.

Field register:

On January 20, 2012, 7-year-old young female Arxi gave birth to her second infant. The lunar calendar day was December 27 in the Year of Rabbit, so we named the newborn infant Yutu (Jade Rabbit).

Seven days later on January 27, 2012, or January 5th in the lunar calendar, we were up on the 8 meter tall $4^{\#}$ filming platform at the bottom of the FJC Small Cave night shelter observing, through the tree canopy slits, the interactions between mother Arxi and its daughter Yutu.

16:13: Arxi was sitting on a branch with Yutu in her embrace. The 7-day-old Yutu with a clean white face looked at us with big bright curious eyes.

16:14: Arxi groomed Yutu's hair for 3 s.

16:14: A juvenile langur came to the side of Arxi and her daughter, but Arxi would not allow it to touch Yutu. After 6 s, Arxi, carrying Yutu in her bosom, moved to another branch and the juvenile langur left them.

16:15: Arxi bent her head to lick the little-yellow haired infant and groomed the infant's hair, while Yutu pleasantly obliged. Seven seconds later, the mother and daughter looked at us, and in the next 15 s Arxi still held Yutu in her embrace, and then it groomed Yutu for another 7 s. The little yellow-haired pup remained peaceful in her mother's embrace. We noted it had been 62 s that the infant was facing her mother's nipple, but we could not see for certain whether it was suckling.

18:38: Arxi continued grooming Yutu for a further 36 s, then the little yellow-haired pup turned round to climb on her mother. The mother checked and licked Yutu's tummy.

18:55: Still holding the nipple in its mouth, Yutu fell asleep. After a while, Yutu clasped tightly the hairs on her mother's abdomen and Arxi leaped away among the branches.

Table 9.10 The probabilities of different activities in each of the six "sex-age groups" in the "Yintangxiaotu family troop" in the three reproduction ecological seasons in 2012

Reproduction ecological season	Activity category	Activity	Resident alpha ♂		Adult ♀		Sub-adult ♂ and adolescent ♂		Sub-adult ♀ and adolescent ♀		Juvenile ♂		Juvenile ♀	
			Times occurred	Probability (%)	Times occurred	Probability (%)	Times occurred	Probability (%)	Times occurred	Probability (%)	Times occurred	Probability (%)	Times occurred	Probability (%)
Parturition season	1	Defense	40	32.52	5	0.56	24	13.33	0	0.00	26	2.44	0	0.00
	2	Foster	0	0.00	319	35.56	1	0.56	78	37.32	98	9.18	20	20.41
	3	Forage	49	39.84	340	37.90	117	65.00	82	39.23	578	54.17	47	47.96
	4	Social	25	20.33	163	18.17	25	13.89	30	14.35	204	19.12	23	23.47
	5	Others	9	7.32	70	7.80	13	7.22	19	9.09	161	15.09	8	8.16
Season total			123	100.00	897	100.00	180	100.00	209	100.00	1067	100.00	98	100.00
Rearing season	1	Defense	90	43.90	29	1.10	32	13.79	14	2.40	3	0.55	4	1.53
	2	Foster	1	0.49	1432	54.08	0	0.00	149	25.51	91	16.55	90	34.48
	3	Forage	75	36.59	649	24.51	152	65.52	255	43.66	264	48.00	106	40.61
	4	Social	24	11.71	328	12.39	33	14.22	112	19.18	140	25.45	38	14.56
	5	Others	15	7.32	210	7.93	15	6.47	54	9.25	52	9.45	23	8.81
Season total			205	100.00	2648	100.00	232	100.00	584	100.00	550	100.00	261	100.00
Mating season	1	Defense	68	43.04	11	1.92	59	11.46	3	3.61	2	0.40	1	2.38
	2	Foster	1	0.63	288	50.26	0	0.00	6	7.23	32	6.39	5	11.90
	3	Forage	76	48.10	176	30.72	394	76.50	50	60.24	379	75.65	26	61.90
	4	Social	0	0.00	72	12.57	24	4.66	21	25.30	42	8.38	5	11.90
	5	Others	13	8.23	26	4.54	38	7.38	3	3.61	46	9.18	5	11.90
Season total			158	100.00	573	100.00	515	100.00	83	100.00	501	100.00	42	100.00

Remarks: Mating season: July, August, September, October; Parturition season: November, December, January, February; Rearing season: March, April, May, June

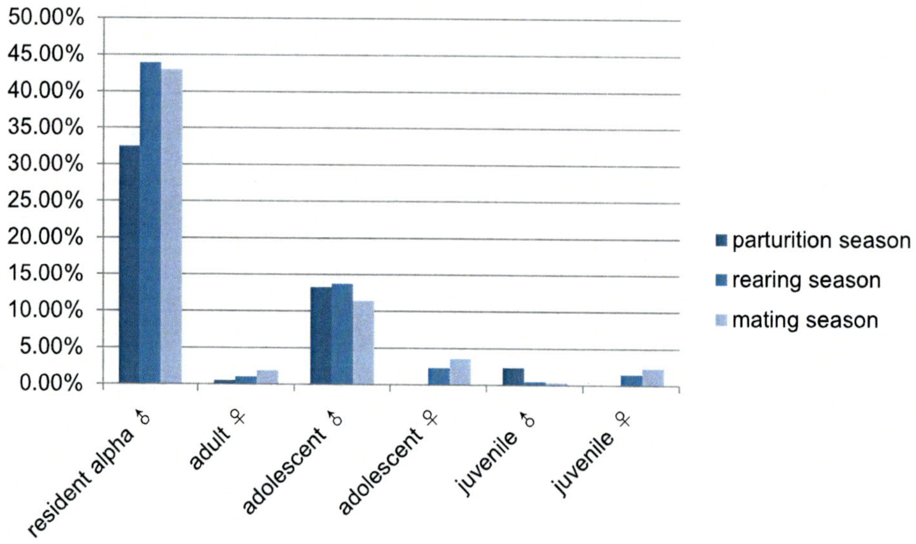

Fig. 9.10 Defense activity percentages by the different sex–age groups in the "Yintangxiaotu family troop" in the three reproduction ecology seasons

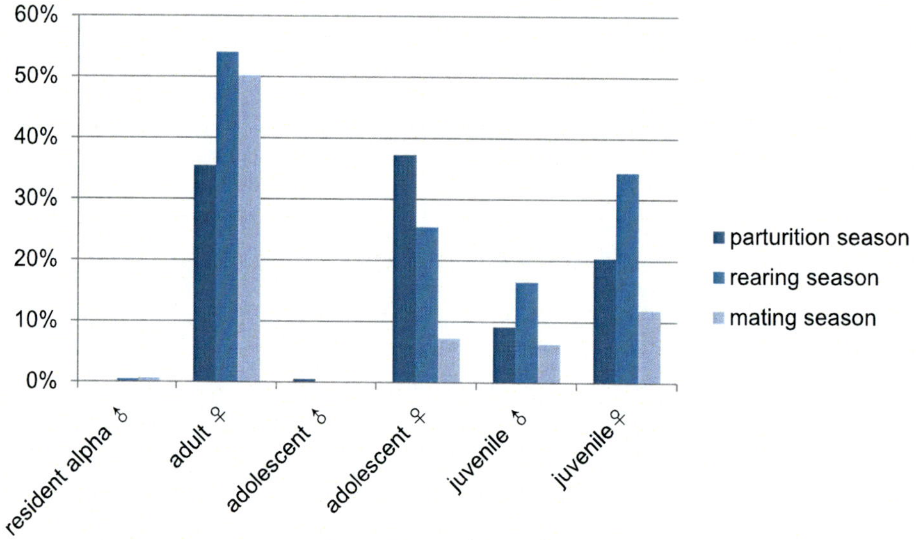

Fig. 9.11 Probabilities of foster activity by the different sex–age groups in "Yintangxiaotu family troop" in the three reproduction ecology seasons

When we compare the time spent on nursing newborn yellow-haired pups by the ♂ groups and ♀ groups in the "Yintangxiaotu family troop" as illustrated in Fig. 9.11, we can clearly see that apart from juvenile ♂ groups which had 10.7% of their time spent on foster activities, adult ♂, sub-adult, and adolescent ♂ almost never took the "role" of nursing.

The statistics have shown that in a white-headed langur family, the roles played by adults, sub-adults, adolescent, and juvenile "sex–age groups" are very different. The following register from the wilderness demonstrates that the most common and important activity in a day of an alpha male langur is defending the family territory.

January 27, 2012, a dark cloudy day with a chilling temperature.

14:21: Yintangxiaotu climbed up from below the Small Cave night shelter to the left side of the precipice and sat on its vigilance position. It half-dropped its lower jaw, silently watching Fangniushan in the south. After a while it turned to watch the direction of Xishan and Big Cave in the north. The entire time its wives and offspring were taking recess under the shade of the dense canopy of trees.

14:24: A 1 year old yellow-haired pup came behind Yintangxiaotu, and with a shriek it leaped onto the back of its father. Yet Yintangxiaotu would not allow itself be disrupted, as it stayed absorbed in vigilance (Fig. 9.12). Not only did it refuse to play, the male langur on duty of defending the territory and protecting its wives and offspring even appeared to be irascible.

Fifteen minutes later, Yintangxiaotu abruptly leaped up. With loud and continuous donkey brays that only adult male langurs could make to show its anger, it dashed up the craggy precipice wall, broke several tree branches with deliberate tramples, and then disappeared into the forest.

Fig. 9.12 The little male langur pup leaped onto its father Yintangxiaotu's back, but the father was not disrupted and stayed absorbed in vigilance. (Photo by Liang Zuhong)

9.7.3 Trait 3: Helper Behavior Is Common Among White-Headed Langurs

The term "helper" refers to an individual who is not the mother of a yellow-haired pup but will actively engage in carrying, looking after, leading, embracing, and nestling it. Helper behaviors were common in the "Yintangxiaotu family troop" at the beginning of each reproduction season, though the behavior took place much more frequently in female groups than in male groups. From Table 9.11 and Fig. 9.13, we can see that in the 4118 monitor recordings on adult females, there were 1756 counts of helping yellow-haired infants, which accounted for 42.64% of all activities. In the 486 monitor recordings on resident alpha male langurs, there was only one instance of it helping to care for an infant, which accounted for only 0.21% of all its activities. The percentage of helper behavior in sub-adult and adolescent females was 19.18%, far higher than the 0.11% in males of the same age group. Juvenile female and yellow-haired pups are strongly inclined toward helper behavior with percentages at 21.45% and 20.08% respectively, which were also far higher than the 8.35% and 8.33% of same age male groups. The probabilities of helper behavior between the sexes are markedly different.

Below is an incident that we observed and registered of a female langur helping her mother to foster her agnate younger brother.

Date: April 19, 2011
Place: "Tree House" at Feijichang

Darkness was about to fall in Feijichang, and we all opened our eyes wide to identify the white-headed langur individuals that were returning late and to register their behaviors in detail.

In particular, we wanted to locate Tiantian and her infant Shishi, who is just a bit over 3 months old. Shishi was bitten by an outlier male langur that snuck into their family territory two nights ago. It was seriously wounded and especially needed care from its mother.

18:49 We saw Tiantian returning to Big Cave alone and went straight to drink from a water puddle. We did not find Shishi in her embrace, which made us really anxious.

18:58 The night got really dark and we had to switch on the light to keep searching.

19:00 At last, we found Shishi with its sister Wenwen in the clump of trees at a location we called Nose Tip above Big Cave (Fig. 9.14a). Wenwen was the oldest child and daughter of Tiantian, and was 1 year and 11 months older than Shishi. As Wenwen was still very young, the task of looking after Shishi was a bit beyond her capabilities. She embraced and carried her younger brother this way and that way, and only after 8 min succeeded moving her brother to a safe place on a branch to wait for their mother Tiantian.

19:10 Shishi held on tight to his sister, and Wenwen carrying her younger brother with great effort started climbing down from Nose Tip. By that time, Tiantian finished her drinking and climbed up to Nose Tip, received her son Shishi from the hands of Wenwen and then climbed over the precipice (Fig. 9.14b). Wenwen followed closely behind her mother and younger brother.

19:15 The three of them, mother and daughter and son, finally reached the night shelter at the Waterfall quarter safely. Wenwen and her brother, who was reclining on their mother, fell asleep in the brink of an eye.

9.7.4 Trait 4: Play Behavior Is of Significant Importance in the Socialization of Young White-Headed Langurs

Every individual in the "Yintangxiaotu family troop" was closely blood-related to each other. While they acted different "roles" relative to their sex–age group in the family, all of them had instinctively fulfilled their obligations in their daily lives to benefit the family. Through their social activities, they had enhanced the closeness and harmony between the family members.

Besides grooming, mounting, and presenting one's anogenital and copulation, the most distinguished behavior in white-headed langurs' social activities is playing. Playing is the predominant behavior of young white-headed langurs. It is a pleasant behavior, and it is nothing else but having fun. In playing, young white-headed langurs learn survival skills and become an accommodating member of the family.

Often we saw groups of yellow-haired pups (age 0.5–1 year) or juvenile white-headed langurs chasing each other among trees or on precipices. They jumped, wrestled,

Table 9.11 Activity percentages of the different "sex–age groups" in "Yintangxiaotu family troop" in year 2012

		Yellow-haired infants		Yellow-haired pups		Juvenile langurs		Adolescent langurs		Adult langurs	
Behavior	Sex	Times occurred	Probability (%)	Times occurred	Probability (%)	Times occurred	Probability (%)	Times occurred	Probability (%)	Times occurred	Probability (%)
Defense	♂	0/665	0.00	0/48	0.00	31/2118	1.46	115/927	12.41	198/486	40.74
	♀	1/957	0.10	0/259	0.00	5/401	1.25	17/876	1.94	45/4118	1.09
Helper	♂	0/665	0.00	4/48	8.33	181/2118	8.55	1/927	0.11	1/486	0.21
	♀	0/957	0.00	52/259	20.08	86/401	21.45	168/876	19.18	1756/4118	42.64
Play	♂	59/665	8.87	20/48	41.67	191/2118	9.02	6/927	0.65	0/486	0.00
	♀	140/957	14.63	54/259	20.85	19/401	4.74	3/876	0.34	1/4118	0.02

Notes: (1) The denominator denotes the total number of monitor recordings on that sex–age group, and the numerator denotes the times occurred of the subject behavior in the recordings. (2) Helper activity includes carrying, looking after, leading, embracing, and nestling

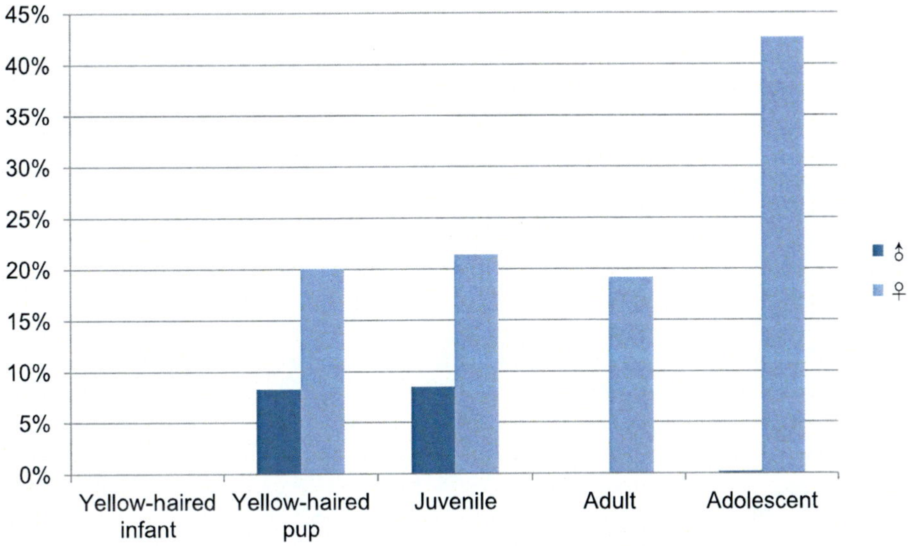

Fig. 9.13 Probabilities of the helper behaviors by the different sex–age groups in "Yintangxiaotu family troop" in 2012

Fig. 9.14 Helper behavior (Photo by Liang Zuhong). (a) Little sister Wenwen was coming down from the clump of trees at Nose Tip carrying very carefully her younger brother Shishi who has a gaping wound on his back. (b) Little sister Wenwen (right) was passing her younger brother Shishi carefully to their mother Tiantian (left)

played fought, and got tangled with each other in a jumble (Fig. 9.15). Once we saw six yellow-haired pups of Yintangxiaotu at the bottom of our work platform 1#, jumping down one after another from the precipice to the tree canopies along the twigs which grow out from crag cervices. After jumping down, they turned to climb up the precipice again and repeated the same process. While playing, some of them pushed and pulled each other or jumbled together. They played again and again in this manner for more than 20 rounds in half an hour while their mothers were taking a rest on the side. We did not detect any goals that the yellow-haired pups might have other than jumping

Fig. 9.15 Yellow-haired pups kept chasing each other and fought playfully while their mothers were taking recess. Playing is the predominant behavior of white-headed langur pups. It is a required course of their growth. (Photo by Liang Zuhong)

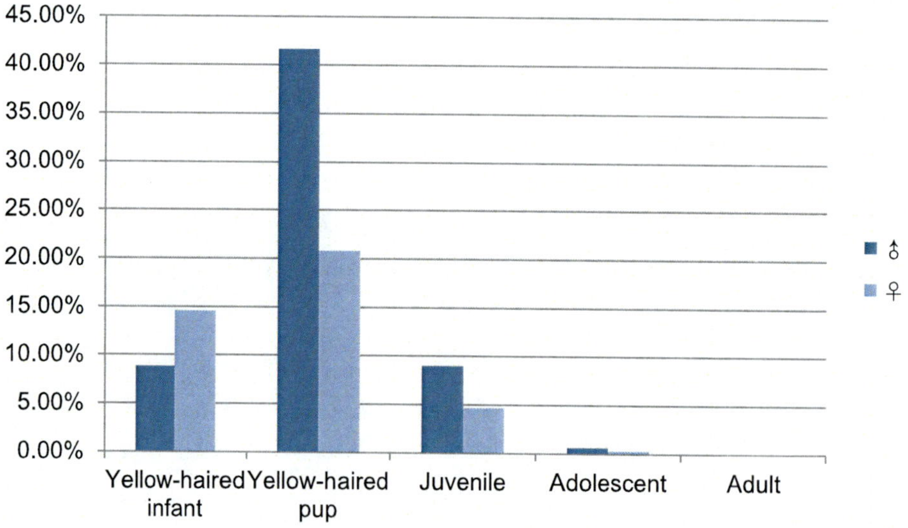

Fig. 9.16 Probabilities of play behavior in the different "sex–age groups" in "Yintangxiaotu family troop" in 2012

down from the precipice. They played just for the sake of playing.

Figure 9.16 shows that the probability of play behavior in yellow-haired infants was around 10%, then drastically increased to 20.85% (♀) and 41.67% (♂) at the age of a pup. After that, the probabilities dropped drastically to 4.72% (♀) and 9.02% (♂) at juvenile ages. Play behavior eventually vanished when the white-headed langurs reach adolescent age and are sexually matured.

9.7.5 Trait 5: All Family Members Forage Simultaneously, While Adult Males Have a Higher Foraging Probability

9.7.5.1 Foraging Probabilities of Adult Males and Adult Females

We found in our observations that in most of the time, white-headed langurs foraged as they moved. The time they spent on foraging practically included their time spent on feeding

Table 9.12 Comparing the foraging probabilities between adult females and the resident alpha male in the "Yintangxiaotu family troop" in year 2012[a]

	Mating season (%)	Giving birth season (%)	Suckling season (%)	Yearly average (%)	The proportion of forage in total activities in 1 day + night (%)[b]
Resident alpha ♂	48.10	39.84	36.59	41.51	19.01
Adult ♀	30.72	37.90	24.51	31.04	14.23
Yearly average	39.41	38.87	30.55	36.28	16.62

[a]Reproduction season defined by daytime activity
[b]Note: The mean 1 day + night activity hours of white-headed langurs in a year (11 h for daytime activities, 13 h for night rests (including slight movements))

and moving. We added the time spent on feeding and moving together in our calculations and compared that of different sex–age groups to find the differences.

As adults are fully grown and mature, their energy requirements and thus their forage activities would be constant in normal circumstances. Hence, we shall take only the total length of time spent on foraging by adult females in the "Yintangxiaotu family troop" in a full year to compare it with that of the resident alpha male. As the forage activities of immature individuals would fluctuate, they were excluded from calculation. Table 9.1 reveals that an adult ♂ has a higher foraging probability than an adult ♀.

9.7.5.2 Foraging Probability Differs in Accord with Sex-Age Groups

Table 9.13 shows the foraging and movement probabilities in a year for respective "sex-age groups" in the "Yintangxiaotu family troop." Figure 9.17a, b reveal that regardless of age, the forage and movement probabilities are higher in males than in females.

In respect of the probability of forage activities, the differences between males and females could be related to the differences between their body weight and growth rate. But at this moment, there are no solid statistics that can be used for making comparisons. In 2015, we placed an electronic scale in the wilderness and registered the body weight of Yintangxiaotu to be 10.2 kg, and its eldest son Wuque weighed 10 kg. In 1998, we obtained the weight of the oldest adult female in "FJC Big Cave troop" Heimama (dead from poaching) to be 8.5 kg. Since the bodies of adults ceased to develop, we may suppose that their body weights would remain relatively steady. In that case, the body weight ratio between adult male and adult female should be:

$$(10.2 + 10)/2 : 8.5 = 1.2 : 1.$$

That means adult males weigh about 15% more than adult females. This naturally leads one to conclude that males would have to expend more time in foraging.

9.7.6 Discussion

1. We shall take the mean of the total activities of Yintangxiaotu and that of the 13 adult females in 2012 (all of them are matured) to verify if our registers from observations are correct and reliable.
   There are behaviors that still need to be determined, as researchers observing in the wilderness could not yet make out the meanings behind behaviors such as climbing, jumping, and very short rests. Those behaviors, categorized as others in Table 9.10, had taken up a certain proportion of the total activities of the whole year. The mean probability of these "others" activities in a year by Yintangxiaotu (resident alpha male) was 7.62% and that of the adult females was 6.76%.
2. Then we sum up the percentages of all activities (defense, foster, forage, social, others) in the reproduction ecological year of 2012 listed in Table 9.10:
   That of Yintangxiaotu being $39.82\% + 0.37\% + 41.49\% + 10.68\% + 7.62\% = 99.98\%$ ($n = 486$)
   That of adult females being $1.19\% + 46.63\% + 31.04\% + 14.38\% + 6.76\% = 100.00\%$ ($n' = 4118$)

> This was indeed an immense amount of sample data ($n + n' = 4604$) that was to be computed; before the result came out, we were so anxious—could we be correct? We could only leave it to the hands of divinity. Now that the result comes out to be flawless, it feels like a heavy stone has been taken off my heart.
> (Pan Wenshi, 2015.12.10 03:45)

In the first half year of 2012, the "Yintangxiaotu family troop" inhabiting at FJC was an integrated polygynous family. In the daytime, the father, the mothers, and the offspring foraged in the forest and had stable food resources, and at night they slept safely high up on the precipitous precipice. The biggest concern in the family was the survival of yellow-haired newborns. All the others in the family took respective "roles" dictated by the activity rhythm of the three

Table 9.13 Foraging and movement probabilities from monitor records in the different "sex-age groups" in the "Yintangxiaotu family troop" in 2012

Activity	Male ♂						Female ♀						Subtotal	
	Resident alpha male		Sub-adult and adolescent		Juvenile		Adult		Sub-adult and adolescent		Juvenile			
	Times occurred	Probability (%)	Times occurred	Probability (%)	Times occurred	Probability (%)	Times occurred	Probability (%)	Times occurred	Probability (%)	Times occurred	Probability (%)	Times occurred	Probability (%)
Forage	104/486	21.40	329/927	35.49	575/2118	27.15	553/4118	13.43	112/876	12.79	64/401	15.96	1737/8926	19.46
Movement	93/486	19.14	329/927	35.49	622/2118	29.37	608/4118	14.76	274/876	31.28	115/401	28.68	2041/8926	22.87

Notes: The denominator denotes the total number of monitor recordings on that sex–age group, and the numerator denotes the times subject activity occurred in the recordings (probability accounted on activities in daytime only)

Fig. 9.17 (**a**) Forage probabilities in the different "sex–age groups" in the "Yintangxiaotu family troop" in 2012. (**b**) Movement probabilities in the different "sex–age groups" in the "Yintangxiaotu family troop" in 2012

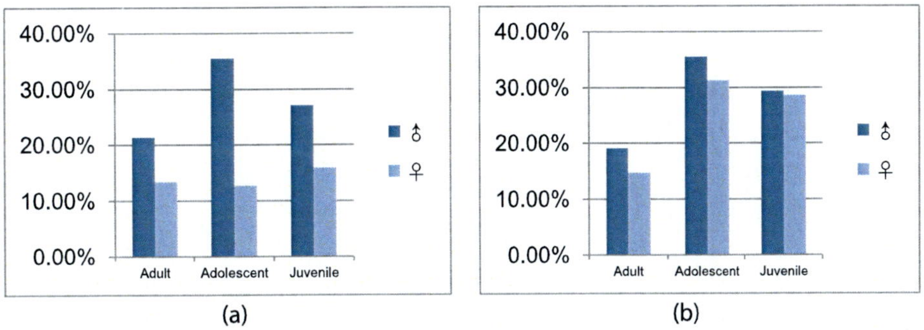

reproduction ecological seasons in the full year. Their activity patterns benefited each other in the family and maintained the birth and survival rate of pups, keeping the mortality rate at the lowest possible level, thereby enhancing the genetic fitness of the population.

White-headed langurs live in troops, and each troop is a society. The activities of each individual are meant for its own survival as well as bringing the most benefit to the continuation of the population. We would like to end this chapter with an excerpt from our observation registers in the wilderness in which we witnessed a rarely seen episode of nursing by a primate species other than humans. It is a living example of white-headed langur midwifery; the 14-year-old mother langur Pingping taking up a role of a midwife in assisting her 5-year-and-2-month-old daughter Yumei in delivering an infant (Fig. 9.18).

Date: 2013.03.11

Hour: 18:45

Place: FJC Big Cave, Tree House

18:45: Mother langur Pingping in the night shelter was sitting at location 4# with the yellow-haired infant pup in her embrace. It was born to her just 2 h ago.

19:00: About one meter to the left of Pingping, a young female langur behaved like she was about to deliver. From the black blotches on the triangular skin patch on her groin we recognized she was Yumei, about to give birth to her first infant. It was a pleasant surprise that the mother and daughter were both giving birth on the same date.

20:00: Yumei appeared to be suffering labor throes. In a moment, she raised her buttock and in another moment she prostrated on the rocks to rest. She kept repeating this over and over. Pingping was resting with her newborn infant by the side, and she did not bother Yumei.

20:05: Pingping, embracing her yellow-haired infant pup, moved about one meter to the right. Her place was taken up by another young female langur, which fell into sleep right away without giving any notice to Yumei's behavior.

21:06: The amniotic sac in Yumei ruptured and fluid were dripping down her groin.

21:27: Yumei raised her buttock; the vagina opening was slightly dilated.

21:39: Yumei lifted her buttock to increase her exertion force. The face of the infant appeared, and it looked a bit purplish under the light beam. After a while, Yumei sat down again to rest.

21:40: Yumei again raised her buttock, and twisted her body to touch with her left hand the crown of the infant, which had come out just a little.

21:45: Yumei raised her buttock and forced. The whole of the infant's head came out but the infant's eyes were tightly shut.

21:46: Yumei prostrated to rest for a while and then lifted up her buttock again. Gripping on the rocks in front with her both hands, she exerted more force and the shoulders of the infant came out. Immediately after, two mini hands came out of the birth canal. Then the infant spread the fingers and slightly opened its eyes.

21:47′20″: Yumei, holding the neck of the infant with her left hand, was ready to pull it out towards her head. At that moment, Pingping on the right, with her yellow-haired infant pup in her embrace, abruptly dashed over the bodies of the sleeping female langurs to reach Yumei.

21:47′26″: Pingping, with the newborn infant of hers still clasping in her embrace, half-squatted behind Yumei, extended both her hands to hold the shoulders of Yumei's infant and pulled. We were all nervous but excited at this sight.

21:47′44″: In 18 s, the infant was pulled out from the birth canal by grandmother Pingping. The young mother Yumei was so exhausted that it fell in a kneeling position with her head rested against a rock. Pingping then had two infants in her embrace, started licking Yumei's infant. After a while, Yumei turned round, picked up the drenched tail of her newborn infant and licked it.

21:50: Now with two infants in her embrace, Pingping rose and pulled to her right. The umbilical cord together with the placenta was yanked out of Yumei's birth canal. The one young female langur that was sleeping behind them woke, turned its head to look, and stretched its hands, wishing to touch the infant

21:52: Pingping, holding two infants on her chest, moved to the middle of location 4# but Yumei did not go after her infant. One old female langur moved to the side of Pingping to take a look and made one touch on the infant with her hand.

21:53: Yumei came to Pingping and her infant, and started eating the placenta. Five or six female langurs came near to take looks and tried to feel the infant.

21:58: Pingping kept holding the two infants until the umbilical cord was broke away from the infant. During that time, a female langur by her side had taken Pingping's infant into her embrace for 4 min before returning it to Pingping.

22:06: Yumei finished eating the placenta and reached her arms to take the infant into her embrace, but she got Pingping's infant. Pingping did not refuse and kept holding Yumei's infant in her embrace. The mother and daughter each holding an infant in their embrace reclined against each other and rested (In the early morning of the next day we found they had returned each other's infant).

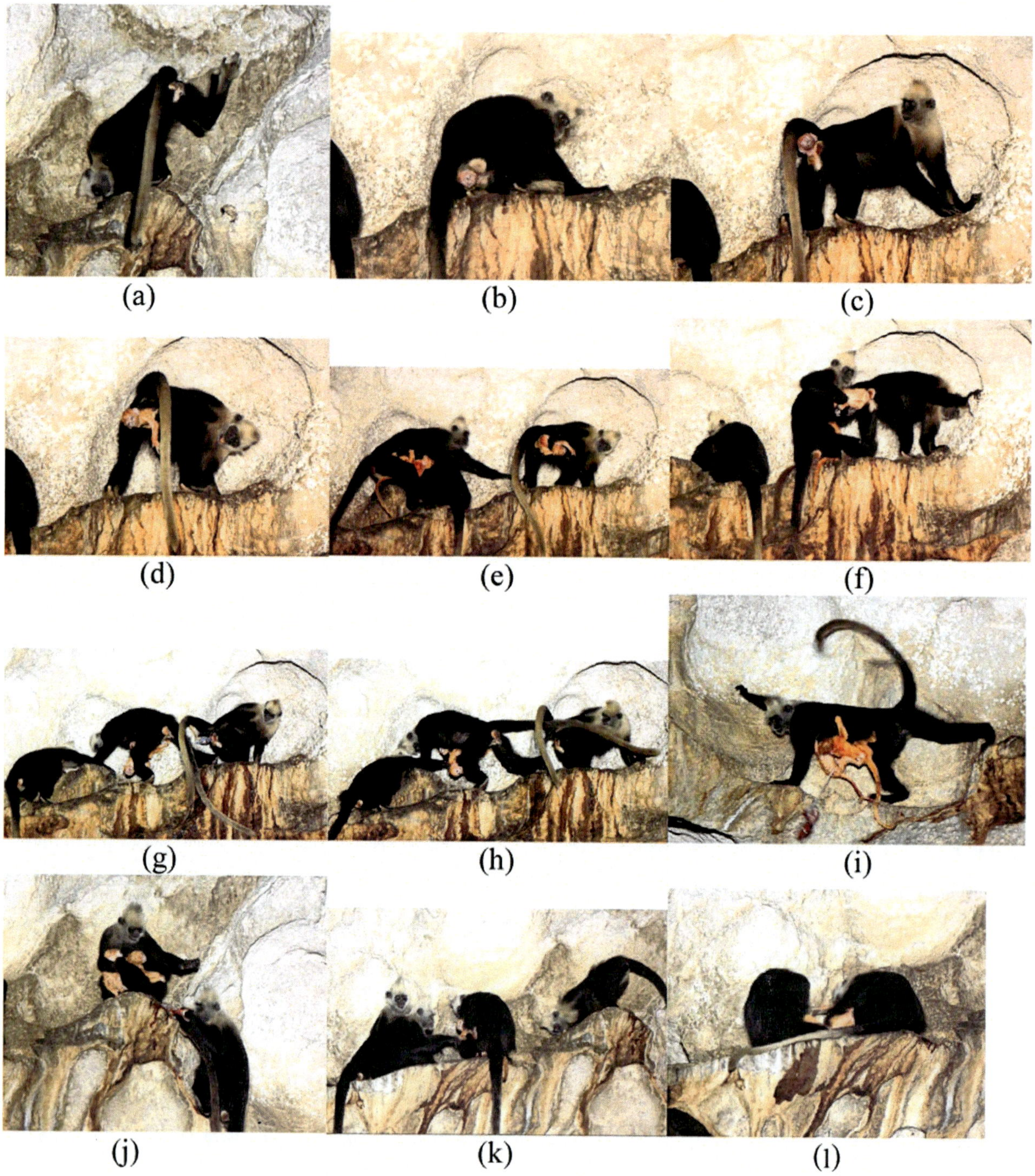

Fig. 9.18 The process of Pingping (mother) taking midwife "role" assisting Yumei (daughter) deliver an infant. (Photo by Liang Zuhong). (**a**) 19:47 Yumei suffering labor throes. (**b**) 21:36 The vaginal opening dilated; face of the infant appeared. (**c**) 21:45 The whole of the infant's head came out, its eyes tight shut. (**d**) 21:46 Forelimbs of the infant came out; its eyes opened a slit. (**e**) 21:47 Yumei holding the infant's neck with her left hand ready to pull the infant out; Pingping dashed to Yumei from her right. (**f**) 21:47′26" Pingping squatted with her own infant in her embrace and then with both hands holding the shoulders of Yumei's infant tugged with force. (**g**) 21:47′44" The infant was tugged out whole from the birth canal of the mother by Pingping. (**h**) 21:50 The placenta, with umbilical cord still attached to the infant, had been tugged out in full from the birth canal. (**i**) 21:52 Pingping holding two infants in her embrace moved to the middle of location $4^{\#}$ in Big Cave night shelter. (**j**) 21:53 Yumei came to Pingping and the infant, started eating the placenta; some old female langurs came over to have a look and felt the infant. (**k**) 21:53 The umbilical cord broke away from the infant; the other female langurs came to see and kiss the infant. (**l**) 22:06 Yumei clasped Pingping's infant in her embrace

Appendix

Table 9.14 ♂ (Male) behavior probabilities by different age groups in "Yintangxiaotu family troop" in the season of mating ecology of year 2012

Ecologic activity	Sequence	Behavior	Resident alpha ♂		Adolescent ♂		Juvenile ♂		Yellow-haired pup ♂		Yellow-haired infant pup ♂		Item total	Item % of total behaviors
			Times occurred	Probability (%)	Times occurred	Probability (%)	Times occurred	Probability (%)	Times occurred	Probability (%)	Times occurred	Probability (%)		
Defense	1	Protection												
	2	Vigilance	39	24.68	25	4.85							64	5.00
	3	In alert	27	17.09	30	5.83							57	4.46
	4	Intimidation	1	0.63									1	0.08
	5	Driving away	1	0.63	4	0.78	2	0.40					7	0.55
Sub-total			68	43.04	59	11.46	2	0.40					129	10.09
Forage	6	Recess	29	18.35	98	19.03	85	16.97			7	6.86	219	17.12
	7	Forage	24	15.19	155	30.10	141	28.14			4	3.92	324	25.33
	8	Drink												
	9	On the move	23	14.56	141	27.38	153	30.54			1	0.98	318	24.86
Sub-total			76	48.10	394	76.50	379	75.65			12	11.76	861	67.32
Foster	10	Carry							1	33.33	9	8.82	10	0.78
	11	Look after					8	1.60			14	13.73	22	1.72
	12	Snatch					1	0.20			1	0.98	2	0.16
	13	Suckle					12	2.40			4	3.92	16	1.25
	14	Lead					2	0.60	1	33.33	5	4.90	9	0.70
	15	Embrace					7	1.40	1	33.33	21	20.59	29	2.27
	16	Nestle	1	0.63			1	0.20			2	1.96	4	0.31
Sub-total			1	0.63			32	6.39	3	100.00	56	54.90	92	7.19
Social	17	Affection												
	18	Groom			11	2.14	6	1.20			3	2.94	20	1.56
	19	Cuddle			10	1.94							10	0.78
	20	Showing teeth in appeasement												
	21	Showing teeth in resentment					9	1.80					9	0.70
	22	Offering place												
	23	Holding place												
	24	Mounting					8	1.60					8	0.63

(continued)

Table 9.14 (continued)

Ecologic activity	Sequence	Behavior	Resident alpha ♂		Adolescent ♂		Juvenile ♂		Yellow-haired pup ♂		Yellow-haired infant pup ♂		Item total	Item % of total behaviors
			Times occurred	Probability (%)	Times occurred	Probability (%)	Times occurred	Probability (%)	Times occurred	Probability (%)	Times occurred	Probability (%)		
	25	Intimate behaviors												
	26	Kissing												
	27	Present anogenital												
	28	Copulation												
	29	Play			3	0.58	19	3.79			20	19.61	42	3.28
Sub-total					24	4.66	42	8.38			23	22.55	89	6.96
Others	30	Climbing	1	0.63	1	0.19	18	3.59			10	9.80	30	2.35
	31	Leaping	3	1.90	23	4.47	19	3.79					45	3.52
	32	Resting	3	1.90	5	0.97	9	1.80			1	0.98	18	1.41
	33	Sleeping	6	3.80	9	1.75							15	1,17
Sub-total			13	8.23	38	7.38	46	9.18			11	10.78	108	8.44
Total			158	100.00	515	100.00	501	100.00	3	100.00	102	100.00	1279	100.00
% by age group			12.35		40.27		39.17		0.23		7.97		100.00	

Note: Total observed in 79 days in the season of mating ecology (July, August, September, and October)

Table 9.15 ♀ (Female) behavior probabilities by different age groups in "Yintangxiaotu family troop" in the season of mating ecology of year 2012

Ecologic activity	Sequence	Behavior	Adult female ♀		Adolescent ♀		Juvenile ♀		Yellow-haired pup ♀		Yellow-haired infant pup ♀		Item total	Item % of total behaviors
			Times occurred	Probability (%)	Times occurred	Probability (%)	Times occurred	Probability (%)	Times occurred	Probability (%)	Times occurred	Probability (%)		
Defense	1	Protection												
	2	Vigilance												
	3	In alert												
	4	Intimidation												
	5	Driving away	11	1.92	3	3.61	1	2.38					15	1.60
Sub-total			11	1.92	3	3.61	1	2.38					15	1.60
Forage	6	Recess	71	12.39	12	14.46	6	14.29	8	6.15	5	4.59	102	10.89
	7	Forage	69	12.04	17	20.48	10	23.81	13	10.00	6	5.50	115	12.27
	8	Drink											0	0.00
	9	On the move	36	6.28	21	25.30	10	23.81	6	4.62			73	7.79
Sub-total			176	30.72	50	60.24	26	61.90	27	20.77	11	10.09	290	30.95
Foster	10	Carry	49	8.55					8	6.15	9	8.26	66	7.04
	11	Look after	92	16.06	3	3.61	1	2.38	20	15.38	29	26.61	145	15.47
	12	Snatch	3	0.524	2	2.41			1	0.77	2	1.83	8	0.85
	13	Suckle	23	4.01					4	3.08	4	3.67	31	3.31
	14	Lead	20	3.49					6	4.62	3	2.75	29	3.09
	15	Embrace	96	16.75	1	1.20	4	9.52	10	7.69	9	8.26	120	12.81
	16	Nestle	5	0.87					1	0.77			6	0.64
Sub-total			288	50.26	6	7.23	5	11.90	50	38.46	56	51.38	405	43.22
Social	17	Affection												
	18	Groom	28	4.89	10	12.05	1	2.38	3	2.31	1	0.92	43	4.59
	19	Cuddle	19	3.32	3	3.61							22	2.35
	20	Showing teeth in appeasement	21	3.66	4	4.82							25	2.67
	21	Showing teeth in resentment	2	0.35									2	0.21
	22	Offering place	2	0.35	2	2.41							4	0.43
	23	Holding place												
	24	Mounting												
	25	Intimate behaviors												
	26	Kissing												
	27	Present anogenital												
	28	Copulation												
	29	Play			2	2.41	4	9.52	28	21.54	24	22.02	58	6.19

(continued)

Table 9.15 (continued)

Ecologic activity	Sequence	Behavior	Adult female ♀		Adolescent ♀		Juvenile ♀		Yellow-haired pup ♀		Yellow-haired infant pup ♀		Item total	Item % of total behaviors
			Times occurred	Probability (%)	Times occurred	Probability (%)	Times occurred	Probability (%)	Times occurred	Probability (%)	Times occurred	Probability (%)		
Sub-total			72	12.57	21	25.30	5	11.90	31	23.85	25	22.94	154	16.44
Others	30	Climbing	9	1.57			2	4.76	16	12.31	16	14.68	43	4.59
	31	Leaping	5	0.87			2	4.76	1	0.77			8	0.85
	32	Resting	10	1.75	1	1.20			5	3.85	1	0.92	17	1.81
	33	Sleeping	2	0.35	2	2.41	1	2.38					5	0.53
Sub-total			26	4.54	3	3.61	5	11.90	22	16.92	17	15.60	73	7.79
Total			573	100.00	83	100.00	42	100.00	130	100.00	109	100.00	937	100.00
% by age group			61.15		8.86		4.48		13.87		11.63		100.00	

Note: Total observed in 79 days in the season of mating ecology (July, August, September, and October)

Table 9.16 ♂ (Male) behavior probabilities by different age groups in "Yintangxiaotu family troop" in the season of parturition ecology of year 2012

Ecologic activity	Sequence	Behavior	Resident alpha ♂ Times occurred	Resident alpha ♂ Probability (%)	Adolescent ♂ Times occurred	Adolescent ♂ Probability (%)	Juvenile ♂ Times occurred	Juvenile ♂ Probability (%)	Yellow-haired pup ♂ Times occurred	Yellow-haired pup ♂ Probability (%)	Yellow-haired infant pup ♂ Times occurred	Yellow-haired infant pup ♂ Probability (%)	Item total	Item % of total behaviors
Defense	1	Protection	12	9.76	4	2.22	1	0.09					17	1.05
	2	Vigilance	19	15.45	7	3.89	1	0.09					27	1.67
	3	In alert	8	6.50	13	7.22	3	0.28					24	1.48
	4	Intimidation	1	0.81									1	0.06
	5	Driving away					21	1.97					21	1.50
Sub-total			40	32.52	24	13.33	26	2.44					90	5.55
Forage	6	Recess	11	8.94	8	4.44	53	4.97	2	4.88			74	4.57
	7	Forage	6	4.88	33	18.33	201	18.84	6	14.63	1	0.48	247	15.24
	8	Drink	1	0.81	4	2.22	23	2.16					28	1.73
	9	On the move	31	25.20	72	40.00	301	28.21	7	17.07	2	0.95	413	25.48
Sub-total			49	39.84	117	65.00	578	54.17	15	36.59	3	1.43	762	47.01
Foster	10	Carry					8	0.75			24	11.43	32	1.97
	11	Look after					21	1.97	1	2.44	4	1.90	26	1.60
	12	Snatch									62	29.52	62	3.82
	13	Suckle					7	0.66	2	4.88	1	0.48	10	0.62
	14	Lead			1	0.56	29	2.72	1	2.44			31	1.91
	15	Embrace					30	2.81			86	40.95	116	7.16
	16	Nestle					3	0.28			5	2.38	8	0.49
Sub-total					1	0.56	98	9.18	4	9.76	182	86.67	285	17.58
Social	17	Affection	9	7.32									9	0.56
	18	Groom	10	8.13	17	9.44	55	5.15			11	5.24	93	5.74
	19	Cuddle	1	0.81			8	0.75					9	0.56
	20	Showing teeth in appeasement			1	0.56	3	0.28					4	0.25
	21	Showing teeth in resentment			1	0.56	1	0.09					2	0.12
	22	Offering place			2	1.11	2	0.19					4	0.25
	23	Holding place												
	24	Mounting	3	2.44	1	0.56	15	1.41	1	2.44			20	1.23
	25	Intimate behaviors			2	1.11	2	0.19					4	0.25
	26	Kissing	2	1.63									2	0.12
	27	Present anogenital												
	28	Copulation												
	29	Play			1	0.56	118	11.06	19	46.34			138	8.51

(continued)

Table 9.16 (continued)

Ecologic activity	Sequence	Behavior	Resident alpha ♂		Adolescent ♂		Juvenile ♂		Yellow-haired pup ♂		Yellow-haired infant pup ♂		Item total	Item % of total behaviors
			Times occurred	Probability (%)	Times occurred	Probability (%)	Times occurred	Probability (%)	Times occurred	Probability (%)	Times occurred	Probability (%)		
Sub-total			25	20.33	25	13.89	204	19.12	20	48.78	11	5.24	285	17.58
Others	30	Climbing			3	1.67	42	3.94	2	4.88	14	6.67	61	3.76
	31	Leaping			5	2.78	71	6.65					76	4.69
	32	Resting	3	2.44	5	2.78	31	2.91					39	2.41
	33	Sleeping	6	4.88			17	1.59					23	1.42
Sub-total			9	7.32	13	7.22	161	15.09	2	4.88	14	6.67	199	12.28
Total			123	100.00	180	100.00	1067	100.00	41	100.00	210	100.00	1621	100.00
% by age group			7.59		11.10		65.82		2.53		12.95		100.00	

Note: Total observed in 63 days in the season of parturition ecology (November, December, January, and February)

Table 9.17 ♀ (Female) behavior probabilities by different age groups in "Yintangxiaotu family troop" in the season of parturition ecology of year 2012

Ecologic activity	Sequence	Behavior	Adult female ♀		Adolescent ♀		Juvenile ♀		Yellow-haired pup ♀		Yellow-haired infant pup ♀		Item total	Item % of total behaviors
			Times occurred	Probability (%)	Times occurred	Probability (%)	Times occurred	Probability (%)	Times occurred	Probability (%)	Times occurred	Probability (%)		
Defense	1	Protection	2	0.22									2	0.15
	2	Vigilance	1	0.11									1	0.07
	3	In alert												
	4	Intimidation												
	5	Driving away	2	0.22									2	0.15
Sub-total			5	0.56									5	0.36
Forage	6	Recess	61	6.80	8	3.83	8	8.16	6	4.72	1	2.08	84	6.09
	7	Forage	37	4.12	7	3.35	7	7.14	14	11.02			65	4.71
	8	Drink												
	9	On the move	242	26.98	67	32.06	32	32.65	28	22.05	3	6.25	372	26.98
Sub-total			340	37.90	82	39.23	47	47.96	48	37.80	4	8.33	521	37.78
Foster	10	Carry	32	3.57	10	4.78	1	1.02	1	0.79	8	16.67	52	3.77
	11	Look after	48	5.35	7	3.35	4	4.08	7	5.51			66	4.79
	12	Snatch	37	4.12	23	11.00	5	5.10			13	27.08	78	5.66
	13	Suckle	23	2.56			2	2.04	3	2.36			28	2.03
	14	Lead	54	6.02	1	0.48	5	5.10	10	7.87			70	5.08
	15	Embrace	110	12.26	35	16.75	3	3.06	7	5.51	15	31.25	170	12.33
	16	Nestle	15	1.67	2	0.96			2	1.57			19	1.38
Sub-total			319	35.56	78	37.32	20	20.41	30	23.62	36	75.00	483	35.03
Social	17	Affection												
	18	Groom	99	11.04	20	9.57	6	6.12	4	3.15	1	2.08	130	9.43
	19	Cuddle	21	2.34	4	1.91	4	4.08	1	0.79			30	2.18
	20	Showing teeth in appeasement	16	1.78	3	1.44	4	4.08					23	1.67
	21	Showing teeth in resentment	15	1.67	2	0.96	4	4.08					21	1.52
	22	Offering place	1	0.11	1	0.48							2	0.15
	23	Holding place												
	24	Mounting	9	1.00									9	0.65
	25	Intimate behaviors												
	26	Kissing	1	0.11			1	1.02					2	0.15
	27	Present anogenital												
	28	Copulation												
	29	Play	1	0.11			4	4.08	25	19.69	4	8.33	34	2.47

(continued)

Table 9.17 (continued)

Ecologic activity	Sequence	Behavior	Adult female ♀		Adolescent ♀		Juvenile ♀		Yellow-haired pup ♀		Yellow-haired infant pup ♀		Item total	Item % of total behaviors
			Times occurred	Probability (%)	Times occurred	Probability (%)	Times occurred	Probability (%)	Times occurred	Probability (%)	Times occurred	Probability (%)		
Sub-total			163	18.17	30	14.35	23	23.47	30	23.62	5	10.42	251	18.20
Others	30	Climbing	42	4.68	2	0.96	6	6.12	16	12.60	3	6.25	69	5.00
	31	Leaping	9	1.00	6	2.87	1	1.02	3	2.36			19	1.38
	32	Resting	1	0.11									1	0.07
	33	Sleeping	18	2.01	11	5.26	1	1.02					30	2.18
Sub-total			70	7.80	19	9.09	8	8.16	19	14.96	3	6.25	119	8.63
Total			897	100.00	209	100.00	98	100.00	127	100.00	48	100.00	1379	100.00
% by age group			65.05		15.16		7.11		9.21		3.48		100.00	

Note: Total observed in 63 days in the season of parturition ecology (November, December, January, and February)

Table 9.18 ♂ (Male) behavior probabilities by different age groups in "Yintangxiaotu family troop" in the season of rearing ecology of year 2012

Ecologic activity	Sequence	Behavior	Resident alpha ♂		Adolescent ♂		Juvenile ♂		Yellow-haired pup ♂		Yellow-haired infant pup ♂		Item total	Item % of total behaviors
			Times occurred	Probability (%)	Times occurred	Probability (%)	Times occurred	Probability (%)	Times occurred	Probability (%)	Times occurred	Probability (%)		
Defense	1	Protection	36	17.56	8	3.45							44	3.27
	2	Vigilance	27	13.17	5	2.16							32	2.38
	3	In alert	22	10.73	9	3.88							31	2.31
	4	Intimidation												
	5	Driving away	5	2.44	10	4.31	3	0.55					18	1.34
Sub-total			90	43.90	32	13.79	3	0.55					125	9.30
Forage	6	Recess	27	13.17	25	10.78	67	12.18	1	25.00	6	1.70	126	9.38
	7	Forage	7	3.41	10	4.31	28	5.09					45	3.35
	8	Drink	2	0.98	1	0.43	1	0.18			1	0.28	5	0.37
	9	On the move	39	19.02	116	50.00	168	30.55	1	25.00	2	0.57	326	24.26
Sub-total			75	36.59	152	65.52	264	48.00	2	50.00	9	2.55	502	37.35
Foster	10	Carry					4	0.73			53	15.01	57	4.24
	11	Look after					18	3.27			46	13.03	64	4.76
	12	Snatch									23	6.52	23	1.71
	13	Suckle					20	3.64			20	5.67	40	2.98
	14	Lead	1	0.49			17	3.09					18	1.34
	15	Embrace					22	4.00	1	25.00	101	28.61	124	9.23
	16	Nestle					10	1.82			17	4.82	27	2.01
Sub-total			1	0.49			91	16.55	1	25.00	260	73.65	353	26.26
Social	17	Affection	3	1.46									3	0.22
	18	Groom	3	1.46	7	3.02	10	1.82			4	1.13	24	1.79
	19	Cuddle			10	4.31	18	3.27			1	0.28	29	2.16
	20	Showing teeth in appeasement	2	0.98	11	4.74	23	4.18					36	2.68
	21	Showing teeth in resentment	1	0.49			6	1.09					7	0.52
	22	Offering place					3	0.55					3	0.22
	23	Holding place	1	0.49	1	0.43	3	0.55					3	0.37
	24	Mounting	5	2.44	2	0.86	23	4.18					30	2.23
	25	Intimate behaviors												
	26	Kissing	6	2.93							1	0.28	7	0.52
	27	Present anogenital	1	0.49									1	0.07
	28	Copulation	2	0.98									2	0.15
	29	Play			2	0.86	54	9.82	1	25.00	39	11.05	96	7.14

(continued)

Table 9.18 (continued)

Ecologic activity	Sequence	Behavior	Resident alpha ♂		Adolescent ♂		Juvenile ♂		Yellow-haired pup ♂		Yellow-haired infant pup ♂		Item total	Item % of total behaviors
			Times occurred	Probability (%)	Times occurred	Probability (%)	Times occurred	Probability (%)	Times occurred	Probability (%)	Times occurred	Probability (%)		
Sub-total			24	11.71	33	14.22	140	25.45	1	25.00	45	12.75	243	18.08
Others	30	Climbing					15	2.73			35	9.92	50	3.72
	31	Leaping	3	1.46	9	3.88	27	4.91			3	0.85	42	3.13
	32	Resting									1	0.28	1	0.07
	33	Sleeping	12	5.85	6	2.59	10	1.82					28	2.08
Sub-total			15	7.32	15	6.47	52	9.45			39	11.05	121	9.00
Total			205	100.00	232	100.00	550	100.00	4	100.00	353	100.00	1344	100.00
% by age group			15.25		17.26		40.92		0.30		26.26		100.00	

Note: Total observed in 65 days in the season of rearing ecology (March, April May, and June)

Table 9.19 ♀ (Female) behavior probabilities by different age groups in 'Yintangxiaotu family troop" in the season of rearing ecology of year 2012

Ecologic activity	Sequence	Behavior	Adult female ♀		Adolescent ♀		Juvenile ♀		Yellow-haired pup ♀		Yellow-haired infant pup ♀		Item total	Item % of total behaviors
			Times occurred	Probability (%)	Times occurred	Probability (%)	Times occurred	Probability (%)	Times occurred	Probability (%)	Times occurred	Probability (%)		
Defense	1	Protection												
	2	Vigilance												
	3	In alert												
	4	Intimidation												
	5	Driving away	29	1.10	14	2.40	4	1.53			1	0.13	48	1.12
Sub-total			29	1.10	14	2.40	4	1.53			1	0.13	48	1.12
Forage	6	Recess	242	9.14	52	8.90	25	9.58			7	0.88	326	7.59
	7	Forage	73	2.76	16	2.74	8	3.07			10	1.25	107	2.49
	8	Drink	4	0.15	1	0.17							5	0.12
	9	On the move	330	12.46	186	31.85	73	27.97	1	50.00	4	0.50	594	13.83
Sub-total			649	24.51	255	43.66	106	40.61	1	50.00	21	2.63	1032	24.03
Foster	10	Carry	279	10.54	28	4.79	6	2.30			112	14.00	425	9.90
	11	Look after	225	8.50	33	5.65	33	12.64			114	14.25	405	9.43
	12	Snatch	33	1.25	40	6.85	7	2.68			35	4.38	115	2.68
	13	Suckle	157	5.93			14	5.36			52	6.50	223	5.19
	14	Lead	38	1.44			13	4.98			3	0.38	54	1.26
	15	Embrace	587	22.17	47	8.05	14	5.36			215	26.88	863	20.09
	16	Nestle	113	4.27	1	0.17	3	1.15			39	4.88	156	3.63
Sub-total			1432	54.08	149	25.51	90	34.48			570	71.25	2241	52.18
Social	17	Affection												
	18	Groom	100	3.78	43	7.36	8	3.07			14	1.75	165	3.84
	19	Cuddle	69	2.61	22	3.77	7	2.68					98	2.28
	20	Showing teeth in appeasement	80	3.02	27	4.62	6	2.30					113	2.63
	21	Showing teeth in resentment	26	0.98	8	1.37	4	1.53					38	0.88
	22	Offering place	28	1.06	6	1.03	1	0.38					35	0.81
	23	Holding place	17	0.64	4	0.68	1	0.38					22	0.51
	24	Mounting	3	0.11							1	0.13	4	0.09
	25	Intimate behaviors												
	26	Kissing	2	0.08	1	0.17					1	0.13	4	0.09
	27	Present anogenital	1	0.04									1	0.02
	28	Copulation	2	0.08									2	0.05
	29	Play			1	0.17	11	4.21	1	50.00	112	14.00	125	2.91

(continued)

Table 9.19 (continued)

Ecologic activity	Sequence	Behavior	Adult female ♀		Adolescent ♀		Juvenile ♀		Yellow-haired pup ♀		Yellow-haired infant pup ♀		Item total	Item % of total behaviors
			Times occurred	Probability (%)	Times occurred	Probability (%)	Times occurred	Probability (%)	Times occurred	Probability (%)	Times occurred	Probability (%)		
Sub-total			328	12.39	112	19.18	38	14.56	1	50.00	128	16.00	607	14.13
Others	30	Climbing	149	5.63	11	1.88	7	2.68			74	9.25	241	5.61
	31	Leaping	14	0.53	10	1.71	9	3.45			1	0.13	34	0.79
	32	Resting	4	0.15							2	0.25	6	0.34
	33	Sleeping	43	0.62	33	5.65	7	2.68			3	0.38	86	2.00
Sub-total			210	7.93	54	9.25	23	8.81			80	10.00	367	8.54
Total			2648	100.00	584	100.00	261	100.00	2	100.00	800	100.00	4295	100.00
% by age group			61.65		13.60		6.08		0.05		18.63		100.00	

Note: Total observed in 65 days in the season of rearing ecology (March, April May, and June)

References

Brown AF, Webb HM. Temperature relations of an endogenous daily rhythmicity in the fiddler crab, Uca. Physiol Zool. 1948;21 (4):371–81.

Carpenter CR. A field study in Siam of the behavior and social relations of the gibbon (*Hylobates lar*). Baltimore, MA: Johns Hopkins Press; 1940.

Fossey D, Harcourt AH. Feeding ecology of free ranging mountain gorillas (*Gorilla gorilla beringei*). In: Clutton-Brock TH, editor. Primate ecology: studies of feeding and ranging behavior in lemurs, monkeys and apes. London: Academic; 1977. p. 539–56.

Gogan PJP. Some aspects of nutrient utilization by Burchell's Zebra (*Equus burchelli bohmi* Matschie) in the Serengeti-Mara Region East Africa. Master thesis. College Station, TX: Texas A & M University; 1973.

Jarman MV, Jarman PJ. Daily activity of impala. East Afr Wildl J. 1973;11:75–92.

Schaller GB. The Serengeti lion: a study of predator-prey relations. Chicago, IL: University of Chicago Press; 1972.

Schaller GB, Hu JC, Pan WS, et al. the giant pandas of Wolong. Chicago, IL: Chicago University Press; 1985.

Stirling I. Midsummer observations on the behavior of wild polar bears (*Ursus maritimus*). Can J Zool. 1974;52:1191–8.

Wyatt JR, Eltringham SK. The daily activity of the elephant in the Rwenzori National Park, Uganda. East Afr Wildl J. 1974;12:273–89.

Yuchang S. Animals ethology (动物行为学). Beijing: Peking University Press; 2005.

Check for updates

10 The Strategy of White-Headed Langur Reproductive Behaviors

Abstract

White-headed langurs have sexual contacts with one another year-round, but there are distinct reproductive peak seasons. Copulations peak between July and August, while the peak season for parturitions is between December and January. The mean age of the female's first parturition is at 5.4 years, with the youngest at 4.8 years and the oldest at 6.2 years. We observed 16 female langurs giving birth to 78 pups, and accurately registered 61 inter-birth intervals of adult females. The average inter-birth interval is 784.3 days or 25.8 months. That is to say, each adult female white-headed langur is able to reproduce 0.93 infants in a year. The length of the suckling period of white-headed langurs is 456 ± 28 days, which is equivalent to 15 ± 0.9 months. Two pieces of data registered from field observations show that the pregnancy period of a female white-headed langur is 138–139 days.

Keywords

Copulation · Parturition · Lactation period · Inter-birth interval · Pregnancy period

The successful restoration of a regional population of a species depends first and foremost on the species' successful reproduction, while the population growth rate is decided by the size of its effective reproduction population. To carry out specific research in this topic, we sorted the reproduction records of the nine white-headed langur families which successively inhabited FJC Big Cave from 1996 to 2015, analyzed the 83 parturition registers by 22 female langurs of FJC Big Cave, kept tracking the growth of their infants and the replacement of their families, and recorded their ecological traits such as their reproduction season, sexual maturity age, sex ratio, interval between births, birth rate, mortality rate, and the age structure in a family troop.

10.1 Peak Seasons for White-Headed Langurs' Copulation and Parturition

It was established in science that in monkeys, apes, and hominids, copulation between males and females may take place at any time of the year. Nevertheless, this knowledge did not make our research of white-headed langur reproduction easier. We had to start with studying their reproductive activities, for if copulations between the different sexes of white-headed langurs do take place at any time, how can we tell which ones are genuine and which ones are not? Are there truly no seasonal influences? We came to note that there were more cases in which females in their estrus period would woo a male than males going after the females. That is quite the opposite to the courting patterns in other mammals.

Since white-headed langur families are polygynous with one single male dominating over a troop of females, male chauvinism is a common trait of males. Hence, we put our attention to documenting the "presenting anogenital area" behaviors initiated by female langurs and the mounting behaviors by male langurs.

10.1.1 Copulation Season

"Presenting anogenital area" is a behavior particular of female white-headed langurs, used to invite a male to mate. We sometimes documented this behavior as "copulation invitation." The behavior may sometimes happen between females, but this was extremely rare. The first account of "copulation invitation" in our registers took place on an early morning in August 1997 at Muzishan.

It was only until after 2008 that our researchers were able to observe the details in white-headed langur copulations at a close distance. Below is one of our field observation registers on how white-headed langurs copulate.

Date: April 18, 2013.

Time: 10:45–12:00 in the morning. Sky was sunny.

W. Pan, *The Population Ecology of White-Headed Langur*, https://doi.org/10.1007/978-981-33-4118-0_10

Fig. 10.1 Copulation between adult male langur Shengdi and female langur Xiaoxiao (Photo by Liang Zuhong). (**a**) Xiaoxiao presenting her anogenital area; viscous secretion was seen dripping down from her groin. (**b**) Xiaoxiao presenting her anogenital area in the face of Shengdi; the sex organ of Shengdi became erect. (**c**) While Shengdi (at the back) was mounting on Xiaoxiao (middle), the other female langurs came to disturb. (**d**) Another female langur presented her anogenital area at the same time as Xiaoxiao (left). (**e**) Shengdi checking on the genital of Xiaoxiao. (**f**) Shengdi copulating with Xiaoxiao

Place: In the bush at mid-hill between FJC Big Cave west and Guoyuan pass.

10:45 After a short recess from forage, the 'Big Cave troop' moved towards the northwest of Big Cave to arrive at Guoyuan Pass. They were about 10 meters from us, and we saw clearly 5–6 female langurs following the male langur Shengdi, staying close by its side and presenting their anogenital area in turns.

11:05 Female langur Xiaoxiao crawled on the floor presenting her anogenital area to the adult male langur sitting behind her. She straightened her back in a horizontal position, and every now and then raised her hind leg to tenderly touch the male langur Shengdi, and swung her head left and right in fast motions. Her tail swayed to the side of her body and she raised her buttock, almost touching Shengdi's face.

11:10 The white triangular shaped area on the groin of Xiaoxiao was moist and viscous secretion was dripping down her right hind leg (Fig. 10.1a).

11:25 The sex organ of male langur Shengdi became erect (Fig. 10.1b).

11:26′18″ Shengdi, holding Xiaoxiao's waist, mounted on her, and came down after 4 seconds. Shengdi then moved about 3 meters to the left. Not long afterwards, Xiaoxiao and another female langur presented their anogenital area respectively on the left and right of Shengdi, but Shengdi paid no attention.

11:29 Xiaoxiao kept her pose. Shengdi looked at it and touched her genital with its left hand, and it appeared like it was checking what state the female langur was in.

11:29′34″ Shengdi again mounted on Xiaoxiao and held Xiaoxiao's calves tight with its hind legs to prevent her from moving about. At that time, 3 other female langurs came to disturb, and one of them scratched Xiaoxiao's face with her hands. After 6 seconds, Shengdi came down from Xiaoxiao's back (Fig. 10.1c).

11:33′48″ Shengdi mounted on Xiaoxiao again for 8 seconds.

11:34 Another female langur presented their anogenital area at the same time as Xiaoxiao (Fig. 10.1d). Shengdi again checked on Xiaoxiao (Fig. 10.1e).

11:35′44″ Shengdi copulated with Xiaoxiao for 6 seconds (Fig. 10.1f).

11:36′40″ Again, Shengdi copulated with Xiaoxiao for 6 seconds.

11:37 Shengdi copulated with Xiaoxiao for 8 seconds. 3 female langurs came to disturb and one of them came in front of Xiaoxiao, held Xiaoxiao's face, and shook it.

Table 10.1 The number of "presenting anogenital area" by female white-headed langurs and "mounting" by male white-headed langurs registered in each month from 2010 to 2015

	January	February	March	April	May	June	July	August	September	October	November	December	Item total
♀ presenting anogenital area	21	15	38	15	6	19	80	73	6	21	43	25	362
♂ Mounting	3	6	6	6	3	17	9	16	6	8	23	3	106
Monthly total and grand total	24	21	44	21	9	36	89	89	12	29	66	28	468

11:37′34″ Shengdi copulated with Xiaoxiao for 8 minutes. One female langur grabbed the hair on Shengdi's head to disturb them.

11:38 In 40 seconds, Shengdi mounted on Xiaoxiao twice, each time for 4 seconds.

11:39′06″ Shengdi held tight Xiaoxiao's waist from behind, with both legs propping behind Xiaoxiao's calves. Shengdi's sex organ inserted into the vagina of the female langur for 12 seconds. One female langur came and held and swayed Xiaoxiao's face while 2 other female langurs watched by the side. Shengdi paused for 6 seconds then again copulated with Xiaoxiao for 7 seconds.

Afterwards, there were 13 female langurs gathered round Shengdi. 3 to 4 of them kept presenting their anogenital area to Shengdi, but Shengdi paid no attention. Soon the troop slowly moved up along the Guoyuan Pass.

By presenting their anogenital area, females in estrus invite the resident alpha male to copulate. But the male does not accept every time. Sometimes the male would have no interest in copulating at all. In fact, experienced male langurs would check meticulously to see if the inviting female is in ovulation. One researcher said that once he had seen a resident alpha male dab the vaginal secretion from a female langur with its finger and taste it in its mouth. If that would be true, would perhaps the fungiform papillae at the tongue tip of white-headed langurs also function to tell the biological reproductive condition of females? African lions and many cloven-hoofed animals such as oxen and sheep have fungiform papillae in their mouth cavities that can catch from the air the ovulation signals of their female individuals. Male white-headed langurs living polygynous lives meeting simultaneous "copulation invitations" from females in estrus must utilize economically and effectively its sperm to ensure the rate of females getting pregnant. For this reason, male white-headed langurs have to pick the females that are truly ovulating to copulate with.

Like all other primates in higher evolutionary ranks, female white-headed langurs are capable of copulation at the prophase of pregnancy, during pregnancy, and even a day before parturition. Hence, we saw all year round in the wilderness love affairs of white-headed langurs. Some were real copulations and some were merely mountings without real copulation.

Between 2010 and 2015, we had captured with HD video cameras in Feijichang 362 counts of females presenting their anogenital areas toward males and 106 counts of males mounting on females. Data in Table 10.1 reveals that white-headed langurs had "copulation" and "presenting anogenital area" behaviors in all the months in a year, but July and August had the most with 178 counts, which accounts to 38% of the total. The data further indicates that both the behaviors had the same peak months, which implies that there is a certain consistent inclination in the reproduction biology of both the sexes of the white-headed langur. It leads us to conjecture that a majority of female white-headed langurs ovulate in July and August every year and that summer is the copulation season for white-headed langurs.

10.1.2 First Parturition Age of Females

From 1998 to 2010, we had 15 registers of female white-headed langurs giving birth for their first time. Among which, the youngest age was 4.8 years and the oldest was 6.2 years, with the mean age of the first parturition at 5.4 years. Table 10.2 lists the details.

10.1.3 Peak Season for Births

Since there is a distinct peak season for white-headed langur copulations, consequently there will be a seasonal peak of births. Such comes as the causal effect regulated by the biology inherent in white-headed langurs.

Table 10.3 provides the 98 births registered in our research in Feijichang over the months from November 1996 to December 2015. In sequence of months, November accounted for 15.3% of births, December had 20.4%, January had 19.4%, February had 16.3%, March had 13.3%, April had 9.2%, and then it dropped to below 4.1% from May onwards. No births were registered in August, September, and October. A total of 70 pups were born from November to February, which amounted to 71.4% of the years' total. It can be seen from Table 10.3 and Fig. 10.2 that winter is the high season for white-headed langur parturitions, and the rate peaks every year in the period between November to February in the next year.

Table 10.2 The first parturition ages of the 15 female white-headed langurs registered in Feijichang from 1998 to 2010

Name of the female langur	Date of birth	Date of first parturition	Name and sex of the first born	Age of the female at first parturition (years)
Yingying	1998.12.15	2004.02.01	Baobao ♂	5.1
Pingping	1998.12.19	2004.12.01	Beibei ♀	5.9
Lanlan	1999.01.01	2004.02.01	Tiantian ♀	5.1
Lanqi	2003.03.05	2007.12	Qiqi ♀	4.8
Tiantian	2004.02.01	2009.01.27	Wenwen ♀	5.0
Beibei	2004.12.01	2010.01.14	Gualai ♂	5.1
Arxi	2005.01.05	2010.01.09	Xiaojiu ♀	5.0
Arlu	2005.02.12	2010.01.18	Hanlin ♂	4.9
Aryu	2005.03.02	2011.01.02	Yiyi ♂	5.8
Mimi	2007.12	2014.02.12	Midzai ♂	6.2
Yumei	2008.01.08	2014.03.11	Xiao-M ♀	6.2
Zhenzhu	2008.03	2013.12.06	Xiaozuan ♂	5.7
Wenwen	2009.01.27	2014.04.29	Wennu ♀	5.3
Xiaojiu	2010.01.09	2015.04.09	Julie ♀	5.3
Qingming	2010.04.05	2015.11.26	Yousheng	5.6
Mean first parturition age of female langur (years)				5.4

10.1.4 Inter-birth Intervals

The term "inter-birth interval" refers to the time between births of infants by a female. The interval is the time span in which an adult female has to go through before giving birth to the next infant, which is the time needed for a full reproduction process from copulation to getting pregnant to suckling to weaning the infant (Fig. 10.3).

From 1996 to 2015, we had observed and registered 16 female langurs in the "FJC troop" giving birth to 78 pups, among which 61 inter-birth intervals had been distinctly registered and are listed in Table 10.4. (Female Pingping gave birth to twins Nannan and Ningning on April 23, 2014, and that was recorded as two infants but only as one inter-birth interval.)

We analyzed the statistics in Table 10.4 and noted the following:

- Inter-birth intervals are not the same every time even with the same female white-headed langur. For instance, Arzhen had inter-birth intervals from 686 to 1538 days.
- We found that the infant's death in the lactation period is a significant factor in shortening the inter-birth interval of a white-headed langur. Pingping had very short inter-birth intervals twice, at 329 and 377 days, respectively, and in both the instances, the preceding pups had been killed in infanticide.
- From the data in Table 10.4, we took the average of the inter-birth intervals of the 16 adult females in the FJC troop under natural circumstances, including circumstances of incidental death, infanticide by intruding males, and miscarriage (see remarks in Table 10.4), and found it to be 784.3 days ($n = 61$) or 25.8 months. That is to say, each adult female white-headed langur is able to reproduce 0.93 infants per year.

While the mean inter-birth interval of the total 16 adult females is 784.3 days, from Table 10.4 we found that this figure was higher than the mean of ten females (62.5% of the total females) among them. That indicates observations in the wilderness should be carried on for a longer period to obtain more truthful results.

10.1.5 Lactation Period

The lactation period is a significant parameter in white-headed langur reproduction.

Observations in the wilderness had allowed us to register distinctly the beginning days of a suckling, for from the day the infant was born, it needed to be suckled by its mother. Yet the length of the lactation period is affected by various environmental factors, social factors, and the biological factors in the mother. The day of weaning is accordingly subjected to influences like social factors, environmental factors, and biological factors in the infant.

In order to precisely estimate the length of the lactation period of the white-headed langur pups, we sort out in Table 10.5 the birth dates, growths, and suckling days of the four similar aged pups in the "Yintangxiaotu family troop" in Feijichang.

Every mother langur would try her best in rearing its pups (Fig. 10.4). Statistics in Table 10.5 shows us that mother white-headed langurs would wean their pups at 15–16 months (we consider white-headed langurs in that age as juveniles or pups). Yet the mother langur did not appear to be in a rush to

Table 10.3 The distribution of white-headed langur pups born in FJC Big Cave and Small Cave areas over the months from 1996 to 2015

	August	September	October	November	December	January	February	March	April	May	June	July	Yearly total
1996.11–1998.04	0	0	0	7	8	2	1	1	1	0	0	0	20
1998.04–2002.07	0	0	0	3	0	5	1	1	0	0	0	0	10
2002.07–2006.09	0	0	0	1	6	1	5	3	1	0	0	0	17
2006.09–2012.07	0	0	0	2	5	9	3	3	4	3	0	0	29
2012.07–2015.12	0	0	0	2	1	2	6	5	3	1	1	1	22
Monthly total and grand total	0	0	0	15	20	19	16	13	9	4	1	1	98
%	0	0	0	15.3	20.4	19.4	16.3	13.3	9.2	4.1	1	1	

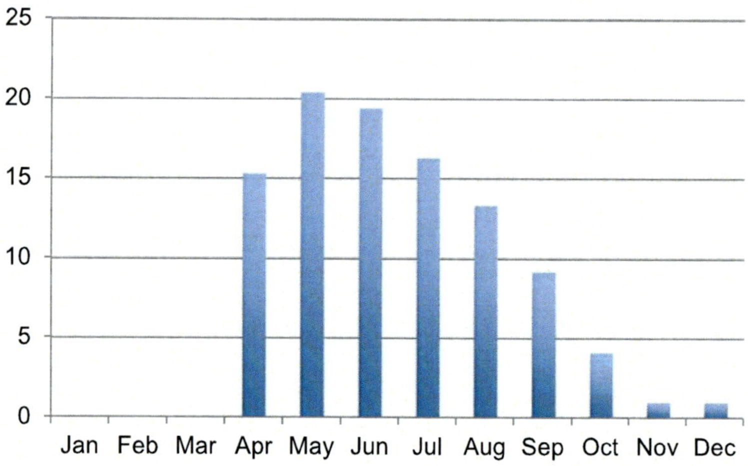

Fig. 10.2 The monthly distribution of white-headed langur births registered in Feijichang from 1996 to 2015

Fig. 10.3 The two inter-birth intervals of adult female Tiantian between her three infants were: 693 days in between Wenwen and Shishi, and 1227 days in between Shishi and Tianniu (Photo by Liang Zuhong). (**a**) Tiantian (left) with daughter Wenwen (right, born 2009.01.27) and son Shishi (middle, born 2010.12.210). (**b**) Tiantian and daughter Tianniu (born 2014.05.02)

bear the next infant, for the weaned pups, though able to feed on their own on tender leaves and shoots, still stuck around their mothers and at times would suck their nipples. Perhaps the seemingly suckling of the pups had aroused the secretion of milk in the mother langur, albeit of a paltry quantity, which stopped her from getting pregnant.

The seemingly suckling had misled researchers to consider a pup mouthing the nipple of its mother to be suckling and hence be mistaken on the length of the lactation period. In 2005, Zhao Qing, Ph.D. (2005) upon his observations on the three pups of the "Shihuiyao (SHY) troop" reported that under natural circumstances the white-headed langur has a lactation period of 19–21 months. Between July 2004 and January 2005, Wang Dezhi, Ph.D. (2004) upon observing two white-headed langur troops in Feijichang discovered that there were two conditions under which pups would wean. One was weaning naturally by themselves when the troop was in normal conditions, while the other type of weaning was involuntary and forced when it was separated from its mother. Wang registered 14 suckling pups and found that the suckling period under natural circumstances lasted 599 $\pm$ 33 days, which equals to 19.7 $\pm$ 1.1 months ($n = 10$), whereas under anomalous circumstances, the period lasted 527 $\pm$ 12 days, which equals to 17.3 $\pm$ 0.4 months ($n = 4$). This provided evidence that the forced separation of the pup from its mother distinctly shortens the suckling period. Under natural circumstances, the average suckling period of female pups is 589 $\pm$ 22 days, equaling to 19.4 $\pm$ 0.7 months ($n = 5$), and for male pups, the average is 620 $\pm$ 41 days, equaling 20.4 $\pm$ 1.3 months ($n = 4$).

We pay high regards to the research done by the postgraduates before 2008. All of them are diligent, their work is carried out scrupulously, and each statistical datum they achieved was valuable. However, there were no high-quality instruments available to them in those years that could tell apart seemingly suckling from real suckling; hence, the length of the suckling period they found was exaggerated.

Table 10.4 Inter-birth intervals registered on the 16 adult females in FJC troop giving birth to 78 pups from 1996 to 2015

Name of the female	No. of births it had given	Interval between each birth (days)	Average interval between births (days)	Remark
Dahei	7	*334; 748; 775; 704; 1091; 870	753.7	*The pup born before that one had been killed in infanticide
Jiajia	8	655; 763; 746; 753; 1010; 1133; 1125	883.6	
Arzhen	8	686; 752; 749; 718; 1053; 862; 1538	908.3	
Bibi	3	757; *368	562.5	*The pup born before that one had been killed in rainstorm
Lingling	4	*379; 722; *474	525	*The pup born before that one had been killed in infanticide (happened twice)
Xiaoxiao	6	828; 673; 1035; 1134; *1455	1025	*Miscarriage before this
Yingying	7	834; 579; 767; 800; 671; *640	715.2	*The pup born before that one had been killed in infanticide
Lanlan	6	*757; *671; 677; 871; 660	727.2	*The pup born before that one had been killed in infanticide (happened twice)
Pingping	7	1126; 732; 835; *329; *377(twins)	679.8	*The pup born before that one had been killed in infanticide
Tiantian	3	693; 1227	960	
Arxi	3	741; 863	802	
Aryu	2	1146	1146	
Lanqi	4	825; 770; 687	760.7	
Arlu	3	800; *664	732	*The preceding pup had been killed in infanticide
Beibei	3	809; **689	749	*One miscarriage, one killed in infanticide
Beike	4	*356; 757; 745	619.3	*The receding pup had been killed in rainstorm
Total	78	($n = 61$)		
The mean inter-birth interval of total 16 females(days)			784.3	

From 2008 onward, the subject in research had been video recorded with HD video cameras, and we were able to analyze the images with repetitive playbacks in the laboratory, which allowed us to exclude seemingly-suckling from true suckling (Fig. 10.5). Thereby, we are able to conclude that the length of the suckling period of white-headed langurs is 456 ± 28 days, which is equivalent to 15 ± 0.9 months.

10.1.6 Length of the Pregnancy Period

The length of the pregnancy period is yet another parameter of significance in the white-headed langur reproduction process.

There is currently no entirely satisfying method for studying the length of the pregnancy period of wild white-headed langurs, since they copulate in the prophase and in the late phase of their pregnancy. For most non-primate animals, the length of the pregnancy period can be determined by the number of days between the date it last copulated and the date it gives birth, but this is not applicable with wild white-headed langurs.

Jin Tong, Ph.D., was the first to work out the number of days of gestation in wild white-headed langurs before 2008. In her doctoral thesis, she calculated the length of the white-headed langur pregnancy period by taking the median date of the period between a female langur's copulations to the date it gave birth. She estimated from observing and registering the copulation dates and the birthdates of four adult females in FJC troop that the average length of the four female white-headed langurs' pregnancy period was 164 ± 34 days which equals to 5.4 ± 1.1 months. See Table 10.6 (Jin Tong 2008).

Since 2008, by using HD monitoring systems, tracking, and in situ video recording in the wilderness, we have registered even more copulations and parturition by the female white-headed langurs in Feijichang.

Case 1: Copulation and Parturition by Adult Female Langur Tiantian

We had authentically registered the date Tiantian gave birth to Shishi on 2010.12.21. Reviewing past video recordings we found Tiantian copulating with Yintangxiaotu on 2010.5.28.

Date and hour: 28 May 2010, 19:00.

Place: On a pit at the left of FJC Big Cave.

Weather/Illumination: Cloudy; light switched on in Tree House

19:26 Male langur Yintangxiaotu climbed straight down the precipice to a level pit west of Big Cave. The 8-year-old Tiantian,

Table 10.5 Suckling day registers of the four pups in "Yintangxiaotu family troop" in Feijichang

Name of the pup and sex	Date of birth	Suckling days of pup (days since birth)																								Name of the mother
		Pup's age in days in the various months in 2011												Pup's age in days in the various months in 2012												
		January	February	March	April	May	June	July	August	September	October	November	December	January	February	March	April	May	June	July	August	September	October	November	December	
Shishi ♂	2010.12.21	+	67	+	134	+	195	+	257	+	318	+	379	+	438	459 #	–	–	–	579*	–	–	–	710*	–	Tiantian
Yiyi ♂	2011.01.02	+	57	+	124	+	185	+	247	+	308	+	369	+	428	+	489	510 #	–	–	–	636 *	–	–	–	Aryu
Gaga ♀	2011.02.07	+	21	+	82	+	143	+	205	+	266	+	327	+	386	+	428 #	–	–	–	565*	–	–	–	–	Jiajia
Chuliu ♀	2011.02.08	+	20	+	81	+	142	+	204	+	265	+	326	+	385	+	428 #	–	–	540*	–	590*	–	–	–	Xiaoxiao

Average length of pup's suckling days: 456 ± 28 days ($n = 4$)

+ denotes month with suckling, – denotes no suckling in that month, # denotes month in which suckling ended, * denotes seemingly suckling observed on that day (the pup mouthing the mother's nipple)

Fig. 10.4 18:45. Four mothers of "Yintang xiaotu family troop" return to the FJC Big Cave night shelter with their young children in their arms: from right to left, the first one on the right is "Tiantian" and "Shishi" of 123 days old; the second one on the right is "Yayu" and "Yiyi" of 113 days old; the third one on the right is "Jiajia" and "Gaga" of 71 days old; the second one on the left is "Xiaoxiao" and "Chuliu" of 70 days old. These babies are carefully nurtured by their mothers and taken care of by their families (Photo by Liang Zuhong)

(a) (b)

Fig. 10.5 True suckling and seemingly suckling (Photo by Liang Zuhong). (**a**) True suckling (1-month-old infant pup); (**b**) Seemingly suckling (18-month-old pup)

which had arrived earlier, immediately raised its buttock presenting her anogenital area towards Yintangxiaotu.

19:26:45 Yintangxiaotu gave Tiantian a slight pat on the back with its hand, then mounted on her and copulated for about 9 seconds. During this, a juvenile male langur sat in front of Tiantian, grabbed the hairs on Tiantian's face with its hands, and shoved its mouth on Tiantian's face and bit her.

19:27 Yintangxiaotu sat down on the near-left position on the level pit facing the Tree House. Tiantian, a yellow-haired pup named Wenwen (her daughter), and a juvenile male langur sat with their backs towards Yintangxiaotu. The juvenile male langur at times patted Tiantian's chin with its hands.

19:28 Tiantian sat with the yellow-haired pup in her embrace.

19:28:30 Yintangxiaotu turned round and gave a tender pat on Tiantian's back, and Tiantian turned its head to have a look but made no response. She kept cuddling the yellow-haired pup in her embrace and Yintangxiaotu remained where it was without making any further moves.

19:29 Yintangxiaotu glanced at Tiantian again and made another pat on Tiantian's back, while Tiantian turned to look and then put down the yellow-haired pup, and then rose up to present its anogenital area to Yintangxiaotu. Yintangxiaotu, holding onto Tiantian's back, mounted on her. At that instance, Wenwen came over to her mother, held her face, bit her, and snatched the hairs on her face while another juvenile male langur seized the hairs on her head and bit her face. From the moment Yintangxiaotu rose to mount, we saw its sex organ was erect and then we saw Yintangxiaotu pulling in and out for 20 seconds. In the end, Yintangxiaotu stopped moving while still mounted on Tiantian's back for a few seconds; could it be ejaculating?

19:35 Yintangxiaotu sat down on the floor but its sex organ remained erect and there were clear traces of viscous secretion on the tip of it.

Table 10.6 The copulation dates, parturition dates, and estimated length of the pregnancy period of the four adult female langurs in FJC troop (Jin Tong, 2008)

Name of the female	Dahei	Arzhen	Jiajia	Xiaoxiao
Copulation dates	2002.8.14	2002.9.25	2002.8.7	2002.7.28
	2002.8.23	2002.9.26	2002.8.8	2002.9.26[a]
	2002.8.27	2002.10.14	2002.10.6	2002.9.28
	2002.8.31	2002.11.10	2002.10.9	2002.10.22
	2002.9.4	2003.1.14	2002.12.13	2002.12.7
Parturition date	2003.2.1	2003.2.25	2003.3.5	2003.4.29
Estimated pregnancy period (days)	158	134	150	213

[a]Date of copulation between Xiaoxiao and Aryuan of FJC III troop

Table 10.7 The copulation dates and parturition dates of the seven female langurs in the FJC troop

Name of the female langur	Name of the male langur	Name and sex of pup born	Copulation date	Parturition date	Period between copulation date and parturition date (days)
Tiantian	Yintangxiaotu	Shishi ♂	2010.05.28	2010.12.21	205
Aryu	Yintangxiaotu	Yiyi ♂	2010.05.09	2011.01.02	238
Jiajia	Yintangxiaotu	Gaga ♀	2010.09.20	2011.02.07	139
Xiaoxiao	Yintangxiaotu	Chuliu ♀	2010.09.22	2011.02.08	138
Arxi	Yintangxiaotu	Yutu ♀	2011.06.25	2012.01.20	209
Lanlan	Yintangxiaotu	Xiaolan ♀	2011.08.05	2012.03.28	236
Yumei	Duanwei	Xiao-M ♂	2012.08.09	2013.03.11	214

Fig. 10.6 Copulation of a female white-headed langur in her lactation period (Photo by Liang Zuhong). (**a**) Pingping and son Xiaozhen (born 2012.04.16). (**b**) Pingping in her lactation period copulated with resident alpha male Yintangxiaotu (2012.07.01)

In this case, it seems that the length of pregnancy of Tiantian was 205 days. But can this be the basis for our calculation of "pregnancy period" of white-headed langurs?

Case 2: We recorded with HD video cameras the copulations of two resident alpha males of the FJC Big Cave troop with the seven female langurs in the troop from 2010 to 2012. We also registered the dates of copulation and parturition by the females afterwards, as listed in Table 10.7.

The time between the copulation and parturition date shown in Table 10.7 varies from 138 to 238 days. If we calculate in the same way as most other researchers did, then the average length of the pregnancy period should be 197 ± 39 days, which is equivalent to 6.6 ± 1.3 months. The data falls within the pregnancy period range concluded by Jin Tong, Ph.D., as well as the pregnancy period range of other langur species registered by international compeer researchers.

However, we still have our doubt. Since the white-headed langur has evolved to a more developed stage than non-primate animals in that the motive of their copulations is not limited to reproduction, but it is also one of their social behaviors. For that reason, a female white-headed langur would copulate during and before their estrus period, in their pregnancy period, and in their lactation period (Fig. 10.6). It is our idea that since their copulation is not in line with the female langur's ovulation and estrus dates, the

Table 10.8 The parturition records of adult female langur Yingying

Name of pup	Sex of pup	Date of birth	Inter-birth interval(days)	Mean inter-birth interval(days)
Baobao	♂	2004.02.18	–	715.2
Arbao	♂	2006.06.01	834	
Zhenzhu	♀	2008.01.08	579	
Xiaoliu	♂	2010.02.06	767	
Xiaoying	♀	2012.04.16	800	
Liushaotou	♀	2014.02.16	671	
Chaosheng	♂	2015.11.18	640	

last registered copulation date cannot be counted as the start of pregnancy.

Thus, we have to find other methods to estimate the length of the pregnancy period in white-headed langurs. We shall review some of the registers from our recent observations in the wilderness. Perhaps it would provide us alternative evidence in estimating the length of the pregnancy period in white-headed langurs.

10.1.6.1 The Shortest Inter-birth Interval Registered

From the parturition registers of Yingying listed in Table 10.8, we will see that Yingying, after giving birth to male pup Arbao on June 1, 2006, again gave birth to female pup Zhenzhu on January 8, 2008. Both the dates were exactly registered, so the inter-birth interval is exactly 579 days or 19.3 months. That inter-birth interval by Yingying under natural circumstances is the shortest inter-birth interval recorded among the 16 registered births by female langurs in the FJC troop. It provides us significant insight in our discussion on white-headed langur reproduction parameters.

10.1.6.2 Resumption Time of Ovulation in Adult Females

From Table 10.5 we noted that after the suckling period ended, there was still a period of time that white-headed langur pups would search for their mother's nipples. In that period, the pups merely mouthed the nipples, and the secretion of milk in the mothers would gradually stop, and ovulation would resume. As behaviors in individual pups differ, the length of that period is difficult to be ascertained.

We had registered in our observations in the wilderness that female langur Yumei had sex just 13 days after its newborn infant was killed in infanticide by the new resident alpha male (Fig. 10.7).

The Yumei Incident

2013.03.11 19:12

The just young adult female langur Yumei gave birth to her first born infant Xiao-M (♀) under the assistance of her mother Pingping at location 4[#] in the FJC Big Cave night shelter.

2013.03.30 19:48

The troop returned to the FJC Big Cave night shelter. It was found that Yumei was holding the dead body of Xiao-M which had injuries all over. The newborn pup had lived for only 20 days.

2013.03.31 06:19

Yumei at last let go of the dead body of Xiao-M.

2013.04.12 09:25–09:50

It was the thirteenth day after her infant was killed. Yumei in the bush in Taohuagu attempted again and again to approach the new resident alpha male Shengdi and present her anogenital area. Video recordings had shown the vaginal opening of Yumei was swollen, red, and slightly moist.

A suckling infant is in a way a contraceptive for its mother, as it curbs ovulation in the mother. Once losing the infant, the physiology in the mother would adjust in order to start a new reproductive cycle. We may regard the period from the date of weaning (Yumei was forced to wean as her infant was killed) to the date when ovulation resumes as the time needed for ovulation resumption. Hence, from the interval between the date Yumei lost her infant and the date she initiatively presented her anogenital area to the resident alpha male, we could find out the resumption time of ovulation in adult female white-headed langurs.

10.1.6.3 Estimation of the Length of the Pregnancy Period

Natural selection endows females with the duty of reproducing progenies, and to ensure that the species thrives in the competition for survival, it must utilize its time efficiently to reproduce a maximum number of offspring.

Under such a premise we focused on Yingying, which had the shortest inter-birth interval registered. Yingying had under normal conditions given birth to Arbao and Zhenzhu respectively at an interval of only 579 days, and this gave evidence that the full process for reproducing the next generation, from ovulation, copulation, pregnancy, suckling, weaning, to the next round of ovulation, can be completed in that period.

Our estimation requires the following conditions to be definite and precise.

- First, the interval between the two births by Yingying was 579 days (Table 10.8).
- Second, we documented the exact suckling periods of four white-headed langur infant pups under normal conditions. The longest was 510 days equaling to 16.8 months and the

Fig. 10.7 The infanticide incident. (Photo by Gu Tieliu, Feng Chunguang). (a) "Yumei" carried the killed infant back to the night shelter in Big Cave. (b) She had to quit the infant

shortest was 428 days equaling to 14.1 months (Table 10.5).

- Third, incidence of Yumei being forced to wean indicates that the days from weaning to having sex again, which signifies that ovulation has resumed, are the days necessary for an inter-birth interval, and this case has shown that the period can be as short as 13 days.

Upon the above, we can estimate the length of the pregnancy period in the white-headed langur. The pregnancy period = inter-birth interval − suckling period − days needed for ovulation resumption. With the specific figures in the above registers, we may estimate that:

Pregnancy period A = 579 − 459 (♂pup) − 13 = 107 (days) or 579–510 (♂pup) − 13 = 56 (days).
Pregnancy period B = 579–428 (♀pup) − 13 = 138 (days).

Looking at the results, the 107 days and 56 days arrived at in pregnancy period A may appear to be too short for an infant ♂ pup to develop in a womb, but would the 138 days in pregnancy period B be enough time to develop a ♀ infant pup?

Let us look at another two pieces of data registered from field observations in Table 10.7:

- On 2010.09.20, Yintangxiaotu copulated with Jiajia, and Jiajia gave birth to her daughter Gaga on 2011.02.07. We estimated that Jiajia had a pregnancy period of 139 days.
- On 2010.09.22, Yintangxiaotu copulated with Xiaoxiao, and Xiaoxiao gave birth to her daughter Chuliu on 2011.02.08. We estimated that Xiaoxiao had a pregnancy period of 138 days.

Is it a coincidence that the number of days in these two pregnancy periods coincides with the above calculation of period B (138 days)? Or in fact, it is possible to conceive a baby in 138 days!

Although we have elaborated to explain the basis of how to calculate the pregnancy period of white-headed langur, and we also refer to the reports of other researchers, we still cannot provide a complete and convincing answer. The fundamental reason is that it is almost impossible to determine the date of copulation, which successfully impregnated the female langur. Under such circumstances, we tried alternative ways of estimating the pregnancy period. We looked for absolute parameters from the copious field observation registers, such as those quoted above. Maybe we will continue to seek solutions to this problem. Some people say that we can measure the changes of their hormone levels, but these tasks are not easy.

10.2 Reproductive Status of White-Headed Langurs in the Core Research Area (FJC), 1996–2006

In November 1996, the number of white-headed langur females living in the core district of our research, FJC, was seven adult females and four immature females. Between April 1998 and June 2002, two drastic events occurred and changed the lives of them, and the initial troop was divided into three family troops each with its own inhabiting territories which we are going to present in detail in the following paragraphs.

The statistics in the following discussion are only related to the white-headed langurs in the core district FJC. We divided the entire district of FJC into three sub-districts (Fig. 10.8) and counted the number of white-headed langurs in each. The three sub-districts are:

Fig. 10.8 Aerial view of Feijichang (FJC), the core district in our research

Sub-district 1: FJC East, which included the activity districts of the white-headed langurs living in FJC Big Cave, Small Cave, and areas from Fanghoushan to Boyuekshan.

Sub-district 2: FJC South, areas from Fangshaoshan to Fangniushan, southwest wall of Research Base.

Sub-district 3: FJC West, which includes the activity districts of those white-headed langurs living in FJC Xishan.

10.2.1 Reproductive Multiplication in FJC Troops

10.2.1.1 Reproductive Multiplication in FJC East

FJC Big Cave and Small Cave in FJC East are the center of the core district (Fig. 10.8). We have been continuously tracking the white-headed langurs there and documented the life history of every one of them. The three adult female langurs and the three female progenies listed in Table 10.9, which stayed behind in "FJC Big Cave troop," were the initiators of the family troops living in FJC Big Cave and FJC Small Cave. Since we began documenting their lives in November 1996, they have stayed in FJC Big Cave and Small Cave until today. Some of the females among them have since reproduced no less than five generations of offspring, Jiajia being one example (see Fig. 8.12, Jiajia lineage graph).

10.2.1.2 Reproductive Multiplication in "Fangshaoshan Troop" (FSS)

"Fangshaoshan troop" was initiated by three young females, Dagege, Ergege, and Sangege. They were daughters of the male langur Queque. In June 2000, they "grafted" to FSS in the neighborhood of FJC and started a family troop centered on the three of them. They started reproducing offspring in June 2001. From their inherent reproductive ability and on the average length of an inter-birth interval of 784.3 days, we projected that the family would be multiplied to include 38 langurs (Table 10.10). In December 2015, we found that there were two family troops living in FSS, in the east and north, and each troop had over 20 langurs. The rate it had multiplied substantiates our projections in Table 10.10.

10.2.1.3 Reproductive Multiplication in the "Xishan troop" (XS)

In 2002, three adult females named Lingling, Bibi, and Beike, and three young females named Arfei, Aryi, and Nianping, "split" from the FJC Big Cave troop and initiated the "XS-β-troop." Structure of this β-troop was quite similar to that of the "α-troop." Based on the innate reproductive capability and on the average 784.3 days of inter-birth interval, we projected that the family troop would be multiplied to 46 langurs from 2004 to 2015 (Table 10.11). In our observations in December 2015, we found the "Xishan troop" had multiplied into three troops, "Xishan troop," "Guoyuan troop," and "South Guoyuan troop," spreading

Table 10.9 Descendants through 1994–2015 of the female initiators from the beginning troop in FJC Big Cave and Small Cave, and those of the female progenies which stayed behind living in FJC Big Cave and FJC Small Cave

Lineage	Lineage initiator and her female progenies	First parturition date	No. of descendants	Name, sex, and number of descendants	Descendants Split away from troop	No of surviving descendants	Survival rate (%)	Died descendants	No. of deaths	Mortality rate (%)
Dahei lineage	Dahei	1998.01	7	♂5: Xiaoarge, Yiyi, Yixi, Zuoque, Xiaohei ♀2: Aryi, Arxi	Aryi, Yixi	6	72.7	Xiaoarge –	3	27.3
	Arxi	2010.01.09	3	♂1: Arweng ♀2: Xiaojiu, Yutu				Yutu		
	Xiaojiu	2015.02.21	1	♀1: Zhuli						
Jiajia lineage	Jiajia	Around 1995.01	9	♂4: Erarge, Arlan, Arqi, Youque ♀5: Dagege, Lanlan, Lanqi, Gaga, Xiaosanba	Dagege, Erarge, Arlan, Lanqi, Gaga	22	91.7		2	8.3
	Lanlan	2004.02	6	♂3: Artian, Mengli, Wangzai ♀3: Tiantain, Mimi, Xiaolan	Mengli, Xiaolan, Wangzai			Artian		
	Lanqi	2007.12.15	4	♂1: 007 ♀3: Qiqi, Qingming, Qinu	Qingming, 007			Qinu		
	Tiantian	2009.01.27	3	♂1: Shishi ♀2: Wenwen, Tiannu	Shishi					
	Mimi	2014.02.21	1	♂1: Mizai						
	Wenwen	2014.04.29	1	♀1: Wennu						
Arzhen lineage	Arzhen	Around 1995.01	8	♂4: Feifei, Feilu, Wuque, Chongsheng ♀4: Sangege. Arfei, Arlu, Zhenni	Sangege, Feifei, Arfei, Feilu, Wuque, Chongsheng	10	90.9		1	9.1
	Arlu	2010.01.18	3	♂2: Hanlin, Xiaolu ♀1: Nuhanzi	Hanlin			Xiaolu		
Xiaoxiao lineage	Xiaoxiao	2001.01.21	6	♂3: Xinxin, Xinyu, Xiaoyuer ♀3: Aryu, Bandazai, Chuliu	Xinxin, Xinyu	6	75	Bandazai, Chuliu	2	25
	Aryu	2011.01.02	2	♀2: Yiyi, Huanghunhou	Yiyi					
Yingying lineage	Yingying	2004.02	7	♂4: Baobao, Arbao, Xiaoliu, Chaosheng ♀3: Zhenzhu, Xiaoying, Liushaotou	Baobao, Arbao, Xiaoliu, Xiaoying	7	87.5	Liushaotou	1	12.5
	Zhenzhu	2013.12.06	1	♀1: Xiaoxuan						
Pingping lineage	Pingping	2004.12	7	♂3: Yiling, Xiaozhen, Nannan ♀4: Beibei, Yumei, Xiao-P, Ningning	Yiling	7	63.6	Xiaozhen, Xiao-P	4	36.4
	Beibei	2010.01.14	3	♂3: Gualai, Xiaobei, Beckham	Gualai			Xiaobei		
	Yumei	2013.03.11	1	♂1: Xiao-M				Xiao-M		
Total			73			61	83.6		12	16.4

Table 10.10 Projected reproduction multiplication by the female white-headed langurs in "Fangshaoshan troop" (FSS) through from 2001 to 2015

Name of the female	First parturition date	Reproductive years since reaching♀ adulthood	Projected number of births (counts)	Projected number. of offspring	Projected number of pups by grown-up ♀ offspring	Total number. of grown-up♀ + third generation offspring according to projection	Projected number of offspring deaths	Projected number of surviving offspring
Dagege	2001.06	14.6	6.8	7 (3♂4♀)	9	16	2.6	13
Ergege	2001.04	14.8	6.9	7 (4♂3♀)	9	16	2.6	13
Sangege	2002.09	13.5	6.3	6 (3♂3♀)	8	14	2.3	12
Total				20	26	46	7.5	38

Remark: (1) Judging from field observations, the death rate was estimated at 16.4%; (2) Judging from field observations, the sex ratio at birth is 1:1 (see Table 11.3) and the same is applied in the projection of the sexes of offspring

Table 10.11 Projected reproductive multiplication by the female white-headed langurs in "XS-β-troop" through from 2001 to 2015

Name of the female	First parturition date	Reproductive years since reaching♀ adulthood	Projected number of births (counts)	Projected number of offspring	Projected number of pups by grown-up ♀ offspring	Total number. of grown-up♀ + third generation offspring according to projection	Projected number of offspring deaths	Projected number of surviving offspring
Bibi	2004.02.13	11.9	5.5	6 (3♂3♀)	7	13	2.1	11
Lingling	2004.04.12	11.8	5.5	6 (3♂3♀)	4	10	1.6	8
Beike	2005.02.03	10.9	5.1	6 (3♂3♀)	4	10	1.6	8
Arfei	2006.08.06	9.5	4.4	5 (3♂2♀)	2	7	1.1	6
Aryi	2006.06.18	9.6	4.5	5 (2♂3♀)	4	9	1.4	8
Nianping	2007.07.17	8.5	4.0	4 (2♂2♀)	1	6	1.0	5
Total				33	22	55	9	46

Remark: (1) Judging from field observations, the death rate was estimated at 16.4%. (2) Judging from field observations, the sex ratio at birth is 1:1 (see Table 11.3) and the same is applied in projection of offspring sexes

out to the west and north. There are 12 langurs, 20 langurs, and more than 10 langurs in the three troops, respectively. This outcome substantiates the projections in Table 10.11 as well.

10.2.2 Estimation of the Total Number of White-Headed Langurs in FJC

The ways male and female white-headed langurs disperse are different. Females generally stay all their lives at their birthplace, whereas males have to travel to far distant places. In view of this predisposition of white-headed langurs, we have to make an account of the movement of males in FJC in the last 20 years.

First, we must record the number of male white-headed langurs that moved out of FJC.

Adult male white-headed langurs are intolerant of other males. Whenever a male (or male pair) successfully invades a family troop, it would kill some of the suckling pups born of the previous male langur and drive out all the males of the former family.

The resident alpha male, after relinquishing the family, would take along its sons and leave. They would lingering around FJC Big Cave for a period and finally depart for other places. In Table 10.12 are listed the ruling periods of four turns of resident alpha male langurs in FJC Big Cave and Small Cave between 1994 and 2015 and the time each langur left the family with a troop of males.

Second, we must find the number of outlier males that came into and settled in the FJC core district.

We take into account only the outlier adult male langurs which came into FJC and successfully invaded a family to be the family's resident alpha male. From our observations on the "Yintangxiaotu all-males troop," we found that invading attacks were usually carried out jointly by several brothers, but after the family was successfully invaded, only one would stay and become its new resident male. Each family troop has the capacity for only one adult male. By this we estimate that there are seven adult male langurs which came from outside FJC, since there are presently seven family troops in FJC (from Leizhai to Fangshaoshan).

Third, we must consider the number of other outlier white-headed langurs coming into the FJC core district.

There would always be all-male troops or transitory troops coming and going in FJC. At present, there are three troops of these kinds frequently spotted in the area. They are respectively the transitory troop with 15 langurs (13♂2♀) which lives in Bokyueshan, the all-male troop with five langurs which appears every day in FJC Xishan and a troop with eight langurs (7♂1♀) which came into FJC from Fangniushan. The sole adolescent female in the last troop had first joined the "Yintangxiaotu transitory troop" but later stayed behind in Small Cave to join the "Laoxie family

Table 10.12 The ruling periods of each of resident alpha male in FJC Big Cave and Small Cave between 1994 and 2015 and the time each turn of family troop split

Resident alpha male	Residency period	Departing from family	Period and places the left home troop roved	Remarks
Queque	1994–1998.04	Queque + 3 sons + 4 daughters + 1 incoming adult ♀	1998.04–2000.06: "Queque transitory troop" successively roved about in Guoyuan, Bokyueshan, and Fangshaoshan 2000.07–2001.11: The three sons remained inhabiting in Bokyueshan	Adult ♀ Weiming died of miscarriage 2006.06: Queque dead from fighting in FSS
Archeng	1998.04–2002.07	Archeng + 5 sons	2002.07–2002.09: The five sons remained inhabiting in Guoyuan	Afterwards dispersed away from FJC
α-Male-Langur	2002.07–2006.09	α-Male-Langur + 4 sons	2006.08–2009.10: All the five males remained inhabiting in Fanghoushan where Research Base locates	Afterwards dispersed away from FJC
Yintangxiaotu	2006.09–2015.07	Yintangxiaotu + 14 sons	From 2015.05 until this day: Yintangxiaotu is leading its 11 sons on their way dispersing to Nonglao in the south of Nongguan Mountains	Three sons had left the troop. Wuque is ascertained to have become the resident alpha male of a troop

troop" when Laoxie invaded the family. (We have also registered other mature or immature single females joining family troops, like sub-adult female Da-S joined "FJC Big Cave troop" in year 2010, and adult female Panma joined "Laoxie family troop" in FJC Small Cave in April 2015.)

The karst hills and wetlands in FJC with white-headed langur habitats have a total area of 1.026 km^2. In June 2015, there were the "Big Cave family troop," the "Small Cave family troop," the "Xishan family troops" (three troops + three resident male langurs), the "Fangshaoshan (FSS) family troops" (two troops + two resident male langurs), the "Yintangxiaotu all-males troop," the "Bokyueshan (BYS) transitory troop," the "Xishan all-males troop," and the "Fangniushan trespassers troop" which had spent half of their time in FJC. To make a conservative estimate, the total number of white-headed langurs comes to:

33 ('Big Cave troop') + 12 ('Small Cave troop') + 46 + 3A♂ (Xishan troops aggregate) + 38 + 2A♂ (FSS troops aggregate) + 14 ('Yintangxiaotu all-males troop') + 15 (BYS transitory troop) + 5 ('Xishan all-males troop') + 4 (half of 'Fangniushan trespassers troop') = 172 langurs.

In June 2015, the "Yintangxiaotu all-males troop" departed from the FJC core district and the total number of white-headed langurs there became 158 after deducting the 14 langurs of the troop (172–14).

Hence, the density of white-headed langurs in the core district comes to 198 langurs/1.026 km^2 = 154 langurs/km^2, which makes the white-headed langur distribution in the core district the highest density population in the Nongguan Mountains.

From 1996 to 2015, the number of sexually mature female individuals in the Big Cave troop of FJC increased continuously and joined the breeding team, which made the number of offspring of each family in the Big Cave troop of FJC increase continuously. Therefore, the reproduction of FJC troop white-headed langurs shows that the white-headed langurs in Nongguan Mountains are entering a period of rapid growth.

References

Dezhi W. A composing unit in wild white-headed langur (*Trachypithecus leucocephalus*) population—the research and tracking of a white-headed langur family troop in Nongguan Mountains, Guangxi (野生白头叶猴社群的结构单元——对广西崇左弄官山区一个白头叶猴家庭的跟踪研究). Doctoral thesis. Beijing: Peking University; 2004.

Qing Z. White-headed langur (Trachypithecus leucocephalus) mother-pup relationship and behavioral development in pups (白头叶猴的母幼关系和幼仔的行为发育). Doctoral thesis. Beijing: Peking University; 2005.

Tong J. Wild female white-headed langur (Trachypithecus leucocephalus) reproduction tactics—its social relationships, mating and reproduction, and surrogate behavior (野生白头叶猴的雌性繁殖策略—社会关系、交配繁殖与拟母亲行为). Doctoral thesis. Beijing: Peking University; 2008.

11 The Number of White-Headed Langurs in the Nongguan Mountains

Abstract

In Nonguan Mountain area, there are 220 hills scattered in the dense jungle. Due to the restriction of the landform, we tracked the white-headed langur troops on fixed spots, along with line transect surveys, to estimate their numbers. From April till December 2015, we surveyed white-headed langur distribution in the Nongguan Mountains and divided it into three districts. According to our survey, we estimated a total of 798 (216 + 425 + 157) individuals. Since 1996, the langur population has been growing rapidly. The sex ratio within phase 1 and 2 was 1:1, but the phase 3 sex ratio after maturity was strongly biased toward females, with a ♀:♂ ratio of 22:3 (overall mean 7.3:1). Old white-headed langurs accounted for 6.5% of the population, adults at reproductive ages accounted for 32.3%, and the pups, juveniles, and adolescents made up 61.3%. Age distribution as such indicates that the age structure of the FJC white-headed langur population is balanced. Meanwhile, the net rate of annual population increase of six families during the last 20 years ranged from 4.0% to 54.5%, with a mean value of 40%.

Keywords

Population estimate · Population growth · Sex ratio · Reproductive age · Age structure

The white-headed langur is a rare and endangered species. Their number is the foremost information that researchers and conservationists must know before implementing conservation measures. In areas where conservation practices are already in place, changes in the population size of the species is an important indicator of how well conservation is working. That information can only be obtained from field observations in the wilderness and through monitoring.

It is not easy to obtain the correct number of white-headed langurs presently in the Nongguan Mountains due to three obstacles:

First, white-headed langurs live in obscure places in densely vegetated tropical forests on karst hills. It is very difficult for researchers to locate and get a clear sight of them.

Second, their distributions are scattered over terrains that are complex and difficult to access. Many of these regions do not even have trails. Researchers must slash with machetes to open a path to the bottom of precipitous karst hills in order to be able to locate the white-headed langurs.

Third, white-headed langurs only appear near their night shelters, which are high up on precipices in the morning and dusk when the light is dim. It is difficult to identify them individually. While the number can be inferred from statistics, getting an accurate tally remains to be extremely difficult.

In the past 20 years, using various advanced electronic equipment, we had conducted continuous research and tracking on the white-headed langurs in FJC Big Cave and Small Cave without interruptions. We were supposed to take the white-headed langur population density in this core district to serve as one quadrant to infer the total number of white-headed langurs in the entire Nongguan Mountains, but now it seems to us that this method is not viable. For the white-headed langurs in the Nongguan Mountains are distributed over no less than 220 karst hills, and the respective populations are growing in various different districts. The research method used in the FJC core district is not applicable to every other district. Hence, we could not estimate the population size by way of density sampling. We must find a practicable way for our research.

W. Pan, *The Population Ecology of White-Headed Langur*, https://doi.org/10.1007/978-981-33-4118-0_11

11.1 Research Method

When observing white-headed langurs in the wild, we always keep in our mind the following four predications (which had been described in earlier chapters):

- White-headed langurs live in karst hill forest habitats in the Northern Tropic for 1.4 million years since their ancestors got there. They have adapted to the environment there and have become an indigenous species of the place.
- White-headed langurs sleep in night shelters high up on karst precipices, but they go out every morning at daybreak to forage for food in the karst hills and surrounding forests. Only at dusk would they return to their night shelters.
- They move and forage at fixed times but not on fixed spots in the forests. It is their routine to have one interval before noon at the place where they forage the most and another interval afterwards, each lasts about 2–3 h. At noon, in between the periods of foraging, they must rest for about 2–3 h. The start and end of the daily forages and rest will be slightly different in different seasons of the year.
- Our tallies in the wild are limited by the dense shades of tropical forests. When tracking the white-headed langurs, researchers were often hindered by lush sprawling vines, which made it difficult to locate and reach the white-headed langurs' resting place. Therefore, it was difficult to reckon their precise numbers.

11.1.1 Limitations of the Belt Transect Method

Since 1930, the line transect method was generally adopted to estimate the population density of wild animals. In the late twentieth century, the belt transect method which was based on the line transect method was developed. It works by recording the number of wild animals (individual or groups) spotted in the natural habitat within a belt that is marked off by equally distanced perpendicular lines on both sides of the transect line, in order to serve as a simple mathematical model which helps to infer the population density of a species. It has once been generally applied in inferring the wild animal population density over extensive areas, but the results were generally overestimated or with large deviations, which makes researchers doubtful about its correctness. We are not applying the belt transect method to infer the white-headed langur population density in the Nongguan Mountains, as some important prerequisites of the method could not be satisfied in the region, such as:

- We could not spot white-headed langurs in every belt section due to craggy karst hills and forest obstacles.
- Belts could not be chosen at random but had to be drawn according to the landscape.
- The requirement that the transect line must be straight could not be met in the region.
- The requirement that the probability of spotting the white-headed langurs not being affected by the population size could not be satisfied. Even in the same area, the number of white-headed langurs could vary at any time.

11.1.2 Testing the Line Transect Method

Prof. Sheng Helin (1992) considered it very difficult to accurately calculate the population of wild animals, yet he regarded the line transect method as one of the best ways to infer their population densities.

We had also thought that selectively adopting the line transect method in our research would be easy and practical, but we also had concerns that the results could be inaccurate, since a researcher could only reach the parts of karst hills that had trails and would not be able to access the precipices where while-headed langurs dwell. In other words, it was not possible to cover the entire white-headed langur distribution with transect lines. To gauge how much the line transact method result would differ from reality, we devised and conducted the following two test cases:

Test case 1: Finding the number of white-headed langurs in the "Bokyueshan transitory troop" by way of the line transect method.

After repetitive "counting at fixed spot" and tracking observations, we have confirmed the home range and exact composition of the "Bokyueshan transitory troop." The home range of the "Bokyueshan transitory troop" is on the isolated karst hill at the outermost northeastern corner of the Nongguan Mountains with limited flatlands surrounding it. It has a hilly terrain, has no swamps, is bounded on the north by sugarcane fields, and covered in the south by Chungzuo Eco-Park. Northern tropical forests grow surrounding the karst hill, and the floras are no different from other places in the Nongguan Mountains. There were 15 langurs in the composition of the Bokyueshan transitory troop, including one adult male, one adult female, eleven adolescent males at ages between 2 and 4 years, one adolescent female, and one yellow-haired pup that is about 1 year old.

Possible transact lines were drawn in the home range of the "Bokyueshan transitory troop." The daily activity areas of the "Bokyueshan transitory troop" on the north included from the hilltop of Bokyueshan to the forests skirting the foothills wherein there were no trails available as the foothills were covered with sugarcane fields. When the surveyor is in the midst of a sugarcane field, his line of sight would be blocked by the sugarcane and would not be able to spot a white-headed langur. On the south, the troop's activity area extended from the hilltop down the hill to reach the Eco-Park, where rows of sidewalk trees such as mango trees, peach trees, banyans, wild kulians (*Melia azedarach*), yinhehuans (*Leucaena leucocephala*), etc. were planted along a footpath. Since November 2015, the white-headed langurs often came to forage for beans of yinhehuans, and they moved along the lines of yinhehuans and Taiwan Acacias (*Acacia confusa*) as they foraged. There, surveyors could observe them and individually identify them at a close distance, sometimes as close as 2 m.

Test period: 2015.12.6–2016.1.6.

Transect line: Drawn on the only available trail down the north side of the hill, which had a length of 13 km.

Test result: Spotted white-headed langurs of the "Bokyueshan transitory troop" 20 times in 19 consecutive days of conducting line transect surveys, averaging at 8.4 langurs in each sighting ($n = 20$). Survey findings are listed in Table 11.1.

Test case 2: Finding the number of white-headed langurs in the "FJC Small Cave family troop" by way of the line transect method.

The survey of the home range of the "FJC Minor Cave family troop," also known as the "Laoxie family troop," covered Xiaodongshan on the south of FJC, the areas that extended eastward and southward until Taohuagu and the research base at Fanghoushan. The southern limit reached Xishan and the western limit reached the opposite point of the FJC wetland.

Test period: 2015.12.6 to 2016.1.7.

Transect line: The trail at the east of Research Base which circles southwestwards around Fanghoushan to FJC was the only viable route, and it had a length of 1.6 km.

Test result: Spotted white-headed langurs of the "FJC Small Cave troop" 24 times in 21 consecutive days of conducting line transect surveys, which averaged to 6.9 langurs in each sighting ($n = 24$). Survey findings are listed in Table 11.1.

11.2 Estimation of the Number of White-Headed Langurs in the Nongguan Mountains

On the whole, the number of white-headed langurs in the Nongguan Mountains is still increasing. Till now, their distributions in the Nongguan Mountains have not reached saturation. Since they are not evenly distributed, it is inappropriate to find the total number of white-headed langurs in the whole area by inferring from the sample plot density found in the belt transect. Hence, we have to divide the white-headed langur distribution in the Nongguan Mountains into three districts according to the localized conditions in each and estimate their numbers with different methods.

Between 2015.4 and 2015.12, we adopted the line transect method to survey the population number of white-headed langurs in the Nongguan Mountains. The transect lines are illustrated in Fig. 11.1, and the findings are listed in Table 11.2.

11.2.1 District 1

District 1, which covered the Fangshaoshan–FJC–Leizhai areas, was the district with the most clearly known number of white-headed langurs. It had an area of 2.2 km^2 with 216 white-headed langurs according to the findings in Table 11.2. With a population density of 98 langurs/km^2, it had the highest density of white-headed langurs in the Nongguan Mountains. In December 2015, there were 12 white-headed langur troops still in the district, with a mean number of 18 langurs per troop.

11.2.2 District 2

District 2 is the area bordering District 1's southern and northern areas in Nonglao with the main peak of the Nongguan Mountain at its center. It had an area of 4.6 km^2

Table 11.1 Statistical numbers of white-headed langurs in "Bokyuehsan troop" and "FJC Small Cave troop" from line transect surveys

Troop	Survey period	Days of survey	Number of surveys conducted	Number of langur spotted	Spotting rate (%)	Average number of langurs in each sighting	Actual number of langurs in the troop	Mean probability of spotting individual langur of the troop (%)
Bokyueshan troop	2015.12.6 to 2016.1.6	19	70	20	28.6	8.4	15	56.0
Small cave troop	2015.12.6 to 2016.1.7	21	87	24	27.6	6.9	12	57.5
Means			78.5	22	28.1	7.65		56.75

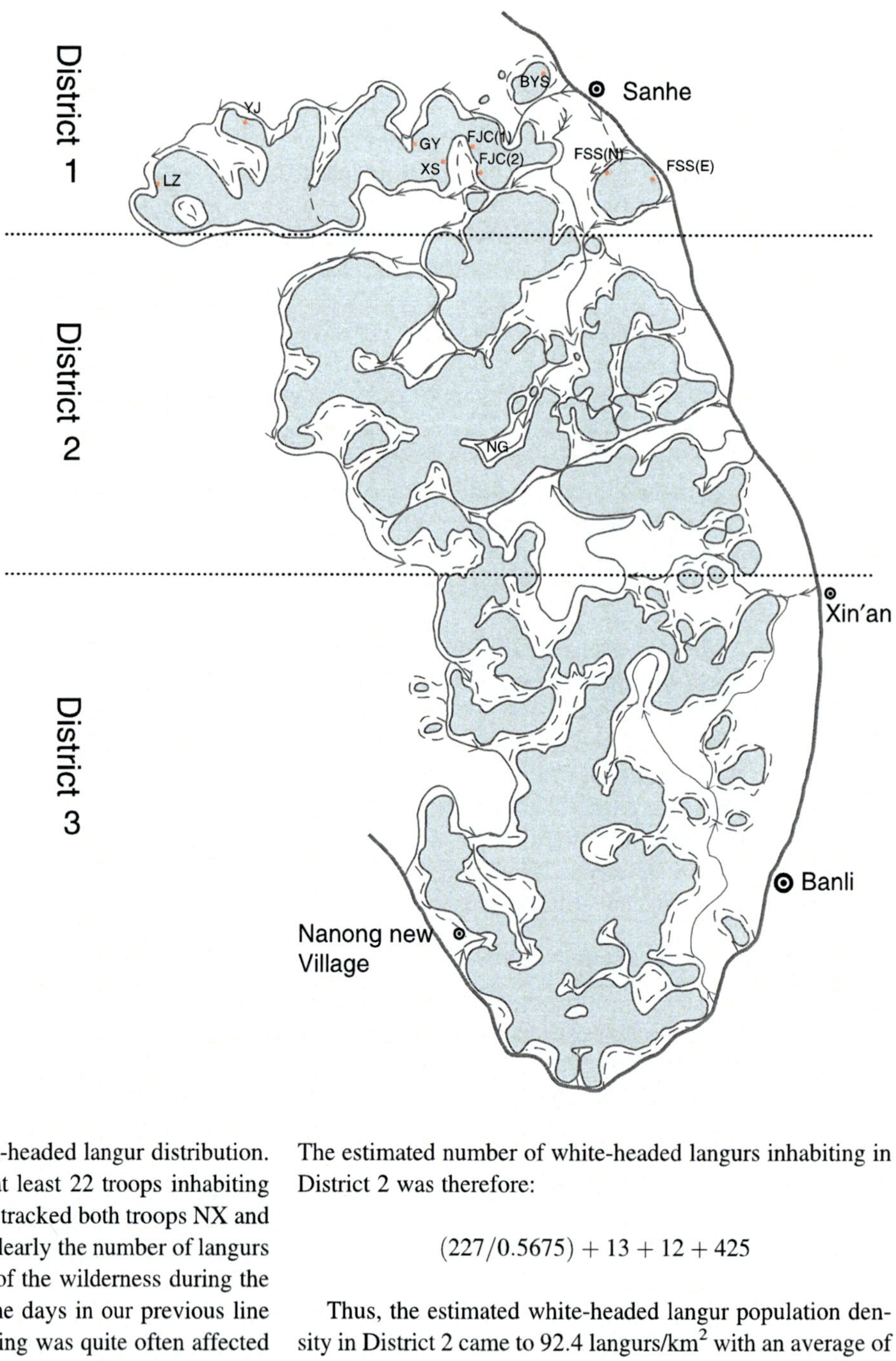

Fig. 11.1 Some of the white-headed langur distributions in the Nongguan Mountains and the surveyed transect lines. Note: The sketched solid and dotted lines, respectively, denote the transect lines and the surveyed areas. Remark: For purposes of protection, some spotting locations had not been flagged

and a comparatively dense white-headed langur distribution. In December 2015, there were at least 22 troops inhabiting this district (see Table 11.2). We tracked both troops NX and NS on fixed spots and obtained clearly the number of langurs in them. The natural conditions of the wilderness during the survey were no different from the days in our previous line transect surveys. While the sighting was quite often affected by weather conditions (cloudy, foggy, etc.), terrain, and vegetation cover which minimized the probability and the actual times of spotting the langurs. In order to get a correct number of the white-headed langurs in the district, we made an estimate with the mean spotting rate of 56.75%, which was obtained in the aforementioned line transect test under natural conditions. We divided the number of white-headed langurs we spotted of the 20 troops other than the NX and BQS troop by the mean spotting rate. Then, we added to it the number of confirmed white-headed langurs of the NX and BQS troops. The estimated number of white-headed langurs inhabiting in District 2 was therefore:

$$(227/0.5675) + 13 + 12 + 425$$

Thus, the estimated white-headed langur population density in District 2 came to 92.4 langurs/km^2 with an average of 19.3 langurs per troop.

11.2.3 District 3

This district extends southwards from Nonglao and crosses a provincial highway to cover the karst hills on the other side. It has an area of 6.4 km^2 and has plenty of traces of white-headed langur night shelters, but recently, white-headed langurs are rarely spotted there. Karst hills in the district are

Table 11.2 Survey findings on the Nongguan Mountain white-headed langur troops between 2015.04 and 2015.12

Sequence	Location	Troop	Number of langurs spotted in survey	Number of night shelters	Troop composition
District 1 (including FJC, the core district in research)					
1	FJC	Big Cave family troop	33	1	1♂; 17♀; 15 pups
2	FJC	Small Cave family troop	12	1	1♂; 9♀; 2 pups
3	XS	Xishan family troop	11	2	1♂; 6♀; 4 pups
4	GY	Guoyuan family troop	20	1	Family troop with yellow-haired pups
5	GYN	Guoyuan South family troop	10	1	Family troop with adolescent female and yellow-haired pups
6	FSS (N)	Fangshaoshan family troop (north)	20	1	Family troop with yellow-haired pups and yellow-haired baby pups
7	FSS (E)	Fangshaoshan family troop(east)	20	2	Transitory troop with yellow-haired pups
8	YJ	Yangjuan	35	2	Family troop with yellow-haired pups and yellow-haired baby pups
9	LZ	Leizhai	28	1	Family troop with yellow-haired pups
10	BYS	Bokyueshan	15	1	Transitory troop with one adult female and one immature female
11	XS-M	Xishan	5	Unknown	All males troop
12	FNS-FJC	Fangniushan-M	7 (8–1)	Unknown	Its one adolescent female had joined Small Cave troop and it became an all-male troop
District 2					
13	FNS (N)	Fangniushan (north)	10	Unknown	Family troop with yellow-haired baby pups
14	FNS (S)	Fangniushan (south)	8	Unknown	Family troop with yellow-haired pups
15	MZS	Muzishan	12	2	Family troop with yellow-haired pups and yellow-haired baby pups
16	NS	Nongshui	10	2	Family troop with at least two yellow-haired pups
17	NQ	Nongqiong	30	2–3	Family troop with yellow-haired pups and yellow-haired baby pups
18	NG	Nonggeng	24	2	Family troop with yellow-haired pups and yellow-haired baby pups
19	NX	Nongxiang	13	2	'Wuque family troop', with yellow-haired baby pups
20	TXD	Taoxindong	8	2	Family troop with yellow-haired pups
21	NA	Nongan	10	1	Family troop, spotted in it two yellow-haired pups
22	NL	Nonglu	8	1	Family troop, spotted in it one yellow-haired pup of about 2 years old
23	NT	Nongtong	20	2	Family troop, spotted in it yellow-haired pup
24	NGK1	Nongguankou (east)	10	2	Family troop, spotted in it one yellow-haired baby pup and at least two yellow-haired pups
25	NGK2	Nongguankou (west)	17	2	Family troop, spotted in it one yellow-haired baby pup and two yellow-haired pups
26	NG-1	Nongguan (west)	16	3–4	Unstable all-male troop, being eyed on by "Yintangxiaotu all-males troop"
27	NG-2	Nongguan (east)	12	2	Family troop, spotted in it two yellow-haired pups
28	NBG-1	Nongbaga (exterior)	10	2–3	Family troop with one yellow-haired pup. "Yintangxiaotu all-males troop" once attempted to invade it but failed
29	NBG-2	Nongbaga (interior)	4	2–3	Family troop with 1♂3♀, spotted no yellow-haired pup in it
30	SHY	Shihuiyao	7	2	Family troop with yellow-haired pups
31	NG-3	Nongguan (middle)	3		Roving all-male troop

(continued)

Table 11.2 (continued)

Sequence	Location	Troop	Number of langurs spotted in survey	Number of night shelters	Troop composition
32	BQS	Beiqishan	12		"Yintangxiaotu all-male troop"
33	NL-M	Nonglao (north)	5		Roving all-male troop
34	NF-1	Nongfeng (north)	3		Roving all-male troop
District 3					
35	NL	Nonglao	6	2–3	Family troop with two baby pups
36	NF-2	Nongfeng (west)	5		Roving all-male troop
37	NF-3	Nongfeng (east)	7		Roving all-male troop
38	NN1-7	Nanong area	Seven family troops with total 71 langurs. In average 6–13 langurs spotted in each troop. Yellow-haired pups and yellow-haired baby pups had been spotted in all the troops		

Note: The troops in the table are named by our research team with reference to conditions found in survey

relatively low and separated, so the distribution density there is low. From April to December in 2015, we spotted eight family troops and two all-male troops in our surveys in this district (see Table 11.2), and they all consisted of only moderate to low numbers of langurs. The population number of white-headed langurs in this district estimated by the 56.75% spotting rate under natural conditions came to:

$$89/0.5675 = 156.8$$

Accordingly, the population density was 24.5 langurs/km^2 with an average of 8.9 langurs per troop.

Summing up the number of white-headed langurs of all three districts, we found that in as late as the end of 2015, in the adjacently distributed and connected tropical karst forests on the karst hills in the Nongguan Mountains, there were at least: 216 + 425 + 156 = 797 white-headed langurs.

The white-headed langurs in the 13.2 km^2 natural habitat in the Nongguan Mountains share a common gene pool. They mate and reproduce naturally. Every year, the period from December to February in the next year is the peak season for white-headed langurs to give birth. The aggregate number of white-headed langurs in December 2015 in the Nongguan Mountains were not at their highest. During the 2015–2016 birth cycle, guess how many new individuals were added to the white-headed langur population there from January to March in 2016?

Based on the number of newborn yellow-haired pups from January to March in 2016, during which the FJC Big Cave troop had three newborns, the FJC Small Cave troop had two newborns, and the Xishan troop had none. Then, possibly 2/3 of the family troops in the Nongguan Mountains would have 2–3 newborn pups added to their families. In that case, out of the 33 family troops in Table 11.2, 22 families would have pups added to it, which would be 44–66 pups added in total. Hence, the white-headed langur population would be increased to 841–863 langurs.

This new round of baby pup births shall add many new faces to the white-headed langur population, all of them with glistening golden yellow hair. Each will be under the protection of its father and will be following its mother, sisters, and brothers while capering about in the tranquil verdant forests in the Nongguan Mountains.

11.3 Ecological Traits of the White-Headed Langur Population in the Nongguan Mountains

In content, a population is more than just a sum of its individuals. The white-headed langur population in the Nongguan Mountains is a close-knit integrated society made up of social units with various compositions, from 2 to more than 43 langurs in each unit, together with a few isolated individuals. In the integrated society, each individual according to its inherited growth development and traits performed different "roles." Coupled with their intricate adaptations to living space, weather cycles, and food resources, the white-headed langur population in the Nongguan Mountains has evolved into a distinct ecology of its own.

In 20 years, we have researched in depth the seven family troops, two transitory troops, and five all-male troops living in FJC Big Cave and Small Cave at the northern margin of the Nongguan Mountains. We had also observed the developments of the two family troops that were initiated by "graft" and "split" at Fangshaoshan and Xishan. While they were located in the FJC district, their behaviors reflected the overall traits of the behavioral ecology of the white-headed langur in the Nongguan Mountains.

11.3.1 Population Size and Density

The karst hills in the Nongguan Mountains are adjacently distributed and white-headed langurs usually act in groups; they live on vegetation on and around the karst hills and almost never venture to the flatlands. As the karst hills

Table 11.3 The sex ratios of white-headed langurs born in the FJC Big Cave and Small Cave troops from 1998 to 2015

Year	Phase 1 sex ratio (♀: ♂)	Phase 2 sex ratio (♀: ♂)	Phase 3 sex ratio (♀: ♂)
1998	4:5	4:3	8:1
2002	6:9	6:5	7:1
2006	7:7	7:5	7:1
Subtotal	17:21	17:13	22:3
♀: ♂	1:1.2	1.3:1	7.3:1
2012	13:16	12:13	13:1
2013*	1:1	–	–
2013.07–2015.12	8:9	8:9	21:2
Subtotal	22:26	20:22	34:3
♀: ♂	1:1.2	1:1.1	11.3:1

Note: 2013* (2012.07–2013.06) In that period, the resident alpha male of the FJC Big Cave troop was successively replaced, during which no single male had successfully resided over the troop

became surrounded by farmlands, they became isolated mountain islands, and their boundaries in the Nongguan Mountains thus became the bounds of white-headed langur distributions in the area. There are broad extensive sugarcane fields on the northern and southern ends of the Nongguan Mountains, while the earthen hills in the west had been exploited for eucalyptus plantations, giving the Nongguan Mountains clear-cut boundaries on three sides. At the eastern side of the Nongguan Mountains is a highway. White-headed langurs can be found on the precipices west of the highway throughout the year, but there would be almost no traces of them on the east of the highway. We occasionally observed individuals or small groups crossing the highway to the east for a brief stop, but they do not stay for long. The highway is therefore effectually the eastern boundary of the white-headed langur population in the Nongguan Mountains. The boundaries on the four sides have encircled an area of 23 km^2 out of which 13.2 km^2 is karst hills with talus, and this area is the actual habitat of the white-headed langurs in the Nongguan Mountains. How many white-headed langurs are living in that area in the Nongguan Mountains?

Since our research team began observations in November 1996, we have found that the white-headed langur population in the Nongguan Mountains is increasing gradually. For example, between 2014 and 2015, in the core district, the 0.5 km^2 karst hill forest at eastern FJC, there were 60 white-headed langurs which made it the area with highest population density (120 langurs/km^2). Additionally, from Fangshaoshan at the east of FJC until Leizhai at the west, there were 12 troops with a total of 216 white-headed langurs inhabiting the karst hills in that 2.2 km^2 district, giving it a density of 98 langurs/km^2. That makes District 1 the most densely populated white-headed langur district in the Nongguan Mountains. District 2, with the chief peak of the Nongguan Mountains at its center, has an area of 4.6 km^2 with at least 22 troops and 425 white-headed langurs, giving it a population density of 92.4 langurs/km^2. The karst hilled areas in District 1 and District 2 are of 6.8 km^2, which is about 51.5% of the total karst-hilled areas in the Nongguan Mountains, yet 80.4% of the total white-headed langur population lived there. In the district from the south side of the Nongguan Mountains to the south side of the highway, there were only small numbers of white-headed langurs. On the 6.4 km^2 of karst hills, there were only ten troops with a total of 156 white-headed langurs, giving a population density of 24.5 langurs/km^2.

At present, white-headed langurs in the Nongguan Mountains mostly inhabit the northern portion of its distribution. We predict that their population and density will keep growing and would continuously disperse southwards to the southern portion of its distribution, for only there is available space for their expansion.

11.3.2 Sex Ratio

The sex ratio is the proportion of male and female individuals in a population. It is also known as the sexual structure of a population. We categorize the sex ratio at birth as the "phase 1 sex ratio," the sex ratio of adolescents before maturity as the "phase 2 sex ratio" (taking into account only the surviving ones), and the sex ratio of matured individuals as the "phase 3 sex ratio" (including outlier females adopted by a family troop).

The figures in Table 11.3 have revealed that the sex ratio of white-headed langurs in FJC had remained steady from 1998 to 2015. The phase 1 sex ratio and phase 2 sex ratio were both at the ♀: ♂ ratio of 1:1. With regard to the phase 3 sex ratio after maturity, we can see from the table a distinct period division. In the three reproduction cycles of the FJC family troops from 1998 to 2006, the ♀: ♂ ratio of adult white-headed langurs in FJC Big Cave was 22:3 (average 7.3:1), whereas in the next three reproduction cycles from 2006 to 2015 in the same FJC family troops, there were a total of 34 wives for the three resident alpha males, meaning the adult ♀: ♂ ratio was 34:3, which averages to 11.3:1. The figures indicated that the reproduction of white-headed langurs in FJC has shifted from the restoring stage to rapid-growth stage. The underlying causes may be:

- The young females in the population had gradually grown up to their reproductive ages, providing more reproduction resources to the robust adult males.
- The vegetation that regenerated over the years provided more abundant food resources to the white-headed langurs, inducing more reproductions.
- It may be an invigorating and balancing act of nature in genetics to benefit the continual survival and adaptation of the species.

11.3.3 Age Structure

The number of individuals in each age group in a wild animal population will vary over time. The distribution of various age groups in a population forms the age structure of the population, which is also called the age composition of the population. Age structure is an important trait in the ecology of a species as it bears direct effects on upcoming reproductions. We have tracked and observed for a long time every white-headed langur in FJC Big Cave and Small Cave, registered their births, their growths, their deaths, and the immigration and emigration of individuals in the troop. We divided the age of a white-headed langur into six stages: the infant stage (0–1.3 years, including yellow-haired newborns and pups), the juvenile stage (1.3–2.5 years), the adolescent stage (2.5–4 years), the sub-adult stage (4–5 years), the adult stage (5–25 years), and the old age stage (♀age > 25 years; ♂age > 13 years).

We analyzed the age structure of the six white-headed langur troops in FJC as an example. The results are listed in Table 11.4.

It can be seen from Table 11.4 that old-aged white-headed langurs accounted for only 6.5% of the FJC population, adults at reproductive ages accounted for 32.3%, wherein the 24 adult females took up 25.8%. The aggregate of infants, juveniles and adolescents regardless of sex made up 61.3% of the population. In the next round of the resident alpha male replacement cycle, the percentage of females at reproductive ages in the three family troops will reach 45.7%. The age distribution as such indicates that the age structure of the FJC white-headed langur population is becoming balanced and that the population size will continue to grow.

Young females have a high reproductive value in a population. There are so many young female white-headed langurs in the Nongguan Mountains that will be reaching sexual maturity in the future. It can be foreseen that the number of white-headed langurs in the Nongguan Mountains is going to increase rapidly, and when the population reaches the saturation phase, they will disperse outwards.

11.3.4 Reproduction Capacity

The reproduction capacity of a population determines its future size. Under natural circumstances, the reproduction capacity is directly affected by factors like birth rate, mortality rate, and the number of individuals joining and departing.

We tabulated in Table 11.5 the reproduction capacity of the family troops in FJC Big Cave and Small Cave over the past 20 years and below are the results of our observation:

We tracked and registered the births and deaths of 40 pups (aged 0–6 months) and found that eight died of various causes within 6 months (Fig. 11.2). The mortality rate was as high as 20% ($n = 40$). In contrast, the mortality rate of female baby pups in the same age group was only 3.1% ($n = 32$). The mortality rate of male pups was 6.5 times that of female pups in the same period, which could be related to the infanticide acts by the incoming outlier male langur.

Our tracking observations on 20 juvenile male white-headed langurs (aged 1–3 years) had revealed a 5% ($n = 20$) mortality rate, while the tracking observations on 22 females of the same age group in the same period revealed a 16% ($n = 22$) mortality rate. The mortality rate of females

Table 11.4 The age structure of the six white-headed langur troops in FJC in December 2015

		OA		A		SA		J		I	
Troop	No. of langurs in the troop	♂	♀	♂	♀	♂	♀	♂	♀	♂	♀
Yintangxiaotu all-male troop	15	1		3		7	.	3		1	
Yuweng family troop	33		3	1	14					15	
Laoxie family troop	12			1	5		2		2	2	
Xishan family troop	11			1	4				2	4	
Bokyueshan transitory troop	15	1			1	8		3	1	1	
Fangniushan all-male troop	7	1				3		3			
Total	93	3	3	6	24	18	2	9	5	23	
Age group proportion in the FJC population (%)		3.2	3.2	6.5	25.8	19.4	2.2	9.7	5.4	24.7	

OA old age, *A* adult, *SA* sub-adult, *J* adolescent and juvenile, *I* infant (yellow-haired newborn and pup)

Table 11.5 Statistical reproduction capacity of the family troops in FJC Big Cave and Small Cave (until 2015.12)

Troop	Duration	Number of langurs	Number of pups born	Birth rate (%)	No. of deaths	Mortality rate (%)	Net increase in number of langurs	Net increase rate (%)	Number of adult and immature females staying hehind in the troop	Number of adult and immature langurs split away from the troop	Remarks
Queque family troop	1994–1998.04	18	4♀ 5♂	50.0	1♀ 2♂	16.7	6	33.3	7	4♂ 4♀	Heimama was killed in poaching
Archeng family troop	1998.04–2002.06	23	6♀ 9♂	65.2	4♂	17.4	11	47.8	7	6♂ 6♀	
α-Gonghou family troop	2002.06–2006.09	22	7♀ 7♂	63.6	2♂	9.1	12	54.5	13	5♂ 1♀	
Yintangxiaotu family troop	2006.09–2012.07	44	13♀ 16♂	65.9	3♀ 3♂	13.6	23	52.3	17	14♂ 7♀	One outlier ♀ came joined the troop
Yuweng family troop	2013.07–(still existed by 2015.12)	32	7♀ 8♂	46.9	–	0	15	46.9	–	–	Until December 2015 family remained intact
Yintangxiaotu transitory troop	2012.07–(still existed by 2015.12)	25	1♀ 2♂	12.0	1♀ 1♂	8.0	1	4.0	8	15♂	One outlier ♀ came joined the troop
Laoxie family troop	2015.03–(still existed by 2015.12)	10	1♀ 1♂	20.0	–	0	2	20.0	–	–	One outlier ♀ came joined the troop Until December 2015 family remained intact

Fig. 11.2 2013.06.12: Pingping kept trying to wake up her baby Xiao-P, which had been killed in infanticide at just 3 months old (Photo by Feng Chunguang)

Fig. 11.3 Liushaotou, the 14-month-old daughter of Yintangxiaotu, died 1 week after being seriously wounded by incoming male langur Laoxie (Photo by Liang Zuhong)

was 3.2 times that of males in this age group, revealing that the juvenile stage for female white-headed langurs could be the most critical period of life or death (Fig. 11.3).

Our tracking on 18 adolescent females that had not yet reached their reproductive ages found that they were safely living in their matrilineal family. While the families had come through brief periods of chaos when the alpha male was replaced, none of them had died in the course of transition. Hence, we deem the mortality rate of female in adolescent age as 0% ($n = 18$).

We had kept tracking 19 almost adolescent and sub-adult-aged male langurs for 0.08–4.25 years. After their family split, following their fathers, they left their families to rove about, which made it difficult to keep on tracking them. But before that, they lived safely in their family just as adolescent females did. No deaths had been registered during that period; hence, the death rate of adolescent males was also at 0% ($n = 19$).

We had kept observing a total of 23 adult female white-headed langurs, and there was only one death registered. The female died of miscarriage while living a tumultuous life in a transitory troop after her family split. The mortality rate of adult females registered is thus 4.3% ($n = 23$).

We only regarded those male white-headed langurs which had successfully invaded and presided over a family as truly reached their adulthood. In this accord, we had registered ten adult male langurs in FJC. None of them had died during their time in family troops; hence, their mortality rate was 0% ($n = 10$).

We have yet to find the age difference between reproductive and post-reproductive stages, because until now, we have not yet observed and registered a male or female white-headed langur completing its life cycle. For the time being, we can only make guesses based on the probabilities found in our following observations in the wild.

11.3.4.1 Females

1. In November 1996, we found an adult female in the FJC Big Cave troop with suckle traces on her body. We named her Dahei for she had more black hair on her body than other langurs did, and estimated that she was over 8 years old. Her first delivery we observed was in January 1998, when she gave birth to male pup Xiaoarge. When she gave birth to Xiaohei on May 20, 2010, she had already given birth to a total of seven pups. In the last 5 years, Dahei did not have any more offspring. We estimated that she had given birth to at least eight pups and surmise that she entered into the post-reproductive period at the age of 23 years.
2. There was another old female Arzhen in the FJC Big Cave troop. From her outward features, we estimated that she was born more or less at about the same time as Dahei. She had given birth to eight pups, and the latest one was a

female pup named Zhenni on July 28, 2014. In May 2015, our HD monitoring system still had images captured of her suckling her infant. We estimate that she was at that time 27 years old and had not yet entered into the post-reproductive period.
3. When we came to FJC in November 1996, we also found in the troop a very old female with suckle traces on her body. As she had black hair almost all over her body, we named her Heimama. She spent the nights in the FJC Big Cave night shelter along the others in the troop, foraged and moved together with her troop, but had no offspring around her. Until she was killed by a poacher in April 1998, she had not given any births, which was a sign that she had entered the post-reproductive period. Her outward features were more or less similar to those of Dahei and Arzhen today.

Considering the above three cases, we conjecture that female white-headed langurs enter into the post-reproductive period at about or after 25 years of age.

11.3.4.2 Males

In our idea, a male white-headed langur enters into the post-reproductive age on the day when it loses his resident alpha male status. Presently, we have data collected only from the behavioral changes of Yintangxiaotu and his three adult sons.

1. Second son Zuoque had left his father and brothers at the age of 5 years to find, alone by itself, his position in the scope of a larger society.
2. Third son Youque at the age of 6 years left the all-male troop where his father and brothers were staying.
3. Eldest son Wuque was the strongest among the brothers in the "Yintangxiaotu family troop." During his life in the transitory troop and the all-male troop, he had effectively helped his father take care of the brothers and younger sisters. We found in our tracking observations that he had successfully invaded a family, became a resident alpha male at the age of 6.75 years and owned 13 wives. He became a father at the end of 2015.

Hence, we suppose that the male white-headed langur enters into the reproductive period only after 6 years old. Yintangxiaotu was the strongest male white-headed langur we found in FJC. It had been a resident alpha male for as long as 6 years (while typically the time to be a resident alpha male is 4 years, $n = 3$), after which it presided over a transitory troop for 3 years before it turned to an all-male troop. Based on these findings, it is appropriate to conjecture that male white-headed langurs enter into the post-reproductive period at the age of 13 years. After reaching 13 years of age, the males which had reproduced their own offspring have almost no more chances to reproduce, though they still take active roles in leading transitory troops and all-male troops. They rove among the habitat, making the Nongguan Mountains one integrated white-headed langur society. This brings new scientific issues to our research on the white-headed langur ethology and leads us to further research on the socio-biology of white-headed langurs. We shall publish a new research report in the next 3 years.

11.3.5 The Innate Capacity of Increase in Bionomic Strategies

The innate capacity for the increase in a population is generally ascertained in laboratory conditions. Since poaching of white-headed langurs had been effectively stopped in the Nongguan Mountains in 2000, the vegetation in their habitats were preserved under conservation policies, and the karst hills had since hardly been damaged, we found in our research over the 20 years that the reproduction capacity of the white-headed langurs in FJC had to a large extent not been restrained. Therefore, we can get the innate capacity for the increase of the white-headed langurs in that period simply by deducting the mortality rate from the birth rate, and it approximately equals the net increase rate shown in Table 11.5. As we found that the number of white-headed langurs in FJC district had increased, we could reasonably conclude that the same has happened with the white-headed langurs in other districts in the Nongguan Mountains.

The white-headed langur population in the Nongguan Mountains is still on an increasing trend. Their bionomic strategy is "K-Selection." Despite that, they are confined in the karst hill tropical forest that provides them abundant food resources and had night shelters high up on cliffs to protect them from predators. Hence, they had a steady, high birth rate and a relatively low mortality rate. The female descendants stay in their matrilineal homes throughout their lives to breed and raise offspring, while their young males would be taken care of by their mothers, aunts, and sisters at home for 1–4 years before the family splits. Then their father would lead the males to explore other areas. The outcome is that the white-headed langur population in the Nongguan Mountains is steadily growing to be a sustainable population for the long term.

When we made the first ever documentation on the FJC Big Cave troop ("Queque family troop") in November 1996, it had 13 langurs (we estimated the troop was initiated in year 1994). In April 1998, it had increased to 18 langurs, but then Queque was replaced by outlier male langue Archeng and the troop became the "Archeng family troop." Each later cycle of resident alpha male replacements had brought changes to the troop, and it successively became the "α-Gonghou family

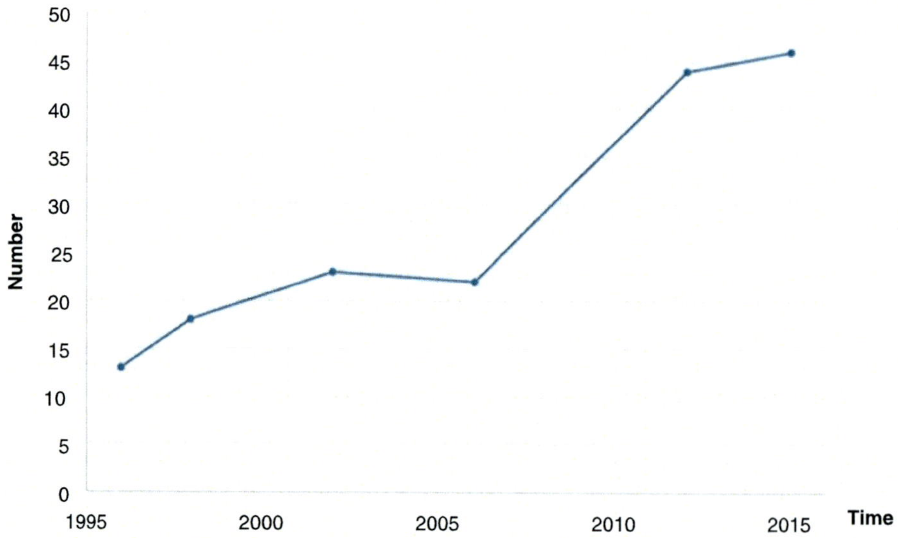

Fig. 11.4 A simple chart showing the increasing number of white-headed langurs in FJC through 1996–2015

troop," the "Yintangxiaotu family troop," the "Yuweng family troop," the "Yintangxiaotu transitory troop," and the "Laoxie family troop."

As time went by, the number of sexually matured females in the FJC Big Cave troop increased, and they took part in producing offspring, increasing the number of descendants in the successive FJC Big Cave troop. As shown in Table 11.5, the "Queque family troop" had successfully raised six offspring and had an average reproduction capacity of 1.5 langurs per year. The "Archeng family troop" successfully raised 11 offspring, and the average reproduction capacity increased to 2.75 langurs per year. The "α-Gonghou family troop" successfully raised 12 offspring and the average reproduction capacity further increased to three langurs per year. The "Yintangxiaotu family troop" successfully raised 23 offspring and the average reproduction capacity came to be 3.8 langurs per year. It can be foreseen that there will be even more descendants from the "Yuweng family troop" and the "Laoxie family troop" in FJC—the core district in our research.

With accord to the increasing number of white-headed langurs in FJC through 1996–2015, we produced a chart (Fig. 11.4) to illustrate that the Nongguan Mountains white-headed langur population may be entering into the period of rapid growth.

11.4 The White-Headed Langur Is a Key Species in the Bio-community

The bio-community of the Nongguan Mountains is one organic unity composed of the correlations between animals, vegetation, and microorganisms. The white-headed langur is a part of that bio-community and has its unique ecologic niche in the community.

Fossils buried in FJC Big Cave provided evidence that when the sea level descended in the first glacial period in the early Pleistocene Epoch, the exposed continental shelf in the Northern Hemisphere gave an opportunity to the ancestors of white-headed langurs to migrate from the Tropic toward places in the Tropic of Cancer. In the 1.4 million years afterwards, the ancestors of white-headed langurs living in the karst hills in the south bank of Zuo Jiang evolved and adapted to the native climate, vegetation, and animals. The correlation between the white-headed langurs and the other species in the Nongguan Mountains bio-community was built over long geologic eras, and they are closely interrelated.

We regard white-headed langurs as a key species in the Nongguan Mountains bio-community, for their activities have made them the pollinator and seed distributor for the native tropical plants. It was found that seeds of shishanrong (*Ficus orthoneura*) had passed through the digestive tracts of white-headed langurs, dropped with their feces on the crevices on the precipices, germinated, and flourished. A large part of the flora growing today in the Nongguan Mountains bio-community was germinated from seeds dropped on the talus by white-headed langurs on their foraging trips. To say that the white-headed langur is a key species in the Nongguan Mountains bio-community does not mean that it dominates or regulates the other species in the bio-community. We are referring to the positive stimulation effects it has on the bio-community. Birds and insects in the bio-community as well have their own stimulation effects.

We shall be more willing to say that the white-headed langur is a charismatic animal and regard it as the flagship species of the Nongguan Mountains. Like a flag, it calls to people to value the survival of all living beings.

Check for updates

A Population Viability Analysis for the White-Headed Langur in Nongguan Mountains

12

Abstract

We used the Vortex Population Viability Analysis (PVA) model to simulate the changes of the white-headed langur population over the past 20 years and verified it with field observation data from 1997 to 2016. We found that the population numbers calculated by the two methods are very close, with a fitting degree of $R^2 = 0.99792$. According to our field investigation, the number of white-headed langurs in the Nongguan Mountains has increased from 105 in 1997 to 797 in 2015, while VORTEX 9.99c, after 500 iterations of simulations, shows that in the 19 years from 1996 to 2016, the population of white-headed langurs would grow from 105 to 858, with an extinction probability of 0% and a survival probability of 100%. Taking into account the effective breeding population size, age distribution, breeding strategy, mortality, population diffusion, environmental capacity, and other factors of the population in 2016, a positive future for the population is predicted through the PVA model. The effective breeding population size will increase with time, and the inbreeding rate will gradually be reduced. When the natural habitat of the whole Nongguan Mountain area of 13.2 km^2 is supposed to sustain a maximum of 1800 white-headed langurs, it is expected to reach carrying capacity in 2027 and, if current conditions persist, remain stable in the future. From the perspective of growth trend, 2003–2021 is the key period for the rapid recovery and growth of this population and thus the golden period that saved the population from extinction.

Keywords

Bottleneck effect · PVA · VORTEX · Population dynamics · Extinction probability · Survival probability

Natural selection is simple and straightforward: it seizes upon the genetic variations that benefit survival and reproduction. We are still unable to describe how the minuscule but beneficial variances that gradually accumulated have defined the evolution of the white-headed langur. Nonetheless, we could be sure that it was due to more than 200,000 generations of mutations, selections, and adaptations since their ancestors arrived at the Nongguan Mountains 1.4 million years ago that they successfully evolved to survive to this date.

12.1 The Small Population of White-Headed Langurs in Nongguan Mountains Is a Typical Example of Bottleneck Effect

Variations are inherent in genetic inheritance. There could be no absolute exactness in any one replication of parental genetic materials, and there could always be deleterious recessive alleles in genes. Whenever two deleterious DNA fragments come together in a gene, the carrier of that gene is subjected to dire consequences. Under the rigorous scrutiny of natural selection, once the defective trait manifests, the carrier would be eliminated by nature. In a sufficiently large population, losses from eliminations of this nature would be negligible, but for a small and isolated population, such losses could be fatal. The smaller the population, the higher the chances of inbreeding and a higher probability of offspring carrying two strands of deleterious DNA, which may induce infertility or early death.

Still, it could be possible that inbreeding depression in a small population may bring about a different consequence, in which case a small population with very few individuals under intense natural selection pressure overcomes the inevitable and breaks through the population bottleneck. Deleterious genes are purged in the breakthrough, and dominant genes remain, which in a way cleanses the gene pool. After a specific time, the inbreeding pressure would gradually be reduced, and the number of individuals in the population will recover.

W. Pan, *The Population Ecology of White-Headed Langur*, https://doi.org/10.1007/978-981-33-4118-0_12

The Nongguan Mountains white-headed langur population is a living example. We had witnessed that humans' ignorance, ruthlessness, and unrestrained greed had in just 10 years (1988–1997) pushed the Nongguan Mountains white-headed langur population to near extinction. Still in the past 20 years (1996–2015), we witnessed this small population in the wild break through the bottleneck. By force of nature, its natural fertility was gradually restored to a thriving vitality. We take that as a typical example exhibiting the bottleneck breakthrough effect.

When the research started in 1996, there were only about 105 white-headed langurs in the Nongguan Mountains. It was really a small and endangered wild population. Generally, only a small number of individuals in a population would be in their reproductive periods. According to the 50/500 rule proposed by Frankel and Soulé (1981), when the number of reproductive individuals becomes less than 500, inbreeding depression will be hazardous, and when the number drops to less than 50, the hazard will become extremely serious. Apart from the genetic hazard, a small population also has the risk of abrupt extinction under the influence of thunderstorms, floods, forest fires, drought, and other natural catastrophes. Hence, it is extraordinary that the Nongguan Mountains white-headed langur population survived and that its numbers are even increasing continuously.

Our research program has avoided interferences such as rearing, artificial breeding, etc. and is based on observations on the natural life cycle of white-headed langurs. We register changes in their population number, social structure, reproduction system, and various other data gathered in situ. Our discussions in this chapter are based on those findings.

We believe, until June 2016, there were 797–841 white-headed langurs living in the Nongguan Mountains, and the population appeared to be increasing continuously at an accelerating rate. Still, in the long run, would the population be affected by genetic drift as a result of the narrow genetic diversity? Could the population continue increasing? What factors might disrupt the population's survival? Could they forever survive in the Nongguan Mountains?

In this chapter, we are going to find the Nongguan Mountains white-headed langur viable population size and its inbreeding depression frequency. By using computer simulations, we conducted a population viability analysis (PVA) with empirical data obtained from our research to simulate and review the restoration process of the small population. We verified the simulation results against the reality to provide references to our work on the conservation of the Nongguan Mountains white-headed langurs.

12.2 Effective Population Size of White-Headed Langurs in Nongguan Mountains

In the early days of our research, we conducted a field survey for 62 days and nights from November 18, 1996, to January 19, 1997, and found only nine white-headed langur troops in Nongguan Mountains with an aggregate of only 105 langurs. They were six family troops comprising of 86 individuals and three all-male troops comprising of 19 individuals. That small population of 105 remnant white-headed langurs is, in reality, the founder of the present-day Nongguan Mountains white-headed langur population. Listed in Tables 12.1 and 12.2 is the data obtained in our research in that period.

From Tables 12.1 and 12.2 and Fig. 12.1, we see that the total of young individuals and mature individuals in reproductive ages accounted for 91.43% of the population. The population appeared to be safe from extinction.

In nature, the total number of individuals regardless of age and sex in a sexual animal population could be large, yet not every individual reaches sexual maturity, and many that appear to be at a reproductive age at one time or another cannot mate or reproduce. Besides, those who participate in reproduction may not have offspring, or their offspring will

Table 12.1 Statistical findings of the Nongguan Mountains white-headed langur in the period November 1996 to January 1997

	Individuals in the troop	Number of heads	Notes
In the six family troops	Adult male	6	All of them were matured individuals in reproductive age
	Adult female	43	There was one post-reproductive old female in each of the six families; all the rest 37 were matured adult females in reproductive age
	Adolescent females	12	Out of the 12 individuals, about ten were sub-adults soon entering into reproductive age
	Yellow-haired juveniles	13	Child age pups, male/female ratio 1:1
	Yellow-haired baby pup	12	Baby pups, male/female ratio 1:1
In the three all-male troops	Adult male	3	All of them were post-reproductive individuals
	Immature male	16	Adolescents and pups
	Total	105	

Table 12.2 The Nongguan Mountains white-headed langur population composition in the period November 1996 to January 1997

Population composing units	Number of heads	Percentage in the population (%)
Post-reproductive individuals	9	8.57
Mature individuals in reproductive ages	43	40.95
Sub-adults and pups	53	50.48
Total	105	100.00

not grow to maturity. No matter what genetic structures those individuals may have, they do not contribute to the genetic structure of the next generation. The effective population size, or "N_e," therefore refers only to those reproductive individuals who contribute to the genetic structure of the next generation. In general, N_e is much smaller than the number of reproductive individuals in a given population of nature (Li 1981).

The Wright–Fisher model proposed by Sewall Wright, R.A. Fisher, and J.B.S. Haldane assumes an "ideal population," which is a fundamental effective population model and follows the following statistical assumptions:

- In a sufficiently large population, the number of individuals participating in reproduction is *N*.
- There is an equal number of male individuals and female individuals, and they mate randomly. The number of sexually matured females is $N_{♀}$, and the number of sexually matured males is $N_{♂}$.
- Every individual has an equal chance of mating and makes an equal contribution to the next generation.
- Every individual has the same number of offspring.
- No generation is overlapping.
- No selections, mutations, migrations, or any factors that disturb the genetic equilibrium of the population come into play.

In this "ideal population," the number of individuals participating in reproduction would be equal to the reproductive population size, meaning:

$$N_e = N = N♀ + N♂ \tag{12.1}$$

Among the 105 white-headed langurs, there were six family troops with a total of 60–61 $N_{♀}$ (37 adult females + 12 sub-adult females + 11–12 female pups), which accounted for about 57.6% of the population. At the same time, there were at least six $N_{♂}$ participating in reproduction (one in each family troop), which accounted for 5.7% of the population. There was one old female in each of the six family troops and one old male in each of the all-male troops making a total of nine individuals, all of whom are likely in post-reproductive age and accounted for 8.57% of the population. The rest were pups and immature individuals, and they made up about 28.13% of the population. Based on the above data, we sort out Table 12.3:

Assuming it was an "ideal population," the Nongguan Mountains white-headed langur effective population size on January 19, 1997, would be:

$$N_e = N = 66.465 \text{ white-headed langurs.}$$

However, in reality, an "ideal population" rarely exists in the wilderness. The white-headed langur is a longevous species (records have shown them living to 29 years). Like most other mammals, the white-headed langur sex ratio at birth is close to 1:1. Within the habitat, any robust adult male (age from 7 to 13) would have a chance to take priority over others in engaging matured females (age from 5 to 25). Hence, without taking into consideration the population of matured white-headed langurs migrating in and out of the Nongguan Mountains during the breeding season, and without considering the complex reproduction system, in a particular breeding season, the number of males and females participating in

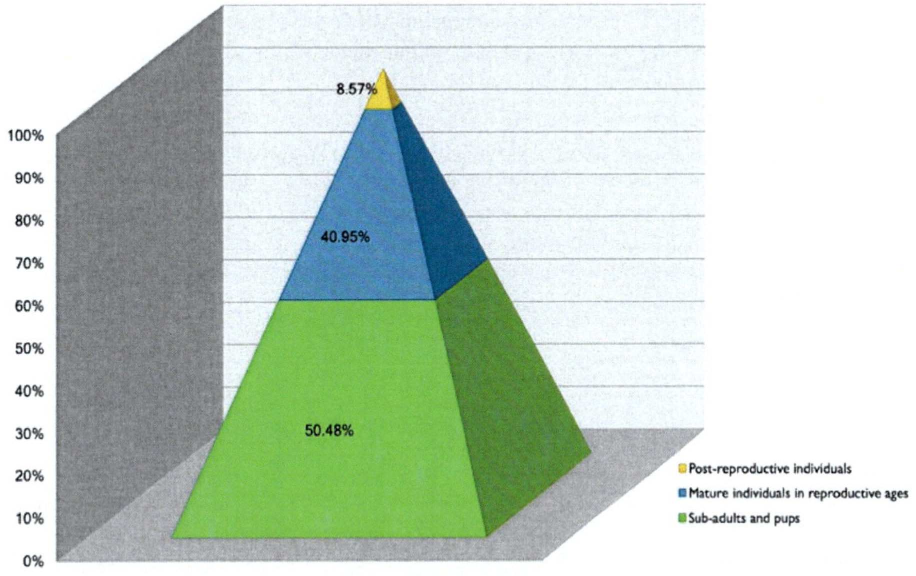

Fig. 12.1 The Nongguan Mountains white-headed langur population composition from November 18, 1996 to January 19, 1997

Table 12.3 Population composition and number of white-headed langurs in Nonguan Mountains from November 1996 to January 1997

Number of heads	Percentage in the population (%)
Total $N_T = 105$	
Post-reproductive individuals $N_o = 9$	$N_o = 9/105 = 8.57\%$
Adults participating in reproduction $N = 60.5 + 6 = 66.5$	$N = 66.5/105 = 63.3\%$
Sub-adults and pups $N_k = 105–9–66.5 = 29.5$	$N_k = 29.5/105 = 28.1\%$

reproduction most likely would not be equal, and generation overlapping could happen.

Say if the ratio of reproductive years between females and males is 20:6 (female: 25 years age span – 5 years pup age = 20 years, male: 13–7 = 6 years):

The proportion of sexually matured female individuals in a population would be 20/(20 + 6) = 76.92%.
The proportion of sexually matured male individuals in a population would be 6/(20 + 6) = 23.08%.

Thus we may infer the numbers of female and male white-headed langurs participating in reproduction in the year 1997 to be:

$$N♀ = \mathrm{N} \times 76.92\% = 66.465 \times 76.92\% \\ = 51.124 \text{ individuals}$$

$$N♂ = \mathrm{N} \times 23.07\% = 66.465 \times 23.08\% \\ = 15.331 \text{ individuals}$$

In a small population with limited N, the assumed conditions in an "ideal population" would be varied. By way of mathematical and statistical modeling, we calculate the appearance frequencies of genes and possible genotypes in different genders, in parents and offspring, and respective family lineages, thereby inferring the relationship between the size of N and the size of N_e in five population models other than the "ideal population" (Guo 1993).

- In the case the numbers of female and male individuals are not the same, the coefficient of the different sexes on the inbreeding frequency would be different. The gender with fewer individuals participating would have a bigger coefficient on inbreeding frequency. In this case, $N_e < N$, and it could be found in the formula that:

$$N_e = \frac{4N_♀ N_♂}{N_♀ + N_♂} = \frac{4 \times 51.45 \times 15}{51.45 + 15} \\ = 46.45 \text{ (individuals)} \tag{12.2}$$

- In the case, different individuals have different fertilities, and there are different numbers of individuals in each family lineage, the number of offspring by different parents could vary considerably, resulting in non-probability lineage sizes. In this case, the effective population size N_e with accord to the actual lineages sizes could be inferred with the following formula:

$$N_e = \frac{8N}{\sigma^2_{k♀} + \sigma^2_{k♂} + 4} \tag{12.3}$$

σ_k^2 = lineage size variance. Suppose the variances of the female and male lineages are both >2, then

$$N_e < \frac{8N}{2 + 2 + 4} < 66.465 \text{ (individulas)}$$

- In the case, the lineage sizes are the same as possible, and with other conditions unchanged, mathematically it could be assumed that the lowest inbreeding frequency is reached. When the lineage size variance is reduced to zero, i.e., if in Formula 12.3

$$\sigma^2_{k♀} = \sigma^2_{k♂} = 0, N_e \approx 2N0 \tag{12.4}$$

then $N_e \approx 2 \times 66.465132.93$ (individuals)

- Population sizes change periodically as the environment keeps changing. It is unlikely that every generation has a constant number; hence, the number of reproductive individuals N changes in every generation. In a population whose size changes in successive generations, the inbreeding frequency (ΔF) must necessarily be increased in the generation with the least individuals. A significant reduction in one generation could drastically alter the gene frequency to the point that it may offset the previous selection and mutation effects and point the gene frequency to a new equilibrium. The smallest population of white-headed langurs in the Nongguan Mountains among our statistics is in 1997 when the number of matured individuals = 105 × 63.2% = 66.465 (individuals), and so far the most significant population was recorded in 2015 with the number of matured individuals = 797× 63.2% = 487.764 (individuals). We put the harmonic mean of the two into the following formula, and the inferred result is:

$$N_e = \frac{t}{\left(\frac{1}{N_1} + \frac{1}{N_2} + \ldots + \frac{1}{N_t}\right)}$$

$$\approx \frac{2}{\frac{1}{66.465} + \frac{1}{487.764}} \approx 116.965 \text{ (individuals)} \qquad (12.5)$$

- In overlapping generations, the number of reproductive individuals in each generation is relative to the number of individuals of the generation, the span of the generation, and the lineage size variance ratio in that generation. The number of reproductive individuals in each generation (N_{et}) can be inferred with the following formula:

$$N_{et} = \frac{4N_c L}{\sigma_k^2 + 2} \qquad (12.6)$$

We found from observations that the oldest reproductive age for female white-headed langurs is at around 25 (E), and the span of a generation with effective reproduction is governed by the average breeding age, which is around 7 (L). N_c represents the total number of individuals reproduced in one generation.

$$N_c = \frac{\text{total number of individuals in one and the same generation } N_r}{\text{theoretical lifespan of an individual } E}$$

with σ_k^2 being the linage variance; hence, we get

$$N_{e_t} = \frac{4 \times \frac{105}{25} \times 7}{2+2} = 29.4 \text{ (individuals)}$$

Applying the white-headed langur population statistics obtained from field observations between 1996 and 1997 into the formulas, we find that the effective population size in those years is in the range between 29.4 and 132.93 individuals. The mean of the above six calculations is 76.61 individuals.

Rich genetic diversity is pivotal in a population's successful adaptation to changing environments and human disturbances. A small population proves very susceptible to many forms of deleterious genetic effects such as random genetic drifts, inbreeding depression, and randomness, which reduces the population size, thereby further losing genetic variance in subsequent generations, increasing the risk of extinction.

Natural selection may reduce the genetic frequency of deleterious genes that harm the population's survival and reproduction and may raise the genetic frequency of genes that benefit survival and reproduction.

Apart from natural selection, genetic variances could also be brought about by stochastic allele mutations. When a particular mutation brings no inferior or beneficial effects to a genetic structure, natural selection will not act upon it, and it is considered a neutral mutation (Motoo Kimura 1968). Stochastic allele mutation is purely accidental and may increase or reduce the genetic frequency of the population. This uncertain and accidental fluctuation in genetic frequency is termed "random genetic drift" (Snustad and Simmons 2011) (Fig. 12.2).

The stochastic allele mutations in the course of random genetic drift may eventually cause some alleles to vanish from a population and some to become fixed in the population, which would reduce the genetic diversity and the heterozygosity of the population as well as cause speciation to increase.

There are two alleles on each locus. The proportion of individuals in a population with two different alleles on the same locus is described as the "heterozygosity" (H) of the population. In a population with N_e effective population, the allele frequency varies stochastically and so does the heterozygosity (Zhuohua et al. 2016). Sewall Wright expressed the relativity of the heterozygosities H and H' between two successive generations mathematically in the formula:

$$H' = \left(1 - \frac{1}{2N_e}\right)H \qquad (12.7)$$

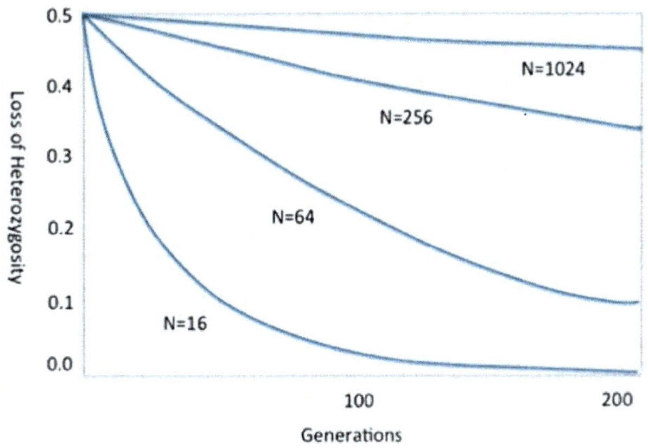

Fig. 12.2 Heterozygosity losses induced by random genetic drift in different N_e populations with initial allele frequencies at 0.5 (Snustad and Simmons 2011)

The formula finds heterozygosity of the next generation. It indicates that in each generation, the heterozygosity is lost a certain rate due to stochastic factors. After (*t*) generations, we may expect heterozygosity at the (*t*) generation to be:

$$H_t = \left(1 - \frac{1}{2N_e}\right)^t H \tag{12.8}$$

where the inbreeding depression rate between each generation is:

$$\Delta F = \frac{1}{2N_e} \tag{12.9}$$

Formula (12.8) allows us to see the accumulated effect of random genetic drifts after many generations. After many generations, the heterozygosity would eventually be reduced to zero when there is only one allele in the gene pool of the population. As a result of stochastic variances of the genetic frequency, genetic drifts would gradually reduce genetic variances, resulting eventually either in the fixation or loss of alleles.

When the size of the effective population is becoming less and less, the inbreeding depression frequency becomes higher and higher. When the effective population size gradually increases, the inbreeding depression frequency gradually gets lower, also reducing the effect of genetic drift.

In the 50/500 rule proposed by Frankel and Soulé (1981), the essential requirement in conserving the genetic diversity of a population is that in any short-term period, the effective population must have no less than 50 individuals and the maximum inbreeding depression frequency must not be over 1% (Frankel and Soulé 1981).

That is to say when $N_e = 50$.

$$\Delta F = \frac{1}{2N_e} = \frac{1}{2 \times 50} = 1\% \tag{12.10}$$

It is clear from the formula that in an effective population of 50 individuals, the heterozygosity in the next generation would be only 99% of the precedent generation losing 1% of allele variance. After ten generations, the heterozygosity would come to only 90% of the first generation. The decreasing heterozygosity is an indication that the genetic diversity of the population is reducing from generation to generation.

A study on 120 bighorn sheep populations in the south-western desert in America, in which some populations had been tracked and studied for more than 70 years, had found that all those populations with less than 50 individuals had come to extinction within 50 years while concurrent populations with over 100 individuals remain (Fig. 12.3). The result of that field research had proven that populations with more than 50 individuals have a chance of surviving for more than 50 years (Berger 1999).

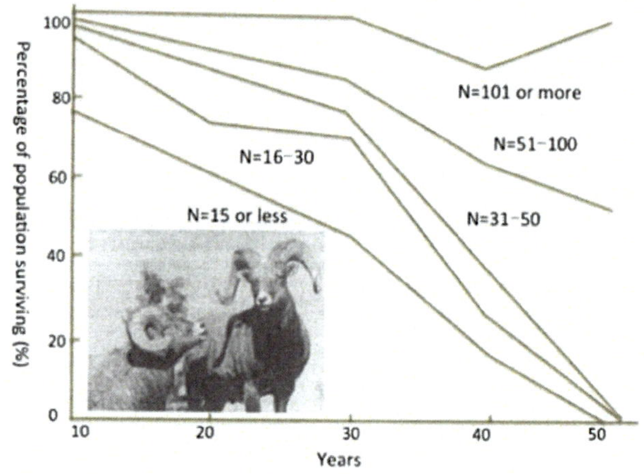

Fig. 12.3 Field researches on 120 bighorn sheep populations in south-western America had proven that populations with more than 50 individuals can survive for more than 50 years

The study has indicated that the smaller the population, the more susceptible it is to influence by environmental and genetic factors. Those influences might further decrease the population, leading it to extinction. When the population is reduced to a particular threshold value, it falls into an extinction vortex and is no longer able to overcome unfavorable factors, and the population would keep decreasing. The extinction vortex often results in the localized extinction of species (Gilpin and Soulé 1986; Guerrant Jr 1992) (Fig. 12.4).

Analysis of the research data obtained from the founding population of white-headed langurs in the Nongguan Mountains in the period from November 1996 to January 1997 found that there were 105 individuals who were above the theoretical threshold value ($N > 100$, Fig. 12.2). The mean effective population N_e was inferred to be 76.33 individuals from calculations in six different genetic statistic models, and the inbreeding depression frequency ΔF at 0.651% was as well below the threshold value ($\Delta F = 1\%$, formula 12.9). 50.48% of the population were sub-adults and pups (Table 12.2).

As our research progressed and conservation measures were put into practice, by April 2015, the population had grown to include 797 individuals. Referring to the population structure illustrated in Chap. 11, we quickly found the inferred mean effective population and inbreeding depression frequency of the white-headed langur in April 2015 by using the calculations in the six different genetic statistics models. The results are listed in Table 12.4 against those of 1997. It shows that the effective population increased along with time, and the inbreeding depression frequency is reducing gradually.

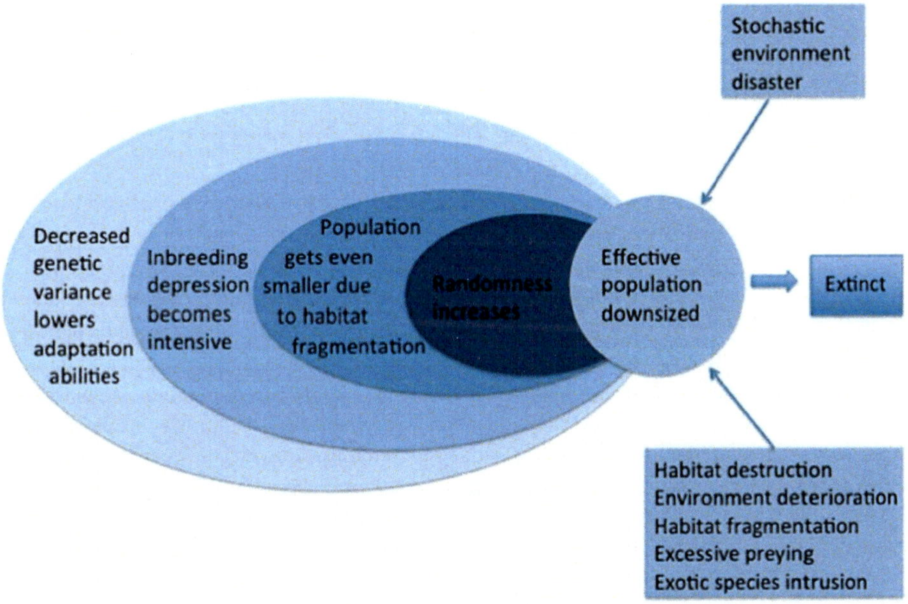

Fig. 12.4 When a population is reduced to a certain extent, it falls into an extinction vortex in which all the unfavorable factors would keep persistently decreasing the population, leading the localized species to extinction (Gilpin and Soulé 1986; Guerrant Jr 1992)

Table 12.4 Population construction, effective population size, and inbreeding depression frequency of the Nongguan Mountains white-headed langur population in April 2015 in contrast to January 1997

Item	Description	January 1997		April 2015	
Population construction	Total number of individuals in the population, N_T	105 (heads)		797 (heads)	
	Number of post-reproductive old age langurs, N_O	8.6%	9.03 heads	6.5%	51.805 heads
	Number of matured adults participating in reproduction, N	63.3%	66.465 heads	32.3%	257.431 heads
	Number of sub-adults and pups, N_K	28.1%	29.505 heads	61.2%	487.764 heads
Effective population calculations	Calculation 1:$N_e = N = N_♀ + N_♂$	66.465 heads		257.431 heads	
	Calculation 2: $N_e = \frac{4N_♀ N_♂}{N_♀ + N_♂}$	46.45 heads		182.808 heads	
	Calculation 3: $N_e = \frac{8N}{\sigma^2_{k♀} + \sigma^2_{k♂} + 4}$	66.465 heads		257.431 heads	
	Calculation 4: $N_e = 2N$	132.93 heads		514.862 heads	
	Calculation 5: $N_e = \frac{t}{\left(\frac{1}{N_1} + \frac{1}{N_2} + \ldots + \frac{1}{N_t}\right)}$	116.965 heads		116.965 heads	
	Calculation 6: $N_{e_t} = \frac{4NcL}{\sigma^2_k + 2}$	29.4 heads		223.16 heads	
	Mean effective population	76.33 heads		258.776 heads	
Inbreeding depression frequency	Inbreeding depression frequency: $\Delta F = \frac{1}{2N_e}$	0.651%		0.193%	

The data in Table 12.4 shows that the remnant wild white-headed langur population in 1997, though small, had not yet fallen into the extinction vortex. Furthermore, it has been a proven fact that given sufficient time and space, an endangered small population could restore itself in nature.

With the knowledge gained on the structure of white-headed langur populations, we further introduced population viability concepts into our study and used computer simulations with past research data to predict and analyze the extinction probability in the future.

12.3 Vortex Model as the Method of Population Viability Analysis (PVA)

Through the five billion years of earth's history, there were asteroid strikes, volcano eruptions, climate changes, sea-level changes, continental drifts, etc., and there were five episodes of "mass extinction of species." Nevertheless, statistics of today has shown us that the contemporary species extinction probability is 100–1000 times higher than that of any other point in geologic history, and the sixth mass extinction of species is approaching as a result of continuous disruptions of nature by humans. In 1963, The International Union for Conservation of Nature launched the IUCN Red List of Threatened Species (IUCN Red List). In 1966, the US Endangered Species Act listed the California condor (*Gymnogyps californianus*), whooping crane (*Grus americana*), the black-footed ferret (*Mustela nigripes*), milu (*Elaphurus davidianus*), and some other species as endangered species. As humans are exploiting natural resources at an ever-increasing pace, natural habitats are fragmented if not destroyed, and the issue of endangered species becomes more and more pressing. Population viability analysis (PVA) thus came into being in the 1970s (Tian Yu et al. 2011).

Population viability analysis refers to the analytic, quantitative research on the survival probability of an endangered species based on ecological knowledge. It puts factors like the population size, distribution, exchanges between individuals, varying amplitudes relative to time, the lifespan of individuals in the population, reproductive rate, etc. into mathematical analysis to simulate the population dynamics. Ever since this tool came up, it received positive responses from biologists. It has provided a means of research and an essential foundation for the conservation of endangered species. Population viability analysis has become a significant subject in biological research since in the past few decades.

These days the Species Survival Commission (SSC) uses PVA as one of its tools to infer the extinction probability of a species by inputting data regarding the species life history, habitat requirements, endangerment factors, particular conservative measures, etc.

The risk of extinction would dramatically increase when the population is reduced to a certain extent. The cause for this is made up of multiple interactive factors. Inbreedings are more prone to happen in small populations, bringing progenies with inborn defects, and small populations are easily impacted by the rise and fall in the number of individuals, and a slight but relatively big fall could lead the population to extinction. The factors enhance each other and become a vortex, drawing a small population into the depths of extinction that is beyond rescue.

White-headed langur is just one step from extinction in the Red List (Bleisch et al. 2008).

After comparing the accuracy of simulation results, we found among the many PVA computing software that four software, GAPPS, ALEX, RAMAS, and VORTEX 4, have similar capabilities and higher accuracies in their simulation results. The VORTEX software, adapting its name from "extinction vortex," was developed by the Chicago Zoological Society. It was first used on mammalian and avian research, and it excels in using a series of events of diploid organisms with sexual reproduction capabilities to simulate a typical life cycle of the diplont.

12.3.1 Specifics of VORTEX Software

- It was designed to work based on the individuals in a population and pay particular attention to the influences of every individual on the population viability.
- It has taken into consideration statistical randomness, environments, genetics, and catastrophes that would have effects on wild animal population dynamics.
- It simulates population dynamics with accord to probabilities of known, isolated, or connected events such as birth, death, mating, catastrophe, etc., and in the model, these could be inputted as constant values or random variables.
- VORTEX provides for abundant data inputs covering parameters of nearly every kind in mainstream ecologic research, requiring the user to possess certain ecological knowledge. It supports "computation with precise parameters input" as well as "stand-alone hypothetical computation." To minimize the effect of randomness on the simulated increase or decrease of a population, the program allows a maximum of 10,000 iterations to get to the nearest value.

Given the above specifics, we, at last, decided to use the software called VORTEX 9.99c in our population viability analysis on white-headed langurs.

12.4 Testing the Reliability of PVA Simulation Computations Against Empirical Population Ecological Data

When we began our research in the Nongguan Mountains, the population of white-headed langurs was at a critical stage. We were fortunate to have witnessed the population at the edge of the extinction vortex miraculously restore itself from the brink of extinction. However, what is more exceptional is the abundant, authentic data we collected in the continuous research through the crucial 20 years which today allow us to conduct a positive step-by-step predictions on the population viability of the Nongguan Mountains white-headed langurs. The events not only provided us real data to verify the reliability of the PVA simulations but also served a meaningful model to wildlife conservation biology all over the world.

For VORTEX 9.99c to recreate in precision the population dynamics of the white-headed langurs in the Nongguan Mountains through 1997 to 2015, we had to sort out the various crucial parameters required for PVA modeling.

Besides those parameters required for mathematical computation, VORTEX also requires inputs of the following ecological parameters:

- Reproductive system and reproductive rate.
- Mortality rate.
- Population dispersal.
- Carrying capacity of the habitat.
- Initial population size and age distribution.
- Prediction of catastrophes.

12.4.1 Reproductive System and Reproductive Rate

1. As illustrated in Sect. 8.4:
   - The Nongguan Mountains white-headed langurs live in female-centered families. Most females live their entire lives in the same matrilineal territory. Every 4 years, the alpha adult male in the family would take along its male offspring and leave their birthplace, and other outlier adult males would come and compete to take over the females, form new conjugal relationships, and begin a new family troop.
   - From November 1996 to December 2015, we had tracked a total of 63 white-headed langur troops of different periods, observed their social structures, and discovered that the structures could be summarized into four basic categories: ante-family troop, family troop, transitory troop, and all-male troop. Apart from all-male troop, all the other three troop states were made up of an adult male and several females in which mating and reproduction could take place. The average conjugal period, which could be temporary or stable, of the three troop forms was 4 years and we take that to be one reproductive cycle.
   - The reproductive system of white-headed langurs is that during the conjugal period, the male will mate with multiple females, whereas a female is polyandrous in her lifetime, for she would mate with successive males who took over the family. Thus, every white-headed langur in its lifetime is polygamous.
2. As illustrated in Chapters 10 and 11:
   - From the first reproduction register of the 15 adult females in "FJC troop" from 1996 to 2015 as listed in Table 10.2, we find that females first reproduce at an average of around 5–6 years.
   - In our tracking registers from early 2008 to 2015, we registered that Wuque, the eldest son of Yintangxiaotu, successfully invaded and took over a family with 13 wives at around age 6.75 years, and by the end of 2015, it had its first offspring. Hence, males have their first children at around the age of 7 years.
   - From our long period of observations on the three adult females in "FJC troop," Heimama, Dahei, and Arzhen, and their reproduction registers, we estimate the mean maximum age for reproduction is 25 years.
   - The female reproduction register of "FJC troop" has documented that adult female Pingping gave birth to twins, Nannan and Ningning, on March 23, 2014. That is the only twin birth that was registered in the 20 years of research, and it accounts for 1.23% of the total 81 registered births.

As shown in Table 11.3, the sex ratio at birth for the white-headed langurs born in FJC Big Cave and Small Cave between 1998 and 2015 was 1:1 (♀:♂).

- From the statistics of inter-birth intervals by the 16 adult females in the "FJC troop" from 1996 to 2015, we calculated the mean inter-birth interval to be 784.3 days, and we estimated the reproductive rate to be 365/783.4 = 46.54%.
- The reproduction statistics from observations of over seven conjugal periods on 21 adult females and seven adult males living in FJC Big Cave and Small Cave in seven various troops (five stable family troops, one ante-family troop, and one transitory troop) has shown that out of 66 conjugal pairings (n = 66), 16 pairings had reproduced zero litter which accounted for 24.24% of the total, 26 pairings had reproduced one litter accounting for 39.39%, 18 pairings had reproduced two litters accounting for 27.27%, and six pairings had reproduced three litters accounting for 9.09% (Table 12.5).

Compiling the statistics above, we sorted out the reproductive system and reproductive rate data in Table 12.6 for inputting into VORTEX.

12.4.2 Mortality Rate

Referring to the illustrations in Sect. 11.3, we sorted out the 1998–2005 register of deaths of white-headed langurs living in FJC Big Cave and Small Cave in Table 12.7.

With the above count of 27 deaths and on the premise of the reproductive capacity of the Nongguan Mountains white-headed langurs illustrated in Sect. 11.3.4, we sorted out the data in Table 12.8 for inputting into the VORTEX model.

12.4.3 Population Dispersal

As illustrated in Sect. 8.4, the white-headed langur in the Nongguan Mountains is almost a "closed" population. They

Table 12.5 Reproduction registers of FJC Big Cave, Small Cave family troops and transitory troop adult females in the conjugal periods from 1994 to 2015

Period	Troop state	Adult female	Offspring	Number of litters reproduced by individual females in the corresponding conjugal period 0	1	2	3	Mate option
1994.10–1998.04	Queque family troop	Jiajia	Dagege; Erarge			✓		1
		Dahei	Xiaoarge		✓			1
		Arzhen	Sangege		✓			1
		Lingling	Sanarge		✓			1
		Bibi	Daarge		✓			1
1998.04–2002.07	Archeng family troop	Lingling	Yingying; Arying; Nianying				✓	1
		Bibi	Pingping; Arping; Nianping				✓	1
		Dahei	Yiyi; Aryi			✓		1
		Jiajia	Lanlan; Arlan			✓		1
		Arzhen	Feifei; Arfei			✓		1
		Xiaoxiao	Xinxin		✓			1
2002.07–2006.06	α-Gonghou family troop	Dahei	Yixi; Arxi			✓		1
		Jiajia	Lanqi; Arqi			✓		1
		Arzhen	Feilu; Arlu			✓		1
		Xiaoxiao	Xinyu; Aryu			✓		1
		Lanlan	Tiantian; Artian			✓		1
		Yingying	Baobao; Arbao			✓		1
		Pingping	Beibei		✓			1
2006.06–2012.07	Yintangxiaotu family troop	Pingping	Zhenzhu; Yiling; Xiaozhen				✓	1
		Lanlan	Mimi; Mengli; Xiaolan				✓	1
		Yingying	Yumei; Xiaoliu; Xiaoying				✓	1
		Lanqi	Qiqi; Qingming; 007				✓	1
		Dahei	Zuoque; Xiaohei			✓		1
		Jiajia	Youque; Gaga			✓		1
		Arzhen	Wuque; Chongsheng			✓		1
		Xiaoxiao	Bandazai; Chuliu			✓		1
		Arxi	Xiaojiu; Yutu			✓		1
		Tiantian	Wenwen; Shishi			✓		1
		Arlu	Hanlin; Xiaolu			✓		1
		Beibei	Gualai; Xiaobei			✓		1
		Aryu	EE		✓			1

(continued)

Table 12.5 (continued)

Period	Troop state	Adult female	Offspring	Number of litters reproduced by individual females in the corresponding conjugal period				Mate option
				0	1	2	3	
2012.07–2013.07	Duanwei-Shengdi family troop	Pingping	Xiao-P		✓			1
		Yumei	Xiao-M		✓			1
		Dahei	–	✓				1
		Jiajia	–	✓				1
		Arzhen	–	✓				1
		Xiaoxiao	–	✓				1
		Arxi	–	✓				1
		Aryu	–	✓				1
		Tiantian	–	✓				1
		Arlu	–	✓				1
		Beibei	–	✓				1
		Xiaojiu	–	✓				1
		Mimi	–	✓				1
		Wenwen	–	✓				1
		Zhenzhu	–	✓				1
		Da-S	–	✓				1
2013.07–2015.04	Yuweng family troop	Jiajia	Xiaosanba		✓			1
		Arzhen	Zhenni		✓			1
		Xiaoxiao	Xiaoyuer		✓			1
		Pingping	Nannan; Ningning. (twin)			✓		1
		Arxi	Arweng		✓			1
		Aryu	Huanghunhou		✓			1
		Tiantian	Tianniu		✓			1
		Arlu	Nuhanzi		✓			1
		Beibei	Beckham		✓			1
		Xiaojiu	Zhuli		✓			1
		Mimi	Mizai		✓			1
		Wenwen	Wennnu		✓			1
		Zhenzhu	Xiaoxuan		✓			1
		Da-S	Xiao-S		✓			1
		Dahei	–	✓				1
		Yumei	–	✓				1
2012.07–2015.04	Yintangxiaotu transitory troop	Lanlan	Wangzai		✓			1
		Yingying	Liushaotou		✓			1
		Lanqi	Qinu		✓			1
Notes	Litters summary /conjugal pairing			16	25	19	6	66
	The proportion of the various number of litters by a female in one conjugal period			24.242%	37.88%	28.79%	9.09%	100%
	Total no. of offspring	81						
	Single pup delivery	80 counts (98.77% of total birth deliveries)						
	Twin pup delivery	1 count (1.23% of total birth deliveries)						

have no easy way to disperse or to mingle with other populations, but can only evolve on their own lineage. Hence, we treat the population as one integral population without inputting a dispersal parameter in VORTEX.

12.4.4 Habitat Carrying Capacity

As illustrated in Sect. 11.2, until December 2015, the number of white-headed langurs in the Nongguan Mountains was increasing continuously, which was an indication that the

Table 12.6 The Nongguan Mountains white-headed langur reproductive system and reproductive rate parameters for VORTEX input

Item	Parameter
Duration of each simulation period	4 years = 4 × 365 = 1460 days This duration was decided concerning the mean conjugal period length which the various reproductive parameters built upon
Reproductive system	Polygyny in a conjugal period Polygamy over the life cycle
Mean first reproduction age of female	5–6 years old
Mean first reproduction age of male	7 years old
Maximum reproduction age of female	25 years old
Maximum number of litters by a female in one conjugal period	3 pups
Maximum number of pups in a litter by a female	2 pups
Sex ratio upon birth	1:1
Reproductive rate of adult female	365/784.3 days = 46.54%
Distribution of the various number of litters	0 litter: 24.24%; 1 litter: 37.88% 2 litters: 28.79%; 3 litters: 9.10%
Distribution of litter with various number pup	1 pup: 98.77% 2 pups: 1.23%
Mate monopolization	5.7%

population had not yet reached a saturation state. The three districts in the Nongguan Mountains that are a part of the white-headed langurs' natural habitat have a total area of 13.2 km^2. In that area, District 1, the Fangshaoshan–FJC–Leizhai district, with an area of 2.2 km^2, had the highest population density at 98 langurs/km^2. If we assume the density to be enlarged 1.5 times and evenly distributes over the entire 13.2 km^2 Nongguan Mountains ecological habitat, the carrying capacity will come to about 1800 langurs.

12.4.5 Initial Population Size and Age Distributions

To better evaluate the prediction accuracy of VORTEX, we pick two cases in the past and input the population size and age distribution of those cases respectively into the program.

12.4.5.1 Model 1

We take the founding population in the core research district FJC and apply the 18 individuals in the "Queque family troop" as the initial population to simulate the population dynamics in the past 18 years (Figs. 12.5 and 12.6).

From our field registers, the "Queque family troop" had 18 individuals in early 1998, which we inputted as the initial

Table 12.7 Register of white-headed langur deaths in research covered districts from 1998 to 2015

	Name	Sex	Birth date	Death date	Age	Cause of death
1	Nianying	♀	2002.07	2002.07	0.3 month	Infanticide
2	Nianqing	♂	2002.07	2002.07	0.3 month	Infanticide
3	Xiao-P	♂	2013.03.11	2013.04	0.7 month	Infanticide
4	Xiaoarge	♂	Early January 1998	1998.03	1.97 months	Infanticide
5	Xiao-M	♂	2013.01.11	2013.03	2.73 months	Infanticide
6	Arying	♂	2001.02	2011.05	2.97 months	Thunderstorm
7	Xiaozhen	♀	2012.04.16	2012.08	3.57 months	Infanticide
8	Xiaobei	♀	2012.04.02	2012.08	4.03 months	Infanticide
9	Xiaolu	♀	2012.03.28	2012.08	4.20 months	Infanticide
10	Siarge	♂	1997.10	1998.03	5.03 months	Infanticide
11	Artian	♂	2006.01	2006.06	5.03 months	Infanticide
12	Yutu	♀	2012.01.20	2012.10	8.50 months	Infanticide
13	Chuliu	♀	2011.08.01	2012.08	12.20 months	Lost during family split
14	Qinu	♀	2014.03.01	2015.04	13.20 months	Dead during family split
15	Liushaotou	♀	2014.02.16	2015.04	13.63 months	Dead during family split
16	–	♂	1999.01.01	2001.01	24.37 months	Disease
17	Bandazai	♀	2008.01.01	2010.03	26.33 months	Lost after wounded
18	Arping	♂	1999.01.01	2001.07	30.40 months	Thunderstorm
19	Half-grown female	♀	Unknown	2009	Below 3 years of age	Wounded in rainstorm, rescued but later died
20	Xiaogonghou 1	♂	Unknown	2012	Below 3 years	Electric shock
21	Xiaogonghou 2	♂	Unknown		Below 3 years	Electric shock
22	Xiaogonghou 3	♂	Unknown		Below 3 years	Killed from crossing highway
23	Xiaogonghou 4	♂	Unknown		Below 3 years	Dead from fighting wounds
24	Weiming	♀	1995.08.01	2001.03	Above 4.5 years	Miscarriage
25	Leizhai Gonghou	♂			Above 13 years	Dead from wounds
26	Queque	♂		2000.06	Above 13 years	Dead from fighting
27	Heimama	♀	1975.01.01	1998.03	Above 23 years	Poaching

Table 12.8 Nongguan Mountains white-headed langur mortality rates in different age groups

	Female ♀		Male ♂		Total tracked number of individuals in the observations
Years of age	No. of deaths	Mortality (%)	No. of deaths	Mortality (%)	
0–1	5	12.5	7	17.5	40
1–2	3	7.1	0	0.0	42
2–3	2	4.7	6	14.3	
3–4	0	0.0	0	0.0	37
4–5	1	3.0	0	0.0	33
5–25	0	0.0	2	0.0	
Above 25	1	25	0	0.0	4

population size for the simulation (Table 12.9). In that same year, the family underwent an invasion by adult male Archeng who replaced Queque as the alpha male and began the family anew with six females that stayed behind. Queque, taking along some individuals, initiated the "FSS family troop" and the "XS family troop" respectively by way of "graft" and "split." Over the years, seven white-headed langur troops had developed from them in the core research district, which comprises Fangshaoshan, FJC Big Cave and Small Cave, and Xishan. According to our research findings as illustrated in Sect. 10.2.2, the number of white-headed langurs in FJC District in June 2015 was 158, and there should be 172 white-headed langurs in June 2016.

We input into VORTEX 9.99c the initial population size and specified age distribution as per Table 12.10.

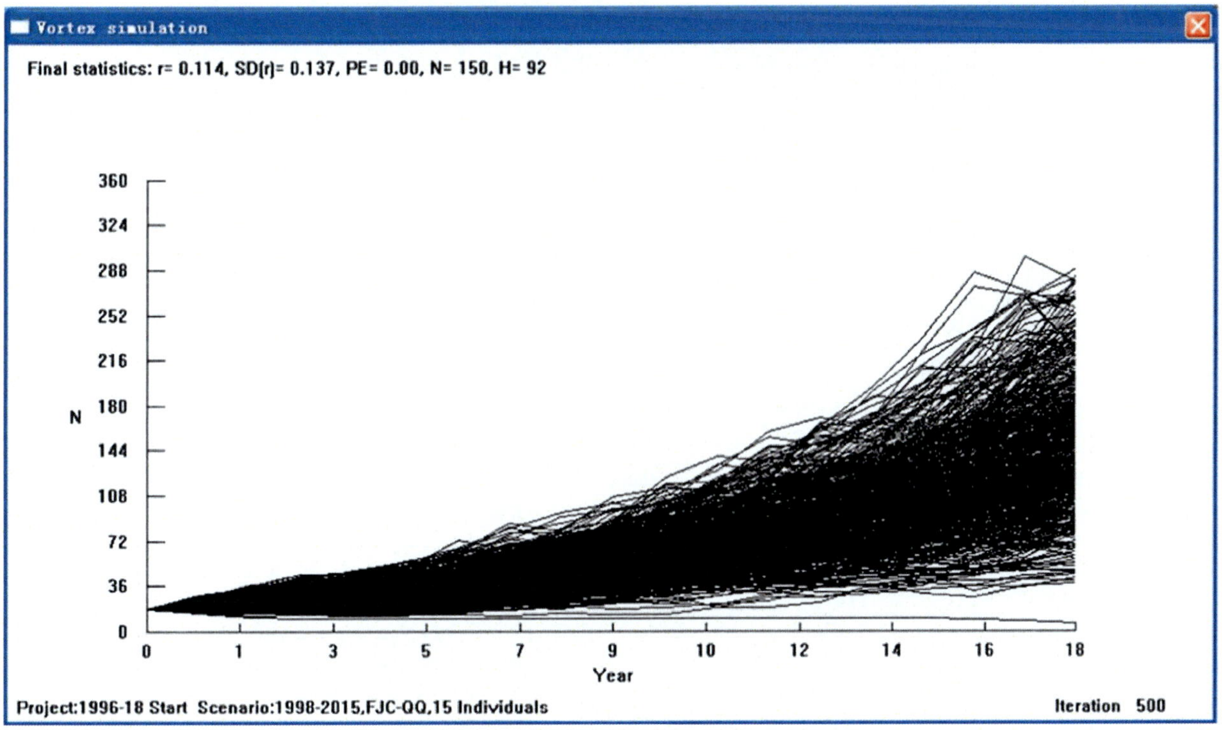

Fig. 12.5 Model 1: Simulations of the population dynamics in the past 18 years of the initial population of the 18 white-headed langurs in the "Queque family troop" in 1988 with the mean age of females when they first give birth set to 6 years. After 500 iterations, the population grows from 18 individuals to 150 individuals, though one iteration shows the population going extinct. The extinction probability is 0.2%, and survival probability is 99.8%

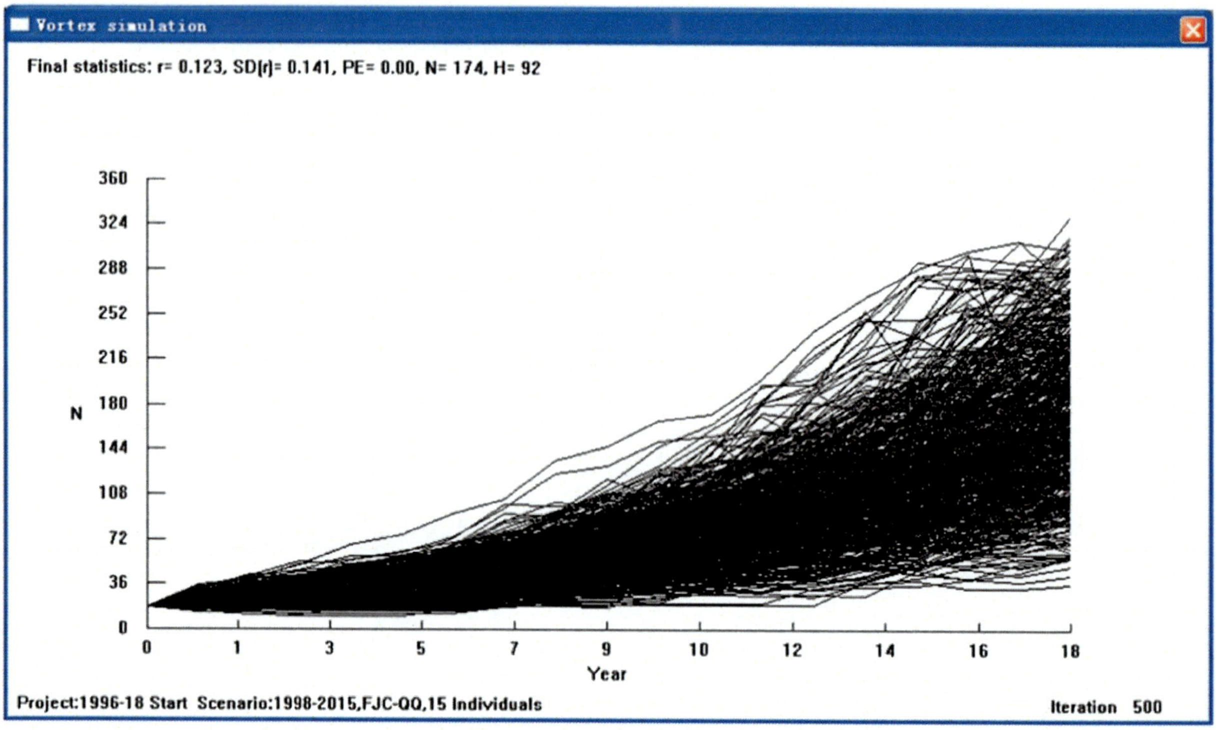

Fig. 12.6 Model 1: Simulations of the population dynamics in the past 18 years of the initial population of the 18 white-headed langurs in the "Queque family troop" in 1988 with the mean age of females when they first give birth set to 5 years. After 500 iterations, the population grows from 18 individuals to 174 individuals, with no iterations predicting extinction. The extinction probability is 0% and the survival probability is 100%

Table 12.9 Age distribution and state of stay/depart of the individuals in "Queque family troop" in 1988

	Individuals	Sex	Birth year	Age (years)	Stay/departed
1	Queque	♂	Before 1987	Around 11	Moved to Queque transitory troop along "split" in March 1998
2	Daarge	♂	Before 1996	2	Moved to Queque transitory troop along "split" in March 1998
3	Erarge	♂	Before 1996	2	Moved to Queque transitory troop along "split" in March 1998
4	Sanarge	♂	1996.12	1	Moved to Queque transitory troop along "split" in March 1998
5	Siarge	♂	1997.12	0.3	Was killed in infanticide in early 1998
6	Xiaoarge	♂	1998.01	0.1	Was killed in infanticide in early 1998
7	Heimama	♀	Before 1975	Above 23	Dead from poaching in April 1998
8	Arzhen	♀	Before 1989	Around 10	Staying in FJC Big Cave troop
9	Dahei	♀	Before 1991	Around 8	Staying in FJC Big Cave troop
10	Jiajia	♀	Before 1991	Around 8	Staying in FJC Big Cave troop
11	Bibi	♀	Before 1991	Around 8	Staying in FJC Big Cave troop
12	Lingling	♀	Before 1994	Around 5	Staying in FJC Big Cave troop
13	Beike	♀	Before 1994	Above 5	Staying in FJC Big Cave troop
14	Xiaoxiao	♀	Before 1996	2	Staying in FJC Big Cave troop
15	Weiming	♀	1995	Above 5	Moved to Queque transitory troop along "split" in March 1998
16	Dagege	♀	Before 1995	3	Moved to Queque transitory troop along "split" in March 1998
17	Ergege	♀	Before 1996	2	Moved to Queque transitory troop along "split" in March 1998
18	Sangege	♀	1997.02	1	Moved to Queque transitory troop along "split" in March 1998

With the input of the above reproductive system, reproductive rate, mortality, dispersal, carrying capacity, and initial population with specified age distribution parameters, VORTEX 9.99c processed 500 iterations of simulation and output as shown in the following graphs and text:

Model Result:

Table 12.10 Initial population size and specified age distribution

Female age (year)	No.	Male age (year)	No.
1	1	1	3
2	2	2	2
3	1	3	0
4	0	4	0
5	1	5	0
6	1	6	0
7	1	7	0
8	2	8	0
9	1	9	0
10	1	10	0
11	0	11	1
12	0	12	0
13	0	13	0
14	0	14	0
15	0	15	0
16	0	16	0
17	0	17	0
18	0	18	0
19	0	19	0
20	1	20	0
21	0	21	0
22	0	22	0
23	0	23	0
Total	12	Total	6

In the 18 years from 1996 to 2015, the population dynamics of the population of 18 wild white-headed langurs in the Queque family troop in Feijichang (FJC) district:

- Was on a steady growing trend.
- The mean deterministic population growth rate was: $r = 0.123 \pm 0.141$–0.114 ± 0.137
- The mean expected end heterozygosity was: $H = 0.9218 \pm 0.0186$–0.9238 ± 0.0164
- Extinct probability 0.2%; survival probability 99.8%.
- On the 18th year, the mean number of individuals in the population was 150–174.

12.4.5.2 Model 2

In Model 2, we take a total of 105 white-headed langurs in the Nongguan Mountains in January 1997 as the initial population to simulate the population dynamics in the past 19 years (Figs. 12.7 and 12.8).

Based on the numbers and age distributions of the white-headed langurs in the Nongguan Mountains from 1996.11.18 to 1997.01.19, we take out the number of post-reproductive females and then sorted the data concerning the initial population size and specified age distribution in Table 12.11 for inputting into the VORTEX 9.99c.

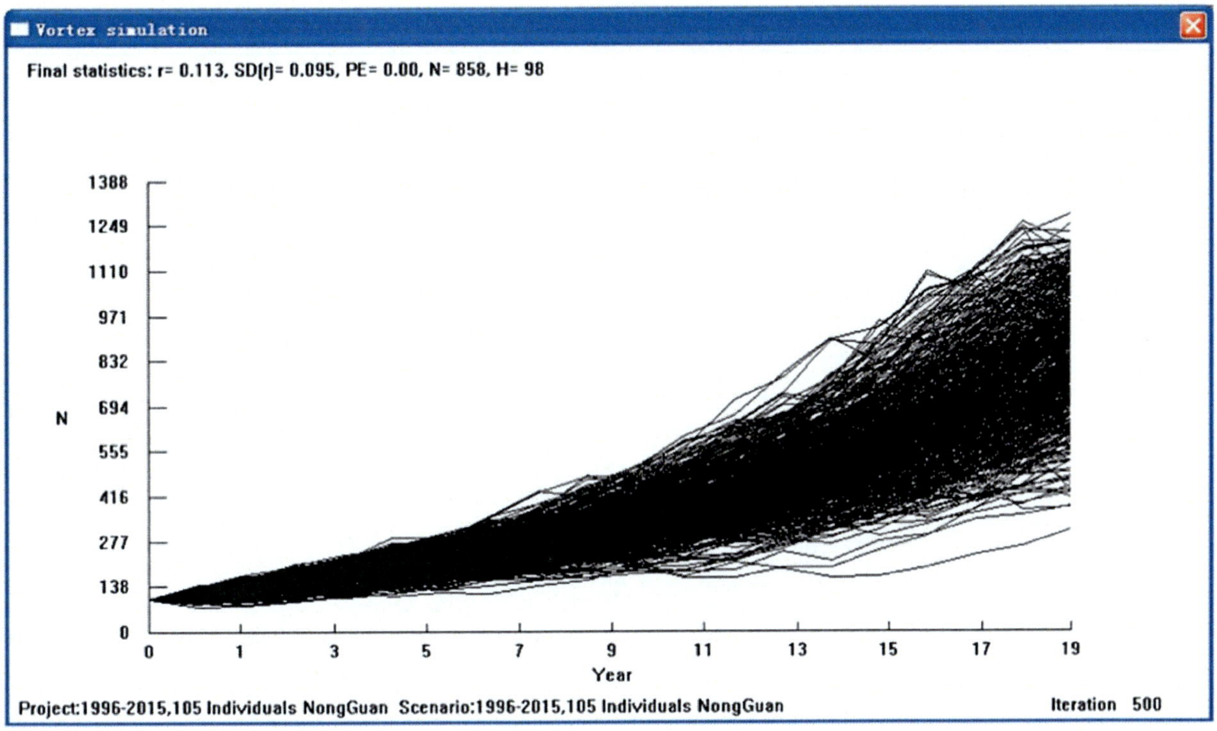

Fig. 12.7 Model 2: Simulations of the population dynamics in the past 19 years of the initial population of 105 white-headed langurs in the Nongguan Mountains in 1997 with the average female age when they give first birth is set to 6 years. After 500 iterations, the population is shown to grow from 105 individuals to 858 individuals with a 0% extinction and 100% survival probability

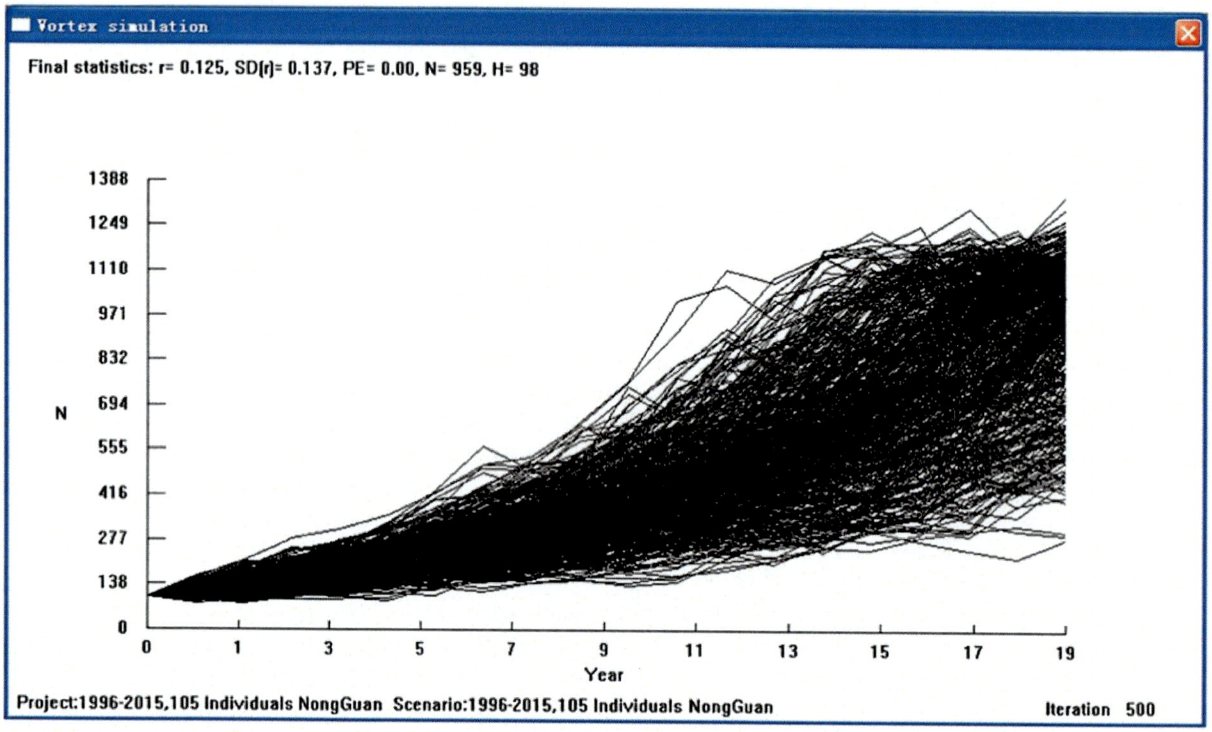

Fig. 12.8 Model 2: Simulations of the population dynamics in the past 19 years of the initial population of 105 white-headed langurs in the Nongguan Mountains in 1997 with the average female age when they give first birth is set to 5 years. After 500 iterations, the population grows from 105 to 959 individuals, with 0% extinction and 100% survival probability

Table 12.11 Initial population size and specified age distribution

Female age (year)	No.	Male age (year)	No.
1	6	1	7
2	6	2	6
3	4	3	4
4	4	4	4
5	5	5	3
6	4	6	3
7	4	7	2
8	3	8	2
9	3	9	2
10	3	10	1
11	3	11	1
12	3	12	0
13	2	13	2
14	2	14	1
15	2	15	0
16	2	16	0
17	2	17	0
18	1	18	0
19	1	19	0
20	1	20	0
21	0	21	0
22	0	22	0
23	0	23	0
24	0	24	0
25	0	25	0
Total	61	Total	38

The initial population size and specified age distribution were inputted along with the same parameters of the reproductive system, reproductive rate, mortality rate, dispersal, and carrying capacity as in Model 1 into VORTEX 9.99c. The following is the output after 500 iterations.

Table 12.12 Comparison of empirical population size statistics and the predictive results by VORTEX

	Population size statistics from field observation	Predictive results by VORTEX 9.99c
Model 1: taking the 18 individuals of the founding troop in FJC core area—"Queque family troop" as the initial population to simulate the population dynamics of the past 18 years	158–172	150–174
Model 2: taking the total of 105 individuals in Nonguan Mountain area in January 1997 as the initial population to simulate the population dynamics in the past 19 years	797–841	858–959

Model Result:

For the past 19 years until 2016, the population dynamics showed that the founding population in the Nongguan Mountains of 105 wild white-headed langurs:

- Was on a steady growing trend.
- The mean deterministic population growth rate was: $r = 0.113 \pm 0.0945–0.125 \pm 0.137$
- The mean expected end heterozygosity was: $H = 0.9804 \pm 0.0037–0.9816 \pm 0.0036$
- Extinct probability 0; survival probability 100%.
- On the 19th year, the mean number of individuals in the population was 858–959.

After going through exponential regression analysis of the above two model results, we found the empirical population size statistics and the predictive results by VORTEX are very close (Table 12.12), with a fitting degree of $R^2 = 0.99792$ (Fig. 12.9):

12.5 Prediction of the Future of the Population by Vortex Model

With detailed and accurate field data, this result not only confirms the reliability of ecological research conclusions, but also confirms the basic prediction ability of the vortex model. We believe that both the number of population simulated by computer and the empirical statistic results are basically close to the truth. On this basis, we continue to predict the future of this population.

When we set the carrying capacity of the whole 13.2 km^2 natural habitat in Nongguan Mountains to 1800 white-headed langurs, it is predicted that saturation state would be reached in 2027, and the population would remain constant from that year onward (Fig. 12.10). At this trend of growth, years 2003–2021 will be the critical years for the population's rapid restoration and growth and as well as the golden years for rescuing the population.

Since conservation biology became a celebrated field of study in the 1980s, it had come to be a discipline that was involved in tackling crises. Sometimes, researchers have to make immediate conservation decisions of threatened species without the benefit of enough research data. The research of the population ecology and conservation biology of the Nongguan Mountains white-headed langurs since 1997 has provided a successful case.

With the rapid development of computer technology, PVA research methods are continuously innovated and upgraded. We have verified the simulation method to be reliable with our empirical data of the white-headed langur population, and we believe that in the future, computer modeling would get results that are even nearer to authenticity. While computer programs may allow various computations as well as deviation inputs, the past is in no way recoverable. The 105 remnant white-headed langurs have steadfastly survived the desperate situations 20 years ago, their descendants have retained their thriving vitality, and the population is still growing today. This is undoubtedly something fantastic and above computer program capabilities.

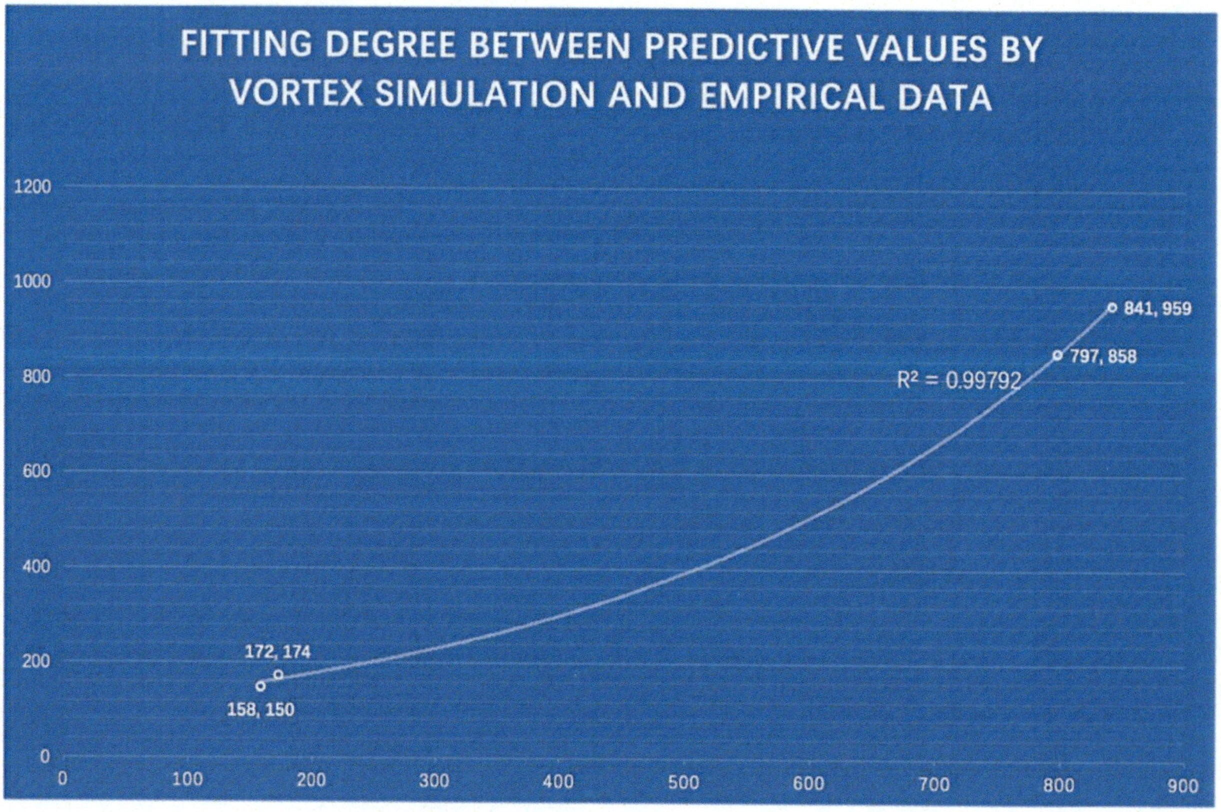

Fig. 12.9 Regression analysis of VORTEX simulation predictive values with empirical data

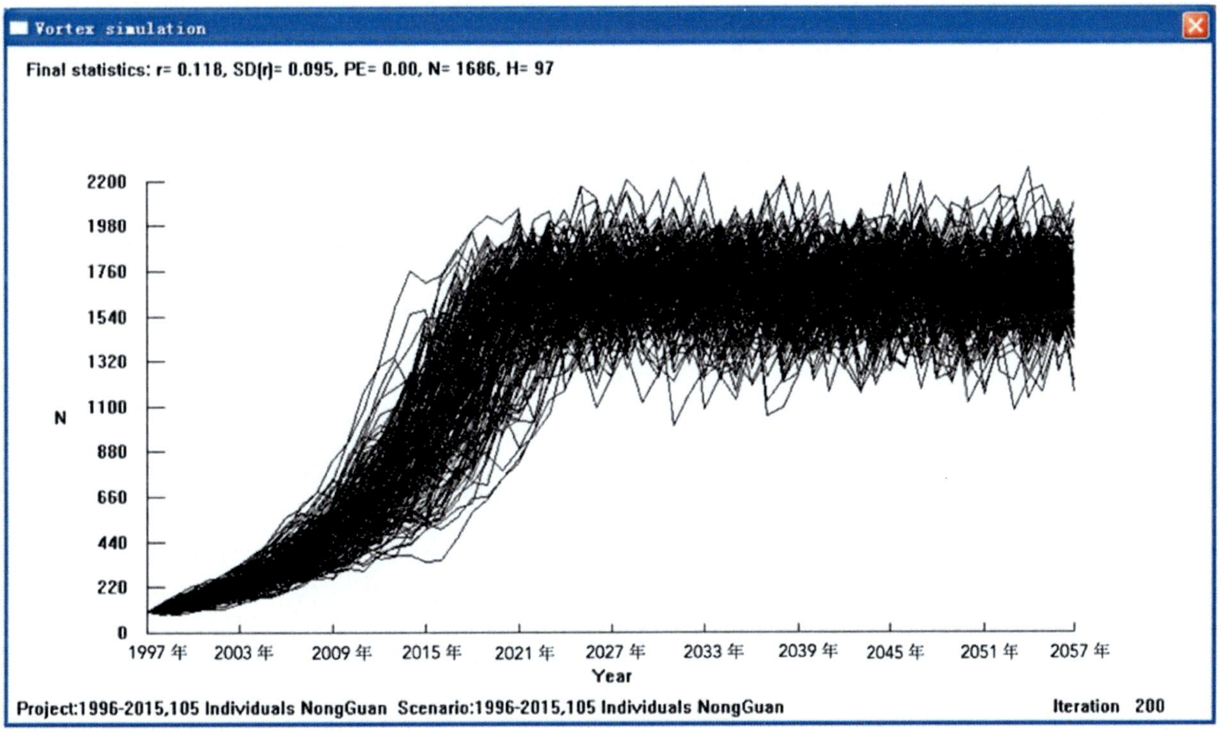

(a)

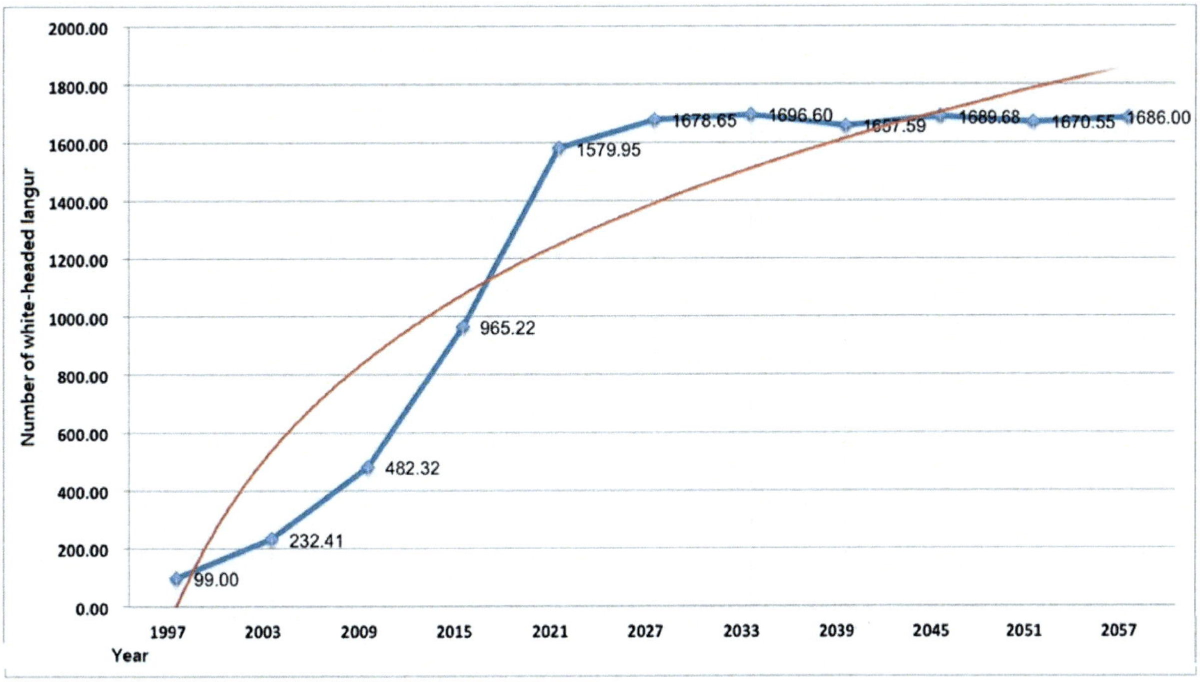

(b)

Fig. 12.10 (**a**) Model 3: Simulation for the 60 years on the population dynamics of the initial population of 105 white-headed langurs in the Nongguan Mountains in 1997. (**b**) Population growth prediction graph of the wild white-headed langur population in the Nongguan Mountains from 1997 to 2057, with habitat carrying capacity set at 1800 langurs

References

Berger J. Intervention and persistence in small populations of bighorn sheep. Conserv Biol. 1999;13(2):432–5.

Bleisch B, Xuan Canh, L, Covert B et al. The IUCN red list of threatened species e. T22045A9351127; 2008.

Frankel OH, Soulé ME. Conservation and evolution. Cambridge: CUP Archive; 1981.

Gilpin ME, Soulé ME. Minimum viable populations: processes of species extinction. Conservation biology: the science of scarcity and diversity. Sunderland, MA: Sinauer Associates; 1986. p. 19–34.

Guerrant EO Jr. Genetic and demographic considerations in the sampling and reintroduction of rare plants. In: Conservation biology. New York. NY: Springer; 1992. p. 321–44.

Guo Z. Introduction to population genetics (群体遗传学导论). Beijing: China Agriculture Press; 1993. p. 158–91.

Kimura M. Evolutionary rate at the molecular level. Nature. 1968;217 (5129):624–6.

Li CC. Population genetics. Translated by Wu Zhongxian. Beijing: China Agriculture Press; 1981. p. 386–9.

Snustad DP, Simmons MJ. Genetics theory. Translated by Zhao Shouyuan, et.al. Beijing: Higher Education Press; 2011. pp. 2, 319, 676, 685–686.

Yu T, Jianguo W, Xiaojun K, et al. Population viability analysis (PVA) and applications (种群生存力分析(PVA)的方法与应用). Appl Ecol J. 2011;1:257–67.

Zhuohua D, Yafu W, Yimin S. Genetics, 3rd edn (遗传学3版). Beijing: Higher Education Press; 2016. p. 494.

The conservation of the white-headed langurs at the same time ensures the existence of biodiversity in the Nongguan Mountains and also the livelihoods of hundreds of thousands people living and working on this piece of land. (Photo by Feng Chunguang)

Check for updates

13 Conservation Suggestions

Abstract

Despite considerable progress, in absolute numbers the white-headed langur still appears to be near extinct due to limited habitat and sparse population in each of the regional distribution centers. In order to conserve this langur species, we have to bear in mind four issues. First, conservation efforts must focus on safeguarding and enhancing the long-term viability of the monsoon rainforests on the karst hills, as core habitat for the langurs. Second, poaching should be banned, and the ban should be strictly enforced. Third, we must draft a conscientious management plan to enable the present population to gradually achieve adequate genetic diversity. Fourth, the purpose of conservation is to ensure the coexistence of both man and nature.

Keywords

Population management, Habitat conservation, Poaching, Viable population size

The white-headed langur lives not only in the karst hills at the south bank of the Zuo Jiang River, China, but also in the hearts of people. Not only are they elegant in appearance, they have had an amazing evolution journey; after the catastrophic damage in the twentieth century, they have been able to live off the leaves of the karst hills without consuming human cultivated crops.

About 1.4 million years ago, when the severe glacial period in the northern hemisphere exposed the Western Pacific continental shelf, their ancestors managed to migrate from the Sunda ancient land. The white-headed langur even survived the Pleistocene Epoch which was marked by repeated glacial cycles which killed off many large species of their contemporaries. It was only recent human history that threatened the species, relegating a little more than 1000 white-headed langurs to a 2000 km^2 region lying between Zuo Jiang River and Ming Jiang River, where since the 1950s, humans have within a short period of time exploited all the flatlands and hill slopes, leaving only the karst craggy hills. They are not wary of the force of nature, but they remain skeptical of human kindness and goodwill.

The white-headed langur is endemic to Chongzuo, Guangxi, China. Therefore, the task of protecting this species naturally falls on the shoulders of the Chinese; the Chinese government and the people of China are now fully aware of their obligation to the white-headed langurs and individuals around the globe who are concerned about the species. We have put in a lot of effort, time, and money in setting up nature reserves, conducting scientific research, and educating the people about environmental protection. Yet, ensuring the bright future of the white-headed langurs still remains an immensely difficult task. In a country with such a large population, the demand for cultivated land, wood, and natural resources is so great that the habitat of the white-headed langur continues to decrease and get divided into smaller, isolated patches of habitats, causing population fragmentation, leading to inbreeding depression, thereby increasing the risk of local extinction.

The primary goal of our research is to conserve the existing populations and restore their once lost habitat. To this end, it is necessary to ensure the security of the natural habitat and the population survival of the white-headed langurs; the effective population size for their survival needs to be determined. At present, the population of the white-headed langur is scarce, their habitat is localized and limited, drawing a picture of near extinct. To draw up a blueprint to ensure their survival, we will need to accept that there will be both morality and scientific challenges to overcome.

There are four factors to focus on in the conservation of white-headed langurs:

W. Pan, *The Population Ecology of White-Headed Langur*, https://doi.org/10.1007/978-981-33-4118-0_13

- First, the "ecological progressive succession" efforts of protecting the karst limestone forest must be central to the conservation efforts.
  The forested pinnacles of Nongguan Mountain and the talus around them are currently the ideal habitat for white-headed langurs and also the last "natural refuge" for their survival. During the coevolution period between humans and white-headed langurs, two distinguishable and closely related ecosystems have been formed in the karst peak forest valleys and peak cluster depressions at the beginning of the twenty-first century—the "sugarcane agro-ecosystem" and the "karst hill ecosystem." The former is an agricultural base on which people in the Nongguan mountainous region have depended on their livelihood. We had witnessed in the past 20 years how farmers here overcame poverty and began to lead sustainable livelihoods by cultivating an infertile land planting sugarcane. The latter is the rocky mountainous monsoon forest which serves as the last refuge of the white-headed langurs. In order to secure the survival of the locals and the white-headed langurs, both the ecosystems need to be protected.
  For the past 2.5 million years, rocks have been falling from the top of the karst hills and accumulated around the foothills into talus. The boundary extending from the karst hills to the farmland formed the boundary between the two ecosystems. Vegetation on the talus not only provides the white-headed langurs with sufficient and steady food resources but also provides a foundation for the biodiversity on the karst hills. If we pull out a small shrub, shake the soil off its roots, and examine with a magnifier, we would see there are many organisms in the black soil, like algae, fungi, nematodes, mites, springtails, little earthworms, and numerous bacteria that cannot be seen. They are a small portion of the Nongguan Mountains ecosystem, yet they are the driving force behind the biodiversity here.
  In the current situation, the deforestation on the talus, the expansion of sugarcane fields, and land reclamation activities continue to threaten the vegetation cover of the talus. Given this, we must take necessary action to prevent the "retrogressive succession" of the vegetation communities on the talus and to ensure a "progressive succession" transforming them to tropical and subtropical monsoon forests.
- Second, poaching should be banned, and the ban should be strictly enforced.
  Although the Big Leap Forward and the Backyard Furnace campaigns in China in the 1950s had completely deforested the forest habitat of the white-headed langurs, there remain six stone mountainous regions sustaining small but viable populations of white-headed langurs. Presently, the six karst regions and the white-headed langur population are scattered between the south of Zuo Jiang River and north of Ming Jiang River, and from Huashan at Ningming County to the Nongguan Mountains, the Nongfeng Mountains, and the Bazhiqi Mountains at the Jiangzhou district, then to the Nonglin Mountains and the Nongdou Mountains at Fusui County. So what was the reason that drove them to extinction? Poachers in the period from the mid-1990s to the beginning of the twenty-first century have eradicated the white-headed langur population from four of their natural habitats!
  The driving force behind poaching was greed from a misguided belief. In spite of a lack of medical evidence, some people believed that tonics made with white-headed langurs could act as an aphrodisiac and pain relief, etc. That created a market demand for white-headed langurs, which aggravated the carnage of this precious and rare species. Despite repetitive affirmations by the government in carrying out wild animal protection decrees, from time to time, there are still white-headed langurs killed by "steel cats" or "loops" planted in the nature reserves. Female white-headed langurs reach reproductive maturity only at age 5 years, and in normal circumstances only bearing one infant every 2 years. Presently the mortality rate of infants in the small populations distributing in the Nongguan Mountains at Jiangzhou district is as high as 16.4%. Add on top mortalities resulting from diseases, old age, and other causes in various age groups, the death rate is even higher. If additional deaths are caused by human poaching, it will have serious consequences for the survival of the population. Hence, deterring poaching by whatever means necessary must be put as a priority in regular work routines.
- Third, we must draft a conscientious management plan to enable the present population to gradually achieve adequate genetic diversity.
  At present, the large tracts of land in Jiangzhou between the Nongguan Mountains and the Nonglin Mountains have been turned into farmland. The isolation of the white-headed langur population in the two regions prevents genetic exchange between them. Even if a nature reserve has been established, it would take a long period for a small population to restore its genetic variation through mutations.
  Since the 1980s, studies on the survival of small populations of wild giant pandas scattered across isolated mountains in Western China have shown that the main factor determining their survival is their genetic fitness. Small isolated endemic populations which lose their genetic variability due to high inbreeding will inevitably lose their viability. Even with adequate space and food in

its habitat and no human interference, it is hard to sustain. An example is the giant panda population in the Mabiandafengding Nature Reserve in Sichuan, which had a rare number of pandas 30 years ago, or today, even in an even more severe condition.

Inbreeding reduces the fertility of the species and the survival rate of their newborns. A scientist researching animal reproduction pointed out that "a fundamental rule in conservation genetics is that the maximum inbreeding in a population shall not exceed 1%." This value is calculated on the premise that animals can mate at will, that adult sex ratios are equal, and that generations do not overlap. However, in the case of the white-headed langur, only the strongest and best males in the same habitat get the chance to mate, so that the sex ratio of adult females and males can be as high as 11.3:1 (see Chap. 11). Under such conditions, to maintain their genetic fitness, the minimum viable population size will need to be quite large.

At present, we do not know how large the viable population size needs to be for the existing white-headed langur population to keep its evolutionary course. Frankel and Soulé (1981) believe that in a population no less than 500 individuals that can randomly mate, the genetic mutations from mating with remotely related partners could offset the loss of genetic variation from inbreeding depression. However, at present, the number of effective breeding white-headed langurs in both the Nongguan Mountains and the Nonglin Mountains is below 500 (the number by the end of 2015 in the Nongguan Mountains was estimated to be 320–360). But Frankel and Soulé (1981) also pointed out that "small populations" could counter the adverse effects of inbreeding by reproducing rapidly. It remains for us to discover how big the minimum viable population for white-headed langurs should be. The probability of survival and extinction of each geographic population can only be logically assessed by calculating the required habitat area and the vital parameters such as their birth and death rates. Hence, a comprehensive maintenance plan for a conservation reserve must be drafted based on scientific research.

- Fourth, we need to acknowledge that all creatures and human beings have the same right to exist. The purpose of conservation is to ensure the coexistence of both man and nature.

  The value of conservation penetrates all aspects of human concern—scientific, economic, political, cultural, aesthetic, and religious. However, at its core is the ethical debate that "all life on the earth has the same right to exist."

  White-headed langurs have formed an indissoluble bond with the karst hills on the south bank of Zuo Jiang River through a course of more than a million years of natural evolution. They could have continued to survive in a natural world through their innate reproduction capabilities; however, human developments had compelled them to face problems that every species including humans must face: the destruction of nature, the increasing scarcity of earth resources (including ground and underground), environmental pollution, excessive hunting, etc. Above all, the most perilous problem is the heedless devastation of nature and ruthless extermination of species by humans for their immediate benefits. All species on Earth, whether they be animals or plants are dependent on each other. The extinction of any single species means the chain of life is broken again. A splendid and unreproducible leaf of history will pass away quietly, and nature's legacy lost to our descendants. The drastic decrease in biodiversity will have catastrophic consequences. Humans must be soberly aware that in order for their future descendants to continue life on this small planet, they will need to restore the fertility of the soil, cultivate new crops, find fossil fuel substitutes, and get inspirations for knowledge and scientific creations.

  Learning from our conservation experience, the goal is to create a win-win situation where the white-headed langur and the local residents coexist harmoniously. The conservation of the white-headed langurs in the karst hills will at the same time have conservation effect on the tens of thousands of other animals, plants, and microorganisms in the biome. It also vouches for the livelihoods of hundreds of thousands of humans who are living and working on this piece of land.

  The white-headed langur is no longer just an ordinary animal, it has become a symbol: protection of the white-headed langur expresses the obligation and confidence of the people of Chongzuo toward themselves and their children, emphasizing the need to create a better world where conserving nature is our core belief. We will treat white-headed langurs with benevolence and compassion and strive with wisdom and hard work to protect their future. Let us work together in this enormous undertaking of conserving nature, so that these beautiful lives will long survive in this beautiful valley of karst hills.

Reference

Frankel OH, Soulé ME. Conservation and evolution. Cambridge: CUP Archive; 1981.